Study Guide

for

Brym and Lie's

Sociology
Your Compass for a New World
Third Edition

Judith Pintar

University of Illinois at Urbana–Champaign

WADSWORTH
CENGAGE Learning™

Australia • Brazil • Japan • Korea • Mexico • Singapore • Spain • United Kingdom • United States

ISBN-13: 978-0-495-00837-8
ISBN-10: 0-495-00837-0

Cover Image: Lawrence Lawry / Getty PictureQuest

Wadsworth
10 Davis Drive
Belmont, CA 94002-3098
USA

Cengage Learning is a leading provider of customized learning solutions with office locations around the globe, including Singapore, the United Kingdom, Australia, Mexico, Brazil, and Japan. Locate your local office at:
www.cengage.com/global

Cengage Learning products are represented in Canada by Nelson Education, Ltd.

To learn more about Wadsworth, visit
www.cengage.com/wadsworth

Purchase any of our products at your local college store or at our preferred online store **www.ichapters.com**

Printed in the United States of America
2 3 4 5 6 7 13 12 11 10 09

TABLE OF CONTENTS

PREFACE

Welcome to the Study Guide for Robert Brym and John Lie's *Sociology: Your Compass for a New World,* Third Edition. The Study Guide is intended to guide on your journey through your course in introductory sociology. The Study Guide chapters correspond to the textbook chapters, and include the following components:

Components of the Study Guide

- Learning Objectives
- Key Terms
- Detailed Chapter Outline
- Study Activities
- Infotrac Exercises
- Internet Exercises
- Practice Fill-In-The-Blank Questions
- Practice Multiple-Choice Question
- Practice True-False Questions
- Practice Short-Answer Questions
- Practice Essay Questions

A SOCIOLOGICAL COMPASS

Student Learning Objectives

After reading Chapter 1, you should be able to:

1. Explain what is unique about sociology in comparison to the other seemingly similar humanities and social science disciplines like economics, political science, philosophy, drama, and psychology.

2. Explain the role of social forces in opening our opportunities and constraining our freedom.

3. Explain what C. Wright Mills referred to as the "sociological imagination,"—and its origins in relation to the three modern revolutions.

4. Understand the connection between theory, research, and values in sociological analyses.

5. Discuss the contribution of Auguste Comte, Herbert Spencer, Karl Marx, Emile Durkheim, and Max Weber to the birth of sociology as a social science and a way of combating social ills.

6. Identify the major sociological theoretical traditions, their central features and key exemplars.

7. Contrast the perspectives of Emile Durkheim, Karl Marx and George Herbert Mead in the context of their respective contributions to functionalism, conflict theory, and symbolic interactionism.

8. Contrast Karl Marx's and Max Weber's perspectives on inequality and conflict.

9. Discuss the contribution of W.E.B. Du Bois to conflict theory and the study of race in the United States.

10. Discuss the contribution of Harriet Martineau, Jane Addams, and feminist theory to sociology, which was largely ignored by other pioneers of sociology.

11. Discuss the features of modern feminist theory.

12. Apply the four sociological perspectives to the issue of fashion.

13. Explain the notion of sociology as "a compass" to examine unresolved issues and dilemmas confronting the postindustrial era and the future.

14. Explain the usefulness of sociology as a practical source of advice to improve society—and as a perspective to enhance your understanding of how we, as individuals, fit in the larger social context of society.

Key Terms

sociology (2)

social context (2)

social solidarity (4)

altruistic suicide (4)

egoistic suicide (4)

anomic suicide (4)

social structures (6)

microstructures (6)

macrostructures (6)

patriarchy (6)

global structures (6)

the Sociological Imagination (7)

the Scientific Revolution (9)

the Democratic Revolution (10)

the Industrial Revolution (10)

theories (12)

research (12)

values (13)

functionalist theories (13)

dysfunctional (14)

manifest functions (14)

latent functions (14)

conflict theory (14)

class conflict (15)

class consciousness (15)

the Protestant ethic (16)

symbolic interactionism (17)

social constructionism (18)

feminist theory (19)

the Postindustrial Revolution (23)

globalization (23)

public policy (27)

Detailed Chapter Outline

I. INTRODUCTION

    A.   Why Robert Brym Decided *Not* To Study Sociology

        1. Sociology: "Thin Soup with Uncertain Ingredients?"

        2. Is Sociology merely like economics, political science, philosophy, drama, and abnormal psychology?

    B. A Change of Mind

        1. Sociology "an altogether new way" to think.

        2. The dilemma of all thinking people ("Life is finite")—study, reflection and the selection of values and goals.

        3. Sociology's Unique Way of Explaining Social Life

            a. **Sociology** is the systematic study of human behavior in social context.

            b. *Social* causes or social forces are distinct from physical and emotional causes.

            c. The **social context** or the organization of the social world opens up some opportunities, but also constrains our freedom.

            d. Understanding the power of *social forces* can help us to know our capabilities and limitations.

    C. The Power of Sociology

II. THE SOCIOLOGICAL PERSPECTIVE

    A. The Sociological Explanation of Suicide
       1. Suicide: a seemingly antisocial, nonsocial, and private act.
       2. Based on his analysis of European government statistics and hospital records, **Emile Durkheim** demonstrated that suicide is more than just an individual act of desperation resulting from psychological disorder, as commonly believed at the end of the 19th century. Durkheim argued that suicide rates vary due to differences in **social solidarity,** which is the degree to which group members share beliefs, and the intensity and frequency of their interaction.
       3. Durkheim's research findings in the late 1800s France:
          a. Married adults are half as likely as unmarried adults to commit suicide due to their social ties.
          b. Women are less likely to commit suicide than men due to their greater involvement in intimate social relations of family life.
          c. Jews are less likely to commit suicide than Christians due to being more socially tight knit in defensive response to centuries of persecution.
          d. The elderly are more prone than the young and middle-aged to commit suicide, because they are more likely to live alone, have lost a spouse, lack a job, and lack a wide network of friends.
       4. The relevance of Durkheim's theory and research to here and now:
          a. Elderly are most likely to commit suicide, especially among the divorced and widowed elderly.
          b. Men are four times more likely than women to commit suicide.
          c. In some parts of the United States, high rates of church membership have low suicide rates, while areas with high divorce rates have high suicide rates.
          d. However, suicide among young people has become more common over the past half century. Why, sociologically?

    B. From Personal Troubles to Social Structures
       1. **Social structures** are stable patterns of social relations that affect your innermost thoughts and feelings, influence your actions, and thus help shape who you are. *Three levels of social structure* that surround and permeate us:
          a. **Microstructures** are patterns of intimate social relations.
          b. **Macrostructures** are patterns of social relations that lie outside and above your circle of intimates and acquaintances.
             i. **Patriarchy**, the traditional system of economic and political inequality between women and men in most societies, is an important macrostructure.
          c. **Global structures** are patterns of social relations that lie outside and above the national level, including international organizations, patterns of worldwide travel and communication, and the economic relations between countries.
       2. One of the sociologist's main tasks is to identify and explain the connection between people's personal troubles and the social structures in which they are embedded. Personal problems are connected to social structures at the micro, macro, and global levels.

    C. The Sociological Imagination

1. The **"sociological imagination,"** coined by C. Wright Mills, refers to the ability to see the connection between personal troubles and social structures.
2. The sociological imagination as a recent addition to the human repertoire, was born in the context of **three modern revolutions** that pushed people to think about society in new ways:

D. Origins of the Sociological Imagination
1. The **Scientific Revolution** which began in Europe around 1550, encouraged the view that sound conclusions about the workings of society must be based on solid evidence, not just speculation.
2. The **Democratic Revolution,** which began about 1750, involved the citizens of the United States, France and other countries broadening their participation in government, and suggested that people organize society, and that human intervention can therefore resolve social problems.
3. The **Industrial Revolution,** which began in Britain in the 1780s, refers to the rapid economic transformation that involved the large-scale application of science and technology to industrial processes, the creation of factories, and the formation of a working class.

III. THEORY, RESEARCH, AND VALUES

A. French social thinker **Auguste Comte** (1798-1857) who coined the term *sociology* in 1838 was eager to adopt the scientific method in his study of society, yet was a conservative thinker, motivated by strong opposition to rapid change in French society – namely the French Revolution, early industrialization, and the rapid growth of cities.

B. British social theorist **Herbert Spencer** (1820-1903), the second founder of sociology, was strongly influenced by Darwin's theory of evolution, and believed he discovered scientific laws that govern the operation of society. According to Spencer, like biological organisms, societies are composed on interdependent parts, evolve from "barbaric" to "civilized."

C. Spencer's ideas became popular and known as "social Darwinism." Today, sociologists have a better understanding of the complex economic, political, military, religious, and other forces that cause social change.

D. Tension between the importance of science and a vision of the ideal society are seen in the work of **Karl Marx** (1818-83), **Emile Durkheim** (1858-1917), and **Max Weber** (1864-1920). They wanted to explain the great transformations of Europe (i.e. transition to industrial capitalism) and suggested ways of improving people's lives and combating social ills. They were also committed to the scientific method of research.

E. The tension between analysis and ideal, diagnosis and prescription, is evident throughout sociology, and becomes clear in the distinction of the terms theories, research, and values.

F. Theory
1. Sociological ideas are usually expressed in the form of theories.
2. **Theories** are tentative explanations of some aspect of social life, that state how and why certain facts are related.

G. Research
1. After the formulation of theories, sociologists conduct research.

2. **Research** is the process of carefully observing social reality, often to "test" a theory or assess its validity.

H. Values
   1. Sociologists must decide which problems are worth studying.
   2. **Values** are ideas about what is considered right and wrong, which help sociologists formulate and favor certain theories over others.
   3. Sociological *theories* may be modified or rejected on the basis of *research,* but they are often motivated by sociologists' *values.*
   4. Durkheim, Marx, and Weber initiated and stood close to the origins of the major theoretical traditions in sociology – functionalism, conflict theory, and symbolic interactionism.

IV. SOCIOLOGICAL THEORY AND THEORISTS

A. Functionalism:
   1. Durkheim
      a. **Functionalist theories** incorporate four features:
         i. Human behavior is governed by stable patterns of social relations, or social structures, usually macro structures.
         ii. Social structures maintain or undermine *social stability*.
         iii. Social structures are based mainly on *shared values*.
         iv. Reestablishment of *equilibrium* can best solve most social problems.
   2. Parsons and Merton
      a. **Talcott Parsons** was the foremost American proponent of functionalism who identified how various institutions work to ensure the smooth operation of society as a whole.
      b. **Robert Merton** proposed that social structures may have different consequences for different groups of people such as disruptive consequences referred to as **dysfunctional.** Merton also noted that some functions are **manifest** (intended and easily observed), while others are **latent** (unintended and less obvious).

B. Conflict Theory:
   1. **Conflict theory** emphasizes the centrality of conflict in social life and incorporates the following features:
      a. Focus on large, macrolevel structures, such as relations between classes.
      b. Shows how major patterns of inequality in society produce social stability in some circumstances and social change in others.
      c. Stresses how members of privileged groups try to maintain their advantages while subordinate groups struggle to increase theirs. Thus social conditions at a given time are the expression of an ongoing power struggle between privileged and subordinate groups.
      d. Suggests that eliminating privilege will lower the level of conflict and increase total human welfare.
   2. Marx
      a. A generation before and radically different from Durkheim, **Karl Marx** originated conflict theory in his observation of the destitution and discontent produced by the Industrial Revolution.

      b. **Class conflict**, the struggle between classes lies at the center of Marx's ideas.

      c. Marx argued that owners or capitalists, in their attempt to improve the efficiency of work and earn higher profits, concentrate workers in larger establishments, keep wages low, and invest little in improving working conditions.

      d. Marx felt that workers would become aware of their exploited class, referred to as **class consciousness,** encourage trade union and labor party organization, and eventually replace private ownership of property with an economic system based on shared property ownership—that is a "communist" society.

  3. Weber

      a. A generation after Marx, **Max Weber** pointed out major flaws in Marx's argument.

      b. Weber noted that growth of the "service" sector with its many nonmanual workers and professionals.

      c. Weber showed that class conflict is not the only driving force of history, but rather politics and religion are also important sources of historical change.

  4. Du Bois and Mills

      a. As the first African American to receive a Ph.D. from Harvard who wrote at the end of the 19th century, **W.E.B. Du Bois** (1868-1963) was an early advocate of conflict theory, and conducted pioneering studies of race in the United States.

      b. He was a founder of the National Association for the Advancement of Colored People (NAACP) and the country's second Department of Sociology at Atlanta University in 1897.

      c. In his best-known work, *The Philadelphia Negro,* Du bois illustrates that poverty and other social problems faced by African Americans were not due to some "natural" inferiority, but to white prejudice.

      d. C. Wright Mills did much to lay the foundations of modern conflict theory in the United States in the 1950s.

      e. In one of his most important books, *The Power Elite,* Mills argued that power is highly concentrated in American society, which is thus less of a democracy than we are led to believe.

C. Symbolic Interactionism

  1. Weber, Mead, and Goffman

      a. In addition to his contribution to conflict theory, Weber also contributed the idea of subjective meanings and motives to sociological theory, which is central to symbolic interactionism. For instance, Weber concluded that capitalism did not develop in the context of economic forces alone, but rather depended on the religious meaning people attached to their work such as the **Protestant ethic** (the belief that religious doubts can be reduced, and a state of grace assured, if people work diligently and live ascetically).

      b. Weber referred to his research emphasis on the importance of understanding people's motives and the meanings they attach to things, as the method of *Verstehen* ("understanding" in German).

      c. Emphasis on subjective meanings found rich soil in the 19th and 20th century United States, therefore much of early American sociology focused on the connection between the individual and the larger society.

d. The work of **George Herbert Mead** (1863-1931) and his colleagues gave birth to symbolic interactionism.

e. **Symbolic interactionism**, coined by Mead's student at the University of Chicago Herbert Blumer (1900-1986), incorporates the following features:

   i. Focuses on face-to-face interaction.

   ii. Emphasizes the subjective meanings people attach to their social circumstances as the basis of social behavior.

   iii. Stresses that people help to create their social circumstances, and do not merely react to them. For instance, **Erving Goffman** (1922-1982) analyzed the many ways people present themselves to others in everyday life to appear in the best possible light, and compared social interaction to a carefully staged play, with a front stage, backstage, defined roles, and props.

   iv. Validates unpopular and nonofficial viewpoints, which increases our understanding and tolerance of people different from us.

2. Social Constructionism

   a. As a variant of symbolic interactionism, **social constructionism** argues that when people interact, they assume things are naturally what they seem to be.

   b. Yet, apparently natural features of life are often sustained by *social* processes that vary historically and culturally.

   c. The study of the subjective side of social life deepens our understanding of how society works, and supplementing the insights of macrolevel analysis.

D. Feminist Theory

1. Although few women figured prominently in the early history of sociology due to the lack of opportunity for women in the larger society in the 19th century, there were a few exceptional women who introduced *gender issues* that were largely ignored by Marx, Durkheim, Weber, Mead, and other early sociologists.

2. Martineau and Addams

   a. **Harriet Martineau** (1802-76), often called "the first woman sociologist," and one of the first feminists, translated Comte into English, wrote one of the first books on research methods, undertook critical studies of slavery and factory laws and gender inequality, and advocated voting rights and higher education for women, and gender equality in the family.

   b. **Jane Addams** (1860-1935), a recipient of the Nobel Prize in 1931, co-founded Hull House (a shelter for the destitute in Chicago's slums) and spent a lifetime focusing on social reform.

      i. Through Hull House, Addams provided a research platform for sociologists at the University of Chicago.

3. Modern Feminism

   a. Since the mid-1960s, feminist theory has had a big influence on sociology. The various strands of **feminist theory** share the following characteristics:

      i. Focuses on various aspects of *patriarchy,* the system of male domination in society, which is at least as important as class inequality in determining a person's opportunities.

      ii. Holds that male domination and female subordination are determined by structures of power and social convention, not biological necessity.

      iii. Examines the operation of patriarchy in both micro and macro settings.

      iv. Contends that existing patterns of gender inequality can and should be changed for the benefit of all members of society.

V. APPLYING THE FOUR THEORETICAL PERSPECTIVES: THE PROBLEM OF FASHION

  A. From "bare midriffs" to "neo-grunge."

  B. *Functionalism* on the ebb and flow of fashion trends.

  C. *Conflict theory* as an alternative view of the fashion world.

  D. Thinking of dress as a form of *symbolic interaction.*

  E. *Feminist* interpretations of Britney Spears.

VI. A SOCIOLOGICAL COMPASS

  A. Sociological founders attempted to solve the great sociological puzzles of their time, the causes and consequences of the Industrial Revolution.
    1. What are the great sociological puzzles of *our* time?
    2. How are today's sociologists responding to the challenges presented by the social settings in which *we* live?

  B. Today, sociology is a heterogeneous enterprise made up of hundreds of theoretical debates.

  C. The Postindustrial Revolution and globalization are perhaps the greatest sociological puzzles of our time.
    1. The **Postindustrial Revolution** is the technology-driven shift from manufacturing to services industries – the shift from employment in factories to employment in offices – and the consequences of that shift for nearly all human activities.
    2. **Globalization** is the process by which formerly separate economies, states, and cultures are becoming tied together and people are becoming increasingly aware of their growing interdependence.

  D. Equality versus Inequality of Opportunity
    1. Optimists suggest that postindustrialism will provide more "equality of opportunity," that is better chances for all people to get a good education, influence government policy, and find good jobs.
    2. The "dark underside" of postindustrialism involves growing inequalities between wealthy and poor Americans, wealthy and poor nations, women and men, and regarding racism and discrimination.

  E. Individual Freedom versus Individual Constraint
    1. Many people are now freer to choose who they want to be with respect to their identities and social relationships.
    2. Increased freedom is experienced only within certain limits and social diversity is limited by pressures of conformity.

  F. Where Do You Fit In?

      1. Rather than being apathetic, people are inclined to look for ways to improve their lives, and this historical period is full of opportunities to do so.

      2. It seems possible to erode some of the inequalities.

G. Careers in Sociology

      1. Sociology offers some useful advice on how to achieve these goals, for sociology is not just an intellectual exercise, but also an applied science, with practical uses, especially in teaching and **public policy** (the creation of laws and regulations by organizations and governments).

      2. Sociology degrees improve one's understanding of the diverse social conditions affecting people, and therefore are well suited to jobs that require good "people skills".

      3. Sociology has benefits for people who do not work as sociologists, such as enabling the connection between the personal and the social-structural.

Student Activity: Applying the Sociological Compass

After reading the chapter, studying the key terms and outline, re-read the section "A Sociological Compass." Identify and write down the key points of the "unresolved issues" and dilemmas. Write down possible questions and solutions from the points of view of functionalism, conflict theory, symbolic interactionism and feminist theory.

Equality Versus Inequality of Opportunity

Key Points:

Pressing Questions:

Functionalism:

1.

2.

3.

Conflict:

1.

2.

3.

Symbolic Interactionism:

1.

2.

3.

Feminist Theory

1.

2.

3.

Potential Solutions:

Functionalism:

1.

2.

3.

Conflict:

1.

2.

3.

Symbolic Interactionism:

1.

2.

3.

Feminist Theory

1.

2.

3.

Individual Freedom Versus Individual Constraint

Key Points:

Pressing Questions:

Functionalism:

1.

2.

3.

Conflict:

1.

2.

3.

Symbolic Interactionism:

1.

2.

3.

Feminist Theory

1.

2.

3.

Potential Solutions:

Functionalism:

1.

2.

3.

Conflict:

1.

2.

3.

Symbolic Interactionism:

1.

2.

3.

Feminist Theory

1.

2.

3.

InfoTrac College Edition Online Exercises

For the following exercises, log on to the online library of InfoTrac College Edition at http://www.infotrac-college.com/. Make note that InfoTrac has implemented a new registration system that will allow easier access to InfoTrac through the use of a personalized username and password. Once you've created your username and password you may proceed directly to the Log On page. To create an account, register your passcode packaged with your textbook, and create a username and password, by following the online prompt. After you are logged in, click on "Infotrac College Edition." You will arrive at a screen that enables you to search topics.

Keyword: **social conflict**. Choose an article that discusses any example of social conflict of your interest. After reading the article, apply the functionalist and social conflict sociological perspectives to the content in the article, by contrasting how a functionalist versus a conflict theorist would explain the sources of conflict in the article.

Keyword: **symbolic interactionism**. Identify two specific topics or issues from at least two articles. Explain how these two topics can be explored from a symbolic interactionist perspective.

Keyword: **feminist sociology**. Identify three challenges feminist sociologists have faced in the past and face in the present, based on at least two of the articles.

Keyword: **theoretical paradigm**. Look for articles that deal with sociology. Why is it important to have theoretical paradigms? How does this relate to the sociological perspectives presented in this chapters. Explain the similarities and differences between those presented in the book and those found on the InfoTrac search.

Keyword: **critical thinking**. Look for articles that discuss what the characteristics of critical thinkers and how to improve or develop critical thinking skills. Do sociologists use critical thinking or just common sense?

Keyword: **nonverbal communication**. Find articles on how nonverbal communication influences our social interpretations. Is nonverbal communication reliable and accurate? Are there any gender differences in nonverbal communication?

Keyword: **virtual reality**. Look for articles that discuss how virtual reality influences our social experiences. How does this relate to sociology?

Keyword: **e-shopping**. There are many articles that deal with attitudes towards e-shopping. How has e-shopping changed the social interaction of shopping in general? For example, is there less social interaction when people e-shop?

Keyword: **ecstasy**. Look for articles that provide information on how ecstasy is used socially. Is there anything specifically geared towards college students? What does this tell us about our methods of social interaction?

Internet Exercises

For a brief introduction to different sociologists and theorists visit the Work and Society: Sociology of Work and Occupation website at:

http://www.was.pe.kr/english1.htm

Click on the links for sociologists. Which sociologists and theorists are discussed. Write a short paper discussing how Internet list compares to the sociologists and theorists that are presented in Chapter 1.

Visit the Annual Review of Sociology official website. This annual publication gives cutting edge@ articles on the different subfields within sociology. Go to: http://www.annurev.org/.

Read three different articles in the most recent Annual Review of Sociology and then write a short paper on how each article uses the sociology theories presented in Chapter 1.

For an overview of how sociology differs from common sense visit these search engines:

http://www.metacrawler.com/ and http://www.megacrawler.com/

Write a small paper on how sociology differs from common sense.

Visit the official website for the American Sociological Association (ASA).

http://www.asanet.org/. ASA is the largest professional organization of sociologists in the United States. What issues does the ASA cover? What resources are available for students of sociology?

Practice Tests

Fill-In-The-Blank

Fill in the blank with the appropriate term from the above list of "key terms."

1. As a variant of symbolic interactionism, _____ argues that when people interact, they assume things are naturally what they seem to be.

2. _____ functions are unintended and less obvious.

3. Disruptive consequences of social structures are referred to as _____ .

4. The _____ which began about 1750, involved the citizens of the United States, France and other countries broadening their participation in government, and that human intervention can thus resolve social problems.

5. _____ is the process of carefully observing social reality, often to "test" theories.

6. _____ are patterns of social relations that lie outside and above your circle of intimates and acquaintances.

7. The _____ which began in Europe about 1550, encouraged the view that sound conclusions about society must be based on solid evidence, not just speculation.

8. The _____ is the belief that religious doubts can be reduced, and a state of grace assured, if people work diligently and live ascetically.

9. According to Durkheim, the degree to which group members share beliefs, and the intensity and frequency of their interaction is referred to as _____ .

10. _____ is the struggle between classes.

11. _____ is the process of carefully observing social reality, often to "test" a theory or assess its validity.

Multiple-Choice Questions

Select the response that best answers the question or completes the statement:

1. Sociology is the social scientific study of human behavior in:
   a. the context of personality development.
   b. a social context.
   c. a psychological context.
   d. the context of human development.
   e. a biological context.

2. Which term refers to the degree to which group members share beliefs and the intensity and frequency of interaction.
   a. social forces
   b. social context
   c. social causes
   d. social solidarity
   e. social conflict

3. According to Durkheim's Theory of Suicide, as the level of social solidarity increases, the suicide rate:
   a. increases
   b. declines
   c. remains constant
   d. All of these are correct.
   e. None of the above.

4. Among males, which age cohort seemed to be at a particularly high risk to commit suicide in 2002?
   a. 85+
   b. 60-64
   c. 20-24
   d. 10-14
   e. 15-19

5. Which of the following is NOT a level of social structure?
   a. microstructures
   b. macrostructures
   c. global structures
   d. innate structures
   e. All of the answer choices are levels of social structure.

6. The sociological imagination was born in the context of the following revolutions EXCEPT:
   a. the Scientific Revolution
   b. the Civil Rights Revolution
   c. the Democratic Revolution
   d. the Industrial Revolution
   e. Women's liberation in the United States

7. Which social revolution is responsible for posing new sets of unresolved issues?
   a. the Scientific Revolution
   b. the Democratic Revolution
   c. the Industrial Revolution
   d. the Postindustrial Revolution
   e. None of the above

8. Who coined the term "sociology?"
   a. Auguste Comte
   b. Emile Durkheim
   c. Karl Marx
   d. Max Weber
   e. W.E.B. Du Bois

9. Functionalists refer to disruptive consequences of society produced by social structures as:
   a. manifest functions
   b. latent functions
   c. equilibrium
   d. dysfunctions
   e. social tensions

10. Which American sociological pioneer played a major role in the study of race and ethnic relations in the United States?
    a. Talcott Parsons
    b. Robert Merton
    c. W.E.B. Du Bois
    d. George Herbert Mead
    e. Emile Durkheim

11. In 2002, the world's developing countries paid nearly _____ times more in interest on loans than they received in official aid.
    a. three
    b. seven
    c. ten
    d. fifteen
    e. twenty-three

12. Which term supports the idea that capitalism did not develop in the context of economic forces alone, but also depended on the meaning people attached to their work?
    a. the Protestant ethic
    b. class consciousness
    c. industrialization
    d. postindustrialism
    e. proletariat

13. Which of the following statements is NOT associated with symbolic interactionism?
    a. focuses on large, macrolevel structures.
    b. focuses on face-to-face interaction.
    c. emphasizes subjective meanings people attach to their social circumstances.
    d. stresses that people help create their social circumstances.
    e. All of the answer choices are associated with symbolic interactionism.

14. Which is NOT a characteristic of feminist theory?
    a. focuses on various aspects of patriarchy.
    b. holds that male domination and female subordination are determined by structures of power.
    c. examines the operations of patriarchy in both micro and macro settings.
    d. contends that existing patterns of gender inequality should be changed exclusively for the benefit of women.

15. Who was among the first early sociologists to translate Comte's writing into English?
    a. W.E.B. Du Bois
    b. Karl Marx
    c. Harriet Martineau
    d. Robert Merton
    e. Emile Durkheim

16. Which sociological theoretical tradition found especially rich soil in the 19th and 20th century United States?
    a. functionalism
    b. conflict theory
    c. symbolic interactionism
    d. feminist theory
    e. All of these are correct.

17. Who noted the growth of the "service" sector with its many nonmanual workers and professionals?
    a. Max Weber
    b. Harriet Martineau
    c. Karl Marx
    d. Emile Durkheim
    e. Herbert Spencer

18. The process by which formerly separate economies, states, and cultures are being tied together is called:
    a. commodification
    b. globalization
    c. consolidation
    d. interdependence
    e. communism

19. Brym and Lie refer to the ability of sociology to sketch the main unresolved issues confronting the postindustrial era as:
    a. thin soup with uncertain ingredients
    b. a sociological stepping stone
    c. a sociological compass
    d. a sociological looking-glass
    e. the sociological imagination

20. Which set of unresolved issues do Brym and Lie address in Chapter 1?
    a. technological innovation versus stagnation
    b. the ongoing conflict of Catholicism and Protestantism
    c. equality versus inequality of opportunity
    d. modernism versus postmodernism
    e. None of the above

21. Which type of sociologists assess the impact of particular policies and programs before and after they go into effect?
    a. clinical sociologists
    b. evaluation researchers
    c. Marxist sociologists
    d. sociological researchers
    e. None of the above

22. Which sociological perspective is especially concerned with the stable patterns of social relations?
    a. functionalism
    b. conflict
    c. symbolic interactionism
    d. feminist theory
    e. All of these are correct.

23. Which types of social patterns involve intimate social relations?
    a. microstructures
    b. macrostructures
    c. mesostructures
    d. global structures
    e. institutional structures

True False

1. Durkheim's research findings on suicide in the 1800s in France are outdated and inapplicable to here and now, since contemporary U.S. society has changed substantially since Durkheim's time.

   TRUE or FALSE

2. Strong social bonds increase the probability that a person will commit suicide if adversity strikes.

   TRUE or FALSE

3. The Democratic Revolutions proved that people could replace unsatisfactory rulers and control society.

   TRUE or FALSE

4. Because theories are part of social science, values have little impact on theoretical development.

   TRUE or FALSE

5. Since sociology has long been change-oriented, women and gender issues have always been at the center of sociological theory.

   TRUE or FALSE

6. Although Max Weber's contribution to conflict theory is big, his ideas also contributed to symbolic interactionism.

   TRUE or FALSE

7. Because women were excluded in the early history of sociology, the work of Harriet Martineau has had little impact on sociology today.

   TRUE or FALSE

8. In contrast to industrialization, sociologists agree that postindustrialism promises abundant opportunities to enhance the quality of life with few social barriers.

   TRUE or FALSE

9. Sociology is not only an intellectual exercise, but also a practical applied science.

    TRUE or FALSE

10. A key contribution of sociological theories is their ability to connect the personal with the social-structural.

    TRUE or FALSE

11. From a sociological perspective, future opportunities and constraints associated with postindustrialism are clear and resolvable.

    TRUE or FALSE

12. Social-structures have a significant impact on the innermost feelings and actions of individuals according to sociologists.

    TRUE or FALSE

Short Answer

1. Briefly explain why sociology is a unique way of explaining social life, in spite of the fact that sociology is in some respects like other disciplines such as economics, political science, philosophy, drama, and abnormal psychology. (p. 1-5, 26-29)

    _____

    _____

2. Briefly explain the difference between microstructures and macrostructures. (p. 6)

    _____

    _____

3. What is the "sociological imagination"? (p. 7)

    _____

    _____

4. Illustrate the difference between manifest and latent functions by providing an example of each function. (p. 14)

    _____

    _____

5. Explain the relationship between theory, research and values. (p. 11−13)

    _____

    _____

6. What do Brym and Lie mean by "a sociological compass"? (p. 23-25)

    _____

    _____

Essay Questions

1. Contrast functionalism, conflict theory, and symbolic interactionism. Be sure to identify the key features of each sociological tradition. (p. 13–20)

2. Explain why suicide is not merely an antisocial, nonsocial, and private act according to Durkheim. (p. 3–5)

3. Compare and contrast Karl Marx's and Max Weber's perspectives on social inequality and conflict. (p. 14–15)

4. Contrast the perspectives of American sociologists Talcott Parsons and George H. Mead. (p. 13-14, 17)

5. Explain the transformation of society from industrialization to postindustrialism. (p. 23–26)

Solutions

Practice Tests

Fill-In-The-Blank

1. social constructionism
2. Latent
3. dysfunctional
4. Democratic Revolution
5. Research
6. Macrostructures
7. Scientific Revolution
8. Protestant Ethic
9. social solidarity
10. Class conflict
11. Research

Multiple-Choice Questions

1. B, (p. 2)
2. D, (p. 4)
3. B, (p. 3-5)
4. A, (p. 5)
5. D, (p. 6-7)
6. B, (p. 9-11)
7. D, (p. 23)
8. A, (p. 11)
9. D, (p. 14)
10. C, (p. 15-16)
11. B, (p. 7)
12. A, (p. 16)
13. A, (p. 17)
14. D, (p. 19-20)
15. C, (p. 19)
16. C, (p. 16-18)
17. A, (p. 15)
18. B, (p. 23-24)
19. C, (p. 23)
20. C, (p. 25)
21. B, (p. 28)
22. A, (p. 13-14)
23. A, (p. 6)

True False

1. F, (p. 3-5)
2. F, (p. 3-5)
3. T, (p. 10)
4. F, (p. 12-13)
5. F, (p. 18-20)
6. T, (p. 16-17)
7. F, (p. 18-20)
8. F, (p. 23-26)
9. T, (p. 27-29)
10. T, (p. 20-23)
11. F, (p. 23-26)
12. T, (p. 20-23)

HOW SOCIOLOGISTS DO RESEARCH

Student Learning Objectives

After reading Chapter 2, you should be able to:

1. Explain the role of objectivity and subjectivity in sociology.

2. Distinguish between scientific and nonscientific thinking.

3. Identify and describe the seven steps of the research cycle.

4. Describe the main methods of collecting sociological data.

5. Explain the advantages and disadvantages of participant observation, experiments, surveys, and the analysis of existing documents in terms of the methodological issues.

6. Explain the issue of causality and association in sociological research.

Key Terms

field research (38)

ethnographic research (39)

participant observation (40)

exploratory research (41)

hypotheses (41)

variables (42)

operationalization (42)

reliability (42)

validity (43)

generalizability (44)

causality (44)

experiment (44)

randomization (44)

experimental group (45)

control group (45)

dependent variable (45)

independent variable (45)

surveys (47)

sample (47)

population (47)

representative sample (47)

probability samples (48)

statistical significance (49)

contingency tables (51)

association (52)   spurious association (53)

control variable (52)

Detailed Chapter Outline

I. SCIENCE AND EXPERIENCE

   A. OTTFFSSENT
1. Experience can prevent people from seeing things.
2. Experience filters perceptions of reality, which is the biggest problem for sociological research. Filtering occurs in four stages:
   a. *Values* or real-life experiences and passions of sociologists motivate research.
   b. Sociologists favor certain *theories* for explaining social problems.
   c. *Previous research* influence sociologists' interpretations.
   d. *Methods* sociologists use to gather data mold our perceptions.
3. Objectivity and subjectivity each play an important role in science, including sociology. Objectivity is a reality check; subjectivity leads us to define which aspects of reality are worth checking.
4. We can never be purely objective, yet we can use techniques to minimize bias, which results in more accurate perceptions of reality.
5. Subjective values and passions are sources of creativity.

   B. Scientific Versus Nonscientific Thinking
1. Everyday Life Versus Science
   a. In science, seeing is believing. In everyday life, believing is seeing, as our biases influence our observations.
   b. Scientists, including sociologists, develop ways of collecting, observing, and thinking about evidence that minimize their chance of drawing biased conclusions.
   c. 10 types of non-scientific thinking:
      i. Tradition
      ii. Authority
      iii. Casual observation
      iv. Overgeneralization
      v. Selective observation
      vi. Exception to the rule
      vii. Illogical reasoning
      viii. Ego-defense
      ix. Premature closure of inquiry
      x. Mystification

II. CONDUCTING RESEARCH

   A. The Research Cycle

1. Six steps of the research cycle:
    a. Formulate a research question.
    b. Review existing literature.
    c. Select a research method.
    d. Collect data.
    e. Analyze the data.
    f. Report/publicize the results.
2. Data may be quantitative or qualitative.
    a. *Qualitative* analysis is usually aimed at understanding patterns of social relationships in small scale social settings.
    b. *Quantitative* analysis associates specific social qualities (types and degrees of knowledge, attitudes, and behaviors) with discreet quantities (numbers).

B. Ethical Considerations
1. Researchers must be mindful of and have respect for their *subjects' rights*:
    a. Right to safety
    b. Right to privacy
    c. Right to confidentiality
    d. Right to informed consent

III. THE MAIN METHODS OF SOCIOLOGY

A. Field Research
1. *Eikoku New Digest* and dinner party rules − an example that by observing others casually, we can get things terribly wrong.
2. Some sociologists undertake **field research,** or research based on the observation of people in their natural settings.
3. *Detached observation* involves classifying and counting the behavior of interest according to a predetermined scheme.
4. Two main problems of direct observation.
    a. *Hawthorne effect -* the presence of the researcher may itself affect the behavior of the people being observed.
    b. The *meaning* of the observed behavior may remain obscure to the researcher.
5. **Ethnographic research** involves spending months and even years living with a people to learn their entire culture, and develop an intimate understanding of their behavior.
    a. In rare cases, ethnographic researchers have "gone native," giving up their research role to become members of the group they are studying.

B. Participant Observation
1. **Participant observation** research involves attempts by researchers to observe a social milieu objectively and take part in the activities of the people they are studying. This allows researchers to achieve a deep and sympathetic understanding of people's beliefs, values, and motives.
2. The Professional Fence
3. Lessons in Method

a. There is a healthy tension between subjectivity and objectivity in participant observation research.
   i. Sociological insight is sharpest when researchers stand both inside (participation) and outside (observation) the lives of their subjects.
   ii. Opting for pure observation and pure participation compromises the researcher's ability to see the world sociologically. Subjectivity can go so far as to compromise objectivity. And objectivity can go too far by compromising the accuracy of inferences about the subjects' behavior.
b. Gaining access:
   i. It is difficult for participant observers to gain access to the groups they wish to study. They must first win the confidence of their subjects.
   ii. *Reactivity* occurs when the researcher's presence influences the subjects' behavior.
c. **Exploratory research** is often the initial step in participant observation studies, - and involves attempts to describe, understand, and develop theory about a social phenomenon in the absence of much previous research on the subject. Researchers at first have a vague sense of what they are looking for.
d. **Hypotheses** are unverified but testable statements about the phenomena in question.
e. Observations constitute sociological data that allow researchers to reject, accept, or modify their initial hypotheses. Thus, researchers often choose their observations, which results in the creation of *grounded theory*, an explanation of a phenomenon based on controlled scrutiny of one's subjects.

C. Methodological Problems
  1. Measurement
    a. Researchers use mental constructs or concepts to think about the world. **Variables** are concepts that can have more than one value.
    b. **Operationalization** involves deciding which observations are to be linked to which variables.
    c. Researchers must establish criteria for assigning values to variables − in other words, make measurement decisions.
  2. Reliability, Validity, Generalizability, and Causality
    a. Since participant-observers usually work alone and investigate one type of group, we must be convinced of three things to accept their findings:
      i. **Reliability** is the degree to which a measurement procedure yields consistent results. *Would another researcher interpret or measure things in the same way?*
      ii. **Validity** is the degree to which a measure actually measures what it is intended to measure. *Are the researcher's interpretations accurate?* Doubts may still exist if the criteria used by the participant observers to assess the validity of their measures are all *internal* to the settings under investigation. Our confidence in the validity of researchers' measures increases if *external* validation criteria are used.
      iii. **Generalizability** exists when research findings apply beyond the specific case examined. **Causality** is the analysis of causes and their effects. Information on how widely or narrowly a research finding applies can help us establish the causes

        of a social phenomenon. Participant observers tend not to think in mechanical, cause-and-effect terms, as they prefer to view their subjects as engaged in a fluid process of social interaction.

    b. In sum, participant observation is useful in exploratory research, constructing grounded theory, creating internally valid measures, and developing a sympathetic understanding of people's perceptions, - and is deficient in establishing reliability, generalizability, and causality.

D. Experiments

    1. An **experiment** is a carefully controlled artificial situation that allows researchers to isolate hypothesized causes and measure their effects precisely. Experiments involve **randomization** (assigning individuals to one of two groups by chance processes) to create two similar groups for comparison.

    2. Research steps of an experiment:

        a. Selection of subjects.

        b. Random assignment of subjects to the **experimental group** (the group exposed to the independent variable) and **control group** (the group that is not exposed to the independent variable).

        c. Randomization and repetition make the experimental and control groups similar, and eliminates bias by allowing a chance process to decide which group each subject is assigned to.

        d. Measurement of the **dependent variable** (the presumed effect in a cause-and-effect relationship) in experimental and control groups.

        e. Introduction of the **independent variable** (the presumed cause in the cause-and-effect relationship).

        f. Remeasurement of the dependent variable in experimental and control groups.

        g. Assessment of experimental effect.

    3. Strengths of experiments:

        a. Precise and repeatable.

        b. High reliability and the ability to establish causality.

    4. Weaknesses of experiments:

        a. Highly artificial situations.

        b. Validity problem.

    5. Experiments on Television Violence

        a. *Hawthorne effect.*

        b. *Power Rangers.*

    6. Field and Natural Experiments

        a. *Field experiments* and *natural experiments* are experiments in natural settings that attempt to overcome validity problems and still retain the benefits of experimental design.

E. Surveys

    1. Sampling

a. **Surveys** ask people questions about their knowledge, attitudes, or behavior, either in face-to-face or telephone interview or in a paper-and-pencil format.

b. Survey studies examine part of a group called a **sample** (the part of the population of research interest that is selected for analysis), in order to learn about the whole group, referred to as the **population** (the entire group which the researcher wishes to generalize).

c. Types of samples:

   i. *Voluntary response samples* involve a group of people who choose themselves, which is unlikely to be representative of the population of interest.

   ii. A **representative sample** is a group of people chosen so their characteristics closely match those of the population of interest.

   iii. *Convenience samples* choose people who are easiest to reach, thus is also highly unlikely to be representative.

   iv. **Probability samples** choose respondents at random. The individual's chance of being chosen must be known and be greater than zero. This enhances generalizability and representativeness within known margins of error.

d. A *sampling frame*, which is a list of all the people in the population of interest, is needed to draw a probability sample.

e. Once a sampling frame is chosen or created, individuals must be selected by a random chance process.

2. Sample Size and Statistical Significance

   a. Large samples give more precise results than small samples.

   b. For most sociological purposes, a random sample of 1,500 people will give acceptably accurate results.

   c. Larger and more precise systematic samples enhance **statistical significance** (when a finding is unlikely to occur by chance, usually in 19 out of every 20 samples of the same size).

3. Types of Surveys and Interviews

   a. Three main ways to conduct a survey:

      i. *Self-administered questionnaire* (i.e. mail questionnaire) is a relatively inexpensive, but sometimes results in unacceptably *low response rates* (the number of people who answer the questionnaire divided by the number of people asked to do so).

      ii. *Face-to-face interviews* involve the presentation of questions and allowable responses to the respondent by the interviewer during a meeting. They are very expensive (i.e. training and paying interviewers).

      iii. *Telephone interviews* elicit relatively high response rates and are relatively cheap to administer.

   b. Survey questions and validity:

      i. A *closed-ended question* provides respondents with a list of permitted answers.

      ii. *Open-ended questions* allow respondents to answer questions in their own words.

   c. Threats to validity in survey research that researchers attempt to alleviate:

      i. Undercounting

      ii. Nonresponse

          iii. Response bias

          iv. Wording effects

          v. The "Palm Beach effect" refers to unclear and ambiguous response categories.

4. Causality

    a. Although a survey is not ideal for conducting exploratory research, survey data are useful for discovering relationships among variables, including cause-and-effect relationships.

    b. Like experiments, the effects of independent variables can be measured. However, the effects of irrelevant variables are removed not by randomization but by manipulating the survey data.

5. Reading Tables

    a. **Contingency tables** (a cross-classification of cases by at least two variables) are one of the most useful tools for manipulating survey data.

    b. **Association** between variables - when the value of one variable changes with the value of the other. Association does not itself prove causality.

    c. Examining the association of two variables within the category of a third variable, referred to as a **control variable**, which allows researchers to manipulate the data to remove the effect of a control variable from the original association.

    d. A **spurious association** exists between an independent variable and a dependent variable when the introduction of control variable makes the initial association disappear. To conclude that the association between an independent variable and a dependent variable is nonspurious or causal, three conditions must hold:

        i. There must be an association between the two variables.

        ii. The presumed cause must occur before the presumed effect.

        iii. When a control variable is introduced, the original association must not disappear.

F. Analysis of Existing Documents and Official Statistics

1. Three types of existing documents sociologists have analyzed:

    a. Diaries

    b. Newspapers

    c. Published historical works.

2. The most frequently used sources of official statistics:

    a. Census data

    b. Police crime reports (i.e. *Uniform Crime Reports*)

    c. Records of key life events.

3. Four main advantages of existing documents and official statistics:

    a. Can save researchers time and money.

    b. Official statistics usually cover entire populations and are collected using rigorous and uniform methods, thus yielding high quality data.

    c. Useful for historical and comparative analysis.

    d. Reactivity is not a problem.

4. One big disadvantage:

    a. Often contain biases that reflect the interests of the individuals and organizations that created the data, - and these data sources are not created with the researcher's needs in mind.

IV. THE IMPORTANCE OF BEING SUBJECTIVE
  A. Subjectivity leads, objectivity follows. Objective sociological knowledge has been enhanced as a result of subjective experiences.
  B. Feminism: A Case in Point
    1. *Subjectivity led* - Feminism as a political movement brought many concerns to the attention of the American public, such as the division of labor in the household, violence against women, the effects of child-rearing responsibilities on women's careers, and the social barriers to women's participation in politics and the armed forces.
    2. *Objectivity followed.* As a result, large parts of the sociological community began doing rigorous research on feminist-inspired issues and greatly refined our knowledge about them. Thus male-centeredness is now less common than it used to be.

Study Activity: Applying the Sociological Compass

After reading the chapter, review the strengths and weaknesses of participant observation, experiments, surveys, and the use of official statistics in terms of operationalization, reliability, validity, generalizability, and their appropriateness for different kinds of research problems. Consider the use of each research method for the following scenario:

The Research Scenario:

It is debatable whether or not delinquency is on the rise or has leveled-off in recent years. Imagine that you are a sociologist interested in studying the prevalence and causes of juvenile delinquency. In light of the debates surrounding delinquency rates and causes of delinquency today, consider each research method mentioned above. Briefly describe how you might carry out a research project and collect data utilizing each method (*design*). Write down the *advantages* and *disadvantages* for each method with respect to the research in question.

<u>Participant Observation</u>

Design:

Advantages:

1.

2.

3.

Disadvantages:

1.

2.

3.

Experiments

Design:

Advantages:

1.

2.

3.

Disdvantages:

1.

2.

3.

Surveys

Design

Advantages:

1.

2.

3.

Disadvantages:

1.

2.

3.

Use of Official Statistics

Design:

Advantages:

1.

2.

3.

Disadvantages:

1.

2.

3.

InfoTrac College Edition Online Exercises

For the following exercises, log on to the online library of InfoTrac College Edition at http://www.infotrac-college.com/. Make note that InfoTrac has implemented a new registration system that will allow easier access to InfoTrac through the use of a personalized username and password. Once you've created your username and password you may proceed directly to the Log On page. To create an account, register your passcode packaged with your textbook, and create a username and password, by following the online prompt. After you are logged in, click on "Infotrac College Edition." You will arrive at a screen that enables you to search topics.

Consider a sociological topic or social problem of your interest, such as child abuse, hate crimes, household division of labor, date rape, attitudes toward sex, etc. Search the topic using infotrac to locate an academic article that involves reporting of social scientific research analyses and findings. Research publications customarily contain the following parts – introduction, theory, method, results/finding, and conclusion. You will have to search a bit, and skim through a number of articles to determine if the article reports research findings. Once you have located a suitable article of interest, read the article with the following questions in mind. Answer the questions for your article.

1. What is the pressing issue, topic, or research question studied in the article?

2. What research method(s) is utilized?

3. What are the steps in the process of data collection?

4. What are the major research findings derived from the method(s) used?

5. What are the strengths of the method(s) in terms of the article?

6. How might you design a stronger research project on the topic in question?

7. Does the author(s) offer any suggestions for subsequent research on this topic?

Keyword: **Hawthorne effect**. Look for articles that deal with this phenomenon. How does this phenomenon relate to sociological research? What is the Hawthorne effect? What type of research articles are there about this topic?

Keyword: **research ethics**. Select an article on research ethics and compare it to the American Sociological Association (ASA) Code of Ethics. Are there any similarities and differences? For example, you will find some things on the American Psychological Association (APA), the area of criminal justice and also for dealing with human subjects.

Keyword: **e-research**. The Internet is a growing source of research. Look for articles dealing with validity and reliability. How reliable is the research information provided. How can the accuracy and reliability of research information be assessed? In your opinion, do you think other students take the Internet as a "very" reliable source?

Keyword: **experiment**. Select one of the experimental studies in the data base to see what steps were used in that experiment. How do the steps in the experimental you chose compare with the discussion of experiments in your textbook?

Internet Exercises

Visit this website http://aspe.os.dhhs.gov/poverty/ and click on the document 2001 HHS Poverty Guidelines to see what operational definition the U.S. government currently uses to ascertain if an individual lives in poverty according to the governmental definition. How does this definition different from your own definition of poverty?

Go to the official website for the popular Gallup Poll at http://www.gallup.com/ and click on how Gallup polls are conducted for an overview of the research method and design the Gallup Poll institution uses. How does the research method and design used by the Gallup Poll compare to the research methods you learned from this chapter? What are the similarities? What are the differences?

Every professional organization that conducts research has a code of ethics or written guidelines on conducting research. Read the Code of Ethics for the American Sociological Association at this site: http://asanet.org/page.ww?section=Ethics&name=Ethics

Do you think there are any ethical concerns that have been left out? If yes, what?

Practice Tests

Fill-In-The-Blank

Fill in the blank with the appropriate term from the above list of "key terms."

1. A (An) _____ group is not exposed to the independent variable in an experiment.

2. Examining the association of two variables within the category of a third variable, is referred to as a (an) _____ .

3. _____ is the degree to which a measurement procedure yields consistent results.

4. A (An) _____ is a group of people chosen so their characteristics closely match those of the population of interest.

5. _____ research involves attempts by researchers to observe a social milieu objectively and take part in the activities of the people they are studying.

6. _____ ask people questions about their knowledge, attitudes, or behavior, either in face-to-face or telephone interview or in a paper-and-pencil format.

7. A (An) _____ is a carefully controlled artificial situation that allows researchers to isolate hypothesized causes and precisely measure their effects.

8. _____ exists when research findings apply beyond the specific case examined.

9. _____ involves deciding which observations are to be linked to which variables.

10. A cross-classification of cases by at least two variables is referred to as _____ .

Multiple-Choice Questions

Select the response that best answers the question or completes the statement:

1. One of the biggest problems for sociological research is that experience filters one's perception of reality, and therefore prevents people from seeing things. Which of the following is NOT one of the sources of filtering identified by Brym and Lie?
   a. values
   b. theories
   c. previous research
   d. methods
   e. All of the answers are sources of filtering according to Brym and Lie.

2. Which of the following is a type of non-scientific thinking?
   a. tradition
   b. authority
   c. casual observation
   d. overgeneralization
   e. All of the answers are types of non-scientific thinking.

3. What is the first step in the research cycle?
   a. Review existing literature.
   b. Formulate a research question.
   c. Select a research method.
   d. Report the results.
   e. Analyze the data.

4. Which research subject right specifies that researchers cannot use data in a way that allows them to be traced to a particular subject?
   a. Right to safety.
   b. Right to privacy.
   c. Right to confidentiality.
   d. Right to informed consent.
   e. Right to a fair trial.

5. Which type of research method involves attempts by researchers to observe a social milieu objectively and take part in the activities of the people they are studying?
   a. participant observation
   b. experiments
   c. surveys
   d. analysis of existing documents
   e. analysis of official statistics

6. _____ is often the initial step in participant observation research.
   a. Review existing literature.
   b. Analyze the data
   c. Select a research method
   d. Report the results
   e. Exploratory research

7. _____ are unverified but testable statements about the phenomenon in question.
   a. Variables
   b. Operationalizations
   c. Hypotheses
   d. Validity tests
   e. Reliability tests

8. _____ is the degree to which a measurement yields consistent results, while _____ is the degree to which a measure actually measures what it is intended to measure.
    a. Validity; reliability
    b. Reliability; validity
    c. Generalizability; validity
    d. Reliability; generalizability
    e. Validity; generalizability

9. In an experiment, which group is created through random assignment?
    a. experimental group
    b. control group
    c. variable group
    d. both 'a' and 'c'
    e. both 'a' and 'b'

10. Which type of variable is the presumed effect in the cause-and-effect relationship?
    a. dependent variable
    b. independent variable
    c. control variable
    d. the third variable
    e. spurious variable

11. Which of the following is NOT a weakness of experiments?
    a. repeatability
    b. highly artificial situations
    c. Hawthorne effect
    d. validity
    e. generalizability

12. Which of the following is a threat to the validity of survey research?
    a. undercounting
    b. nonresponse
    c. response bias
    d. wording effects
    e. All of these are correct

13. Which type of sample chooses people who are easiest to reach?
    a. representative sample
    b. convenience sample
    c. probability sample
    d. random sample
    e. sampling frame

14. _____ occurs when the value of one variable changes with the value of the other.
    a. association
    b. causality
    c. contingency
    d. spuriousness
    e. control

15. Which of the following is NOT an official statistic?
    a. census data
    b. police crime reports
    c. diaries
    d. records of key life events.
    e. All of the above are official statistics.

16. Perhaps the first major advance in modern medicine took place when doctors:
    a. stopped relying on religious faith to treat patients.
    b. stopped using unproven interventions in their treatment of patients.
    c. began considering sociological researching their treatment of patients.
    d. All of these are correct.
    e. None of the above.

17. Examining the association of two variables within the category of a third variable involves:
    a. the analysis of a contingency table
    b. the disappearance of a spurious association
    c. the removal of the effect of a control variable from the original association
    d. the manipulation of data to avoid the "Palm Beach effect."
    e. None of the above.

18. The social institution of science makes ideas _____, and subjects them to careful scrutiny.
    a. unknown
    b. private
    c. public
    d. abstract
    e. None of the above.

19. As the saying in sociological research goes, "_____ leads, _____ follows."
    a. objectivity; subjectivity
    b. subjectivity; objectivity
    c. objectivity; common sense
    d. common sense; subjectivity
    e. None of the above.

20. Which effect involves changing the research subject's behavior by making them aware they are being studied?
    a. field effect
    b. control effect
    c. Palm Beach effect
    d. validity effect
    e. Hawthorne effect

21. A _____ of scholars thinks it is possible to examine data without preconceived notions and then formulate theories.
    a. minority
    b. majority
    c. plurality
    d. consensus
    e. All of these are correct.

22. Which step is first in the research cycle?
    a. report results
    b. collect data
    c. select method
    d. formulate question
    e. review existing literature

23. Which is a main advantage of using existing documents and official statistics in sociological research?
    a. can save researchers money
    b. useful for historical and comparative analysis
    c. reactivity is not a problem
    d. None of the above.
    e. All of these are correct.

24. Which is a main way to administer a survey?
    a. self-administered
    b. face-to-face interviews
    c. telephone interviews
    d. None of the above.
    e. All of these are correct.

25. Lillian Rubin's acclaimed book, *Families on the Faultline*, incorporates:
    a. participant observation research.
    b. experimental research.
    c. analysis of existing crime data.
    d. All of these are correct.
    e. None of the above.

26. Researchers collect information using _____ by asking people in a representative sample as set of identical questions.
    a. field experiments
    b. laboratory experiments
    c. participant observations
    d. ethnographies
    e. surveys

27. If homeless people are not counted in the census, public policy:
    a. will ignore them.
    b. will consider their needs nonetheless.
    c. will require a recollection of data to include them.
    d. All of these are correct.
    e. None of the above.

28. One of the most useful tools for manipulating survey data is the:
    a. computer
    b. internet
    c. contingency table
    d. graph
    e. rubric

True False

1. Subjectivity has no place in social science, since the goal of science is objectivity.

   TRUE or FALSE

2. In conducting social scientific research, we can never be purely objective, yet we can use techniques to minimize bias.

   TRUE or FALSE

3. Albert Einstein sometimes ignored evidence in favor of pet theories.

   TRUE or FALSE

4. A major weakness of participant observation research is that this method lacks depth of analysis.

   TRUE or FALSE

5. Variables are concepts that have only one value.

   TRUE or FALSE

6. Experiments are always conducted in laboratory settings, otherwise it is not an experiment.

   TRUE or FALSE

7. Surveys and questionnaires often incorporate both closed-ended and open-ended questions.

   TRUE or FALSE

8. In survey research, smaller and more precise samples enhance statistical significance.

   TRUE or FALSE

9. Open-ended questions allow respondents to answer questions in their own words.

   TRUE or FALSE

10. A sample is the entire group in which the researcher wishes to generalize.

    TRUE or FALSE

11. Research involves taking the plunge from speculation to testing ideas against evidence.

    TRUE or FALSE

12. The methods used by researcher Alfred Kinsey in the film *Kinsey* were primitive and biased by modern sociological standards.

    TRUE or FALSE

Short Answer

1. Briefly explain the difference between scientific and nonscientific thinking. (p. 34-36)

   _____

   _____

2. Draw the figure from Chapter 2 of the Brym and Lie textbook that illustrates the six steps of the research cycle. (p. 36, fig. 2.2))

   _____

   _____

3. List and state the rights of research subjects. (p. 37-38)

   _____

   _____

4. Briefly explain the difference between validity and reliability. (p. 42-43)

   _____

   _____

5. List four advantages of analyzing existing documents and official statistics for sociological researchers. (p. 54-58)

   _____

   _____

Essay Questions

1. Compare and contrast participant observation, experiments, and surveys. In so doing, be sure to identify the major strengths and weaknesses of each research method. (p. 38-58)

2. Although association does not itself prove causality, explain the role of association, control variables, and spurious association in determining causality. (p. 51-54)

3. Explain the statement "Subjectivity leads, objectivity follows," by describing the impact of feminism on sociological research. (p. 58-59) Explain the stages of research involved in investigating a social issue or problem that has some interest to you. (p. 32-59) Describe the groundbreaking research and research methods of Alfred Kinsey and explain why modern sociologists consider some of his methods primitive and biased. Describe the effects of research projects such as Kinsey's on today's society and sociological research. (p. 56-57)

Solutions

Practice Tests

Fill-In-The-Blank

1. control
2. control variable
3. Reliability
4. representative sample
5. Participant observation
6. Surveys
7. experiment
8. Generalizability
9. Operationalization
10. contingency tables

Multiple-Choice Questions

1. E, (p. 34)
2. E, (p. 34-36)
3. B, (p. 36)
4. C, (p. 37-38)
5. A, (p. 40)
6. E, (p. 41)
7. C, (p. 41)
8. B, (p. 42-43)
9. E, (p. 45)
10. A, (p. 45)
11. B, (p. 44-47)
12. E, (p. 50-51)
13. B, (p. 47-48)
14. A, (p. 52)
15. C, (p. 54-58)
16. B, (p. 35)
17. C, (p. 52)
18. C, (p. 36)
19. B, (p. 58)
20. E, (p. 39)
21. A, (p. 36-37)
22. D, (p. 36)
23. E, (p. 54-58)
24. E, (p. 47-54)
25. A, (p. 39)
26. E, (p. 47)
27. A, (p. 48)
28. C, (p. 51)

True False

1. F, (p. 58-59)
2. T, (p. 33-34)
3. T, (p. 36)
4. F, (p. 40-42)
5. F, (p. 42)
6. F, (p. 44-47)
7. T, (p. 50)
8. F, (p. 48-49)
9. T, (p. 50)
10. F, (p. 47)
11. T, (p. 32-36)
12. T, (p. 56-57)

CULTURE

Student Learning Objectives

After reading Chapter 3, you should be able to:

1. Identify functions of culture to solve real-life problems.

2. Describe the origins of culture in terms of the three main tools of human survival.

3. Critique biological and evolutionary psychological perspectives on culture from a sociological point of view.

4. Discuss the problems of language in terms of nature and nurture.

5. Explain the Sapir-Whorf thesis.

6. Explain the role of ethnocentrism, insider and outsider cultural perspectives on our sociological understanding of culture.

7. Contrast the notions of culture as freedom and culture as constraint.

8. Explain increasing cultural diversity in the contexts of globalization and postmodernism.

9. Explain the phenomena of rationalization and consumerism as sources of cultural constraint.

10. Identify aspects of postmodern culture that distinguish postmodernity from modernity.

11. Describe the cases of heavy metal and hip-hop as examples of the negative consequences of consumerism that limits expressions of freedom and change.

Key Terms

culture (64)

high culture (64)

popular or mass culture (64)

society (64)

abstraction (65)

symbols (65)

cooperation (65)

norms (65)

production (65)

material culture (65)

Detailed Chapter Outline

I. CULTURE AS PROBLEM SOLVING

   A. **Culture** broadly defined refers to all the ideas, practices, and material objects that people create to deal with real-life problems.

   B. Examples of real-life problems that culture helps us deal with:
      1. Ease anxiety to maintain focus on tasks at hand (i.e. professional baseball players, college students writing a final exam).
      2. How to build houses and plant.
      3. Facing problems of death gives meaning to life.

   C. Shared culture is *socially* transmitted. **Society** is a number of people who interact, usually in a defined territory, and share a culture.

   D. Origins of Culture
      1. Three main tools in human cultural survival kits:
         a. **Abstraction** refers to the capacity to create general *ideas*, or ways of thinking that are not linked to particular instances, - and enables humans to learn and transmit knowledge in a way no other animal can.
            i.   **Symbols** are a type of idea that refers to things that carry particular meanings (i.e. languages, mathematical notations, and signs).
         b. **Cooperation** is the capacity to create a complex social life, which is accomplished by establishing **norms**, or generally accepted ways of doing things.
         c. **Production** involves making and using tools and techniques, known as **material culture**, that improve our ability to take what we want from nature. **Nonmaterial culture** is composed of symbols, norms, and other nontangible elements of culture.
      2. **Sanctions** are the rewards and punishments aimed at ensuring conformity to cultural guidelines. The system of **social control** is the sum of sanctions in society. **Taboos**, **mores** and **folkways** are norms that define cultural guidelines.

II. CULTURE AND BIOLOGY

    A. The Evolution of Human Behavior

        1. Although every sociologist recognizes that biology sets broad human limits and potentials, including the potential to create culture, the majority of sociologists disagree with the claim that genes account for specific behaviors and social practices.

        2. Misconceptions of evolutionary psychology:

            a. The starting point of evolutionary psychology is *Charles Darwin's theory of evolution*, which argues that the species members who are best adapted to their environments ("fittest") are most likely to live long enough to have offspring, and the species characteristics that endure are those that increase survival chances.

        3. Male Promiscuity, Female Fidelity, and other Myths

            a. Contemporary evolutionary psychologists make similar arguments:

                i. They identify a supposedly universal human behavioral trait.

                ii. They offer an explanation as to why this behavior increases survival chances.

                iii. The behavior in question is assumed to be unchangeable.

            b. Criticisms of the reasoning of evolutionary psychologists by most sociologists and many biologists and psychologists:

                i. First, some behaviors discussed by evolutionary psychologists are not universal and some are not even that common. For example, according to 2000 GSS data, there is no *universal* propensity to male promiscuity. .

                ii. Second, the arguments that specific behaviors and social arrangements are associated with specific genes, has never been verified.

                iii. Finally, even if researchers discover an association between particular genes and particular behaviors, it would be wrong to conclude that variations among people are due just to their genes, because genes never develop without environmental influence.

    B. Language and the Sapir-Whorf Thesis

        1. Is Language Innate or Learned?

            a. A **language** is a system of symbols strung together to communicate thought.

            b. MIT cognitive scientist Steven Pinker concludes that language is not so much learned as it is grown. Should we believe him?

        2. The Social Roots of Language

            a. What is sociologically interesting is not whether we are biologically wired to acquire language and create speech patterns, but rather how the social environment gives these predispositions form. Our biological potential must be unlocked by the social environment to be fully realized.

            b. All language is learned even though our potential is rooted in biology.

        3. The Sapir-Whorf Thesis and its Critics

            a. The proposition that experience, thought, and language interacts is known as the **Sapir-Whorf thesis.**

            b. Whorf saw speech patterns as "interpretations of experience."

            c. In what sense does language *in and of itself* influence the way we experience the world?

    i. By the 1970s, researchers concluded that speakers of different languages did not perceive color differently.

    ii. In the 1980s and 1990s, researchers found some effects of language on perception.

  d. Biological thinking about culture has benefits and dangers.

    i. Helps see the broad limits and potentials of human creativity.

    ii. "Biological straightjacket" – the failure to appreciate the impact of the social environment.

III. CULTURE AND ETHNOCENTRISM: A FUNCTIONALIST ANALYSIS OF CULTURE

  A. Culture is often invisible, because people tend to take their own culture for granted. Yet people tend to be startled when confronted by other cultures. In other words, culture is clearly visible from the margins.

  B. **Ethnocentrism** is judging another culture exclusively by the standards of one's own culture (i.e. Western views of cow worship among Hindu peasants in India). Ethnocentrism impairs sociological analysis. If you refrain from taking your own culture for granted and ethnocentrism, you will have taken the important first steps towards a sociological understanding of culture.

IV. THE TWO FACES OF CULTURE: FREEDOM AND CONSTRAINT

  A. Two faces of culture:

    1. Culture as freedom

    2. Culture as constraint

  B. Culture as Freedom

    1. Cultural Production and Symbolic Interactionism

      a. Culture is like an *independent* variable. People do not just accept culture passively as previously argued by many sociologists until the 1960s.

      b. We actively produce and interpret culture, creatively fashioning it to suit our own needs, shaping our cultural *environments*.

    2. Cultural Diversity

      a. American society, like most societies in the world, is undergoing rapid cultural diversification, partly due to a high rate of immigration, which is evident in every aspect of life.

      b. Cultural diversity has become a source of political conflict, especially in the debates concerning curricula in the American educational system.

        i. Until recent years, the American educational system stressed the common elements of American culture, history, and society, as told from the perspective of European settlers overcoming great odds.

        ii. School curricula typically neglected the contributions of non-whites and non-Europeans to America's historical, literary, artistic, and scientific development, - and neglected the less savory aspects of American history, many of which involved the use of force to create a strict racial hierarchy that persists today.

    3. Multiculturalism

    a. Advocates of **multiculturalism** argue that the curricula of America's public schools and colleges should present a more balanced picture of American history, culture, and society that reflects the country's ethnic and racial diversity and recognizes the equality of all cultures.

    b. Three criticisms of multiculturalism:

       i. Belief that multicultural education hurts minority students by forcing them to spend too much time on noncore subjects.

      ii. Belief that multicultural education causes political disunity and results in more interethnic and interracial conflict.

     iii. Complaint that multiculturalism encourages the growth of **cultural relativism**, which is the belief that all cultures and all cultural practices have equal value. This criticism lies in the assumption that some cultures oppose the most deeply held values of most Americans.

4. The Rights Revolution: A Conflict Analysis of Culture

    a. Social life is an ongoing struggle between more and less advantaged groups.

    b. The process by which socially excluded groups have struggled to win equal rights under the law and in practice is referred to as the **rights revolution.**

    c. The rights revolution was in full swing by the 1960s.

    d. The rights revolution fragments American culture by:

       i. Legitimizing the grievances of groups that were formerly excluded from full social participation and;

      ii. Renewing their pride in their identity and heritage.

5. From Diversity to Globalization

    a. In *tribal societies*, culture tends to be homogeneous and organized based on **rites of passage** - cultural ceremonies that mark the transition from one stage of life to another.

    b. *Preindustrial Western Europe and North America* were culturally fragmented by artistic, religious, scientific, and political forces.

    c. *Industrialization* further fragmented culture, as the variety of occupational roles grew and new political and intellectual movements evolved.

    d. In the *postindustrial era*, cultural fragmentation is quickening in pace due to globalization, the process by which formerly separate economies, states, and cultures are being tied together.

    e. Roots of globalization:

       i. Expansion of international trade and investment.

      ii. Migration and sustained contact among different racial and ethnic groups.

     iii. Transnational organizations are multiplying.

     iv. Globalization of mass media.

6. The Rise of English and the Decline of Indigenous Languages

    a. Spread of English language as a good indicator of the influence and extent of globalization.

    b. "Japlish" in Japan.

7. Aspects of Postmodernism

a. **Postmodernism** is characterized as an eclectic mixing of cultural elements and the erosion of consensus.

b. Some sociologists believe that this new term is needed to characterize the increasing cultural fragmentation and reconfiguration of our times.

c. Postmodernity is a distinct era following *modernity* – the era marked by the last half of the 19th century and the first half of the 20th century, and characterized by the belief in the inevitability of progress, respect for authority, and consensus around core values.

d. Aspects of postmodernism:

   i. Blending cultures: An eclectic mixing of elements from different times and places.

   ii. The erosion of authority.

   iii. Instability of American core values.

   iv. Voting patterns.

   v. Big Historical Projects.

C. Culture as Constraint

1. Value Change in the United States and Globally

a. The World Value Survey (WVS) is the premier source of information on patterns, persistence and change in values.

b. Two Value Dimensions:

   i. *The traditional/modern value dimensions.*

   ii. *The materialist/postmaterialist value dimension.*

c. Two sets of social forces – economic and religious – influence where countries are located on the two value dimensions.

d. Level of economic development and direction of economic change are the main factors along the materialist-postmaterialist value dimension; and religion is the main factor that distinguishes them along the traditional-modern value dimension.

e. Value Persistence

   i. Values are not randomly distributed across populations, and people are not free to choose whatever values they want; instead clustering occurs along identifiable dimensions.

   ii. Values cluster because they are influenced by powerful social forces.

f. The WVS findings say something about **cultural lag**, or the tendency of symbolic culture to change more slowly than material culture.

   i. Cultural lag is less of an explanation, than many sociologists have suggested, and rather an invitation to search for additional variables that might account for the lag.

2. Regulation of Time

a. According to Max Weber, **rationalization** refers to the application of the most efficient means to achieve given goals, and the unintended, negative consequences of doing so.

b. In the 14th century, *Werkglocken* marked the beginning of the workday, the timing of meals, and quitting time.

3. Consumerism

    a. **Consumerism** is the tendency to define ourselves in terms of the goods we purchase. (i.e. the style of clothing and shoes you wear, and the display of clothing labels according to hip apparel advertisements).

    b. We can choose to purchase items that define us as members of a particular **subculture**, adherents of a set of distinctive values, norms, and practices within a larger culture.

    c. Regardless of our individual tastes, nearly all of us are good consumers.

  4. From Counterculture to Subculture

    a. Unintended negative consequences of consumerism:

      i. Excessive consumption tames countercultures. **Countercultures** are subversive subcultures.

      ii. They oppose dominant values and try to replace them.

      iii. Countercultures rarely pose a serious threat to social stability.

    b. Two examples from popular music subcultures:

      i. Ozzy Osbourne - from the "Prince of Darkness" to *The Osbournes*

      ii. The development of Hip-Hop.

Study Activity: Applying the Sociological Compass

After reading the chapter, re-read the last section of the chapter, "From Counterculture to Subculture", and review the functionalist and conflict sociological perspectives in Chapter 1. Accomplish the following tasks:

List the major social and political conditions in which heavy metal and hip-hop music were born.

1.

2.

3.

4.

Based on the conflict perspective, list the aspects of heavy metal and hip-hop as an expression of subcultural revolt and freedom.

1.

2.

3.

Based on the functionalist perspective, list the social forces that "tamed" the rebellious edge of heavy metal and hip-hop, and transformed each music subculture into a lucrative mainstream commodity.

1.

2.

3.

InfoTrac College Edition Online Exercises

For the following exercises, log on to the online library of InfoTrac College Edition at http://www.infotrac-college.com/. Make note that InfoTrac has implemented a new registration system that will allow easier access to InfoTrac through the use of a personalized username and password. Once you've created your username and password you may proceed directly to the Log On page. To create an account, register your passcode packaged with your textbook, and create a username and password, by following the online prompt. After you are logged in, click on "Infotrac College Edition." You will arrive at a screen that enables you to search topics.

Keyword: **multicultural education** and/or **multiculturalism**. Identify at least one article that supports multiculturalism, and at least one article that critiques multiculturalism. After reading the articles, summarize the pros and cons of multiculturalism based on the articles. Identify the arguments from the articles that are consistent with Chapter 3 of the textbook.

Keyword: **deaf culture**. Look at the article called Transcending Revolutions: The Tsars, The Soviets, and Deaf Culture by Susan Burch. What issues does she present about deaf culture?

How similar or different are your comparisons of deaf culture and non-deaf culture?

Keyword: **tattoo parlor**. The tattoo has increased in popularity over recent years. Is there a specific culture that exists in the tattoo parlor? How does this culture compare to the larger American culture?

Keyword: **dowry**. What are the issues and concerns for being against the dowry? Research these by looking through the articles presented. What type of culture supports the notion of the dowry?

Keyword: **Holy War**. Recent events in the United States has caused there to be an increase awareness of Jihad: The Holy War for Muslims. Research the origins of this social phenomenon from a cultural standpoint. What sociological perspectives can you apply to this social phenomenon?

Internet Exercises

Go to: http://www2.uchicago.edu/jnl-pub-cult for access to the home page of Public Culture: Study for Transnational Culture. You can browse and surf endlessly about culture. Look up some of your favorite topics. What types of topics are included in public culture? Are there some topics that you think should be included? If yes, what are they?

Read an article titled Jihad vs. McWorld at: http://www.theatlantic.com/politics/foreign/barberf.htm written by Benjamin R. Barber. It discusses how the global culture is becoming more homogeneous. What points does Barber make to support his hypothesis? How does this relate to material covered in Chapter 3?

Link to: http://www.h-net.org/~pcaaca/pca/pcahistory.htm the official website of the Popular Culture Association. You can research the history of the popular culture movement, read about the history of the American Culture Association and even link to academic discussion groups on culture. Make a list of five things that you learned from this short exercise.

Practice Tests

Fill-In-The-Blank

Fill in the blank with the appropriate term from the above list of "key terms."

1. The system of _____ is the sum of sanctions in society.

2. _____ is the tendency to define ourselves in terms of the goods we purchase?

3. A _____ is a system of symbols strung together to communicate thought.

4. _____ is a number of people who interact, usually in a defined territory, and share a culture.

5. Advocates of _____ argue that the curricula of America's public schools and colleges should present a more balanced picture of American history, culture, and society that reflects the country's racial and ethnic diversity.

6. _____ broadly defined refers to all the ideas, practices, and material objects that people create to deal with real-life problems.

7. Cultural ceremonies that mark the transition from one stage of life to another is referred to as _____ .

8. _____ is judging another culture exclusively by the standards of one's own culture.

9. The belief that all cultures and practices have equal value is referred to as _____ .

10. The process by which socially excluded groups have struggled to win equal rights under the law and in practice is referred to as the _____ .

Multiple-Choice Questions

Select the response that best answers the question or completes the statement:

1. Which of the following is NOT an example of real-life problems that culture helps people deal with in society?
   a. culture helps ease anxiety to maintain focus on tasks at hand.
   b. culture teaches how to build houses and plant.
   c. culture helps people face the problem of death.
   d. culture gives meaning to life.
   e. All of the answers are examples of real-life problems that culture helps people deal with.

2. Which is NOT a main tool of the human cultural survival kit?
   a. abstraction
   b. cooperation
   c. production
   d. consumerism
   e. All of these are correct.

3. What is a sanction?
   a. The use of tools and techniques to improve our ability to take what we want from nature.
   b. The capacity to create complex social life.
   c. Generally accepted ways of doing things.
   d. A type of idea that carries particular meaning.
   e. A reward or punishment aimed at ensuring conformity to cultural guidelines.

4. The majority of sociologists _____ with the claim that genes account for specific behaviors and social practices.
    a. agree
    b. disagree
    c. are neutral
    d. None of the above.
    e. All of these are correct.

5. The species members who are best adapted to their environments are most likely to live long enough to have offspring, and the species characteristics that endure are those that increase the survival chances of that species. This argument is central to which theory?
    a. Darwin's theory of evolution.
    b. Durkheim's theory of suicide.
    c. Weber's theory of rationalization.
    d. Marx's theory of social class.
    e. None of the above.

6. Which is NOT a criticism of evolutionary psychology from a sociological perspective?
    a. Evolutionary psychologists suggest that biology sets broad limits on human behavior.
    b. Many behaviors discussed by evolutionary psychologists are not universal and some are not even that common.
    c. The arguments that specific behaviors and social arrangements are associated with specific genes, has never been verified.
    d. Even if researchers discover an association between particular genes and particular behaviors, it would be wrong to conclude that variations among people are due just to their genes.
    e. Genes never develop without environmental influence.

7. The fact that people tend to get startled when confronted by other cultures suggests that:
    a. People tend to take their own culture for granted.
    b. People tend to be ethnocentric.
    c. Culture is visible from the margins.
    d. All of these are correct.
    e. None of the above.

8. Which of the following is NOT a characteristic of evolutionary psychology?
    a. They identify a supposedly universal human behavioral trait.
    b. They offer an explanation as to why behavior increases survival chances.
    c. They suggest that behavior is best explained in terms of cognition in a social context.
    d. The behavior in question is assumed to be unchangeable.
    e. All of the answers are characteristics of evolutionary psychology.

9. Jim traveled to Cambodia for vacation and became angry and judgmental toward the "foreign" practices. Which term best describes Jim's reaction?
   a. cultural relativism
   b. enculturation
   c. ethnocentrism
   d. passive ethno-bias
   e. subcultural revolt

10. Which is NOT a critique of multicultural education?
   a. Belief that multicultural education hurts minority students by forcing them to spend
   b. too much time on noncore subjects.
   c. Belief that multicultural education causes political disunity.
   d. Belief that multicultural education results in more interethnic and interracial conflict.
   e. Complaint that multiculturalism discourages cultural relativism.
   f. All of the answers are critiques of multicultural education.

11. In the postindustrial era, cultural fragmentation is increasing in pace due to:
   a. globalization.
   b. tribalization.
   c. increased homogeneity.
   d. the incline of consensus.
   e. None of the above.

12. _____ is characterized as an eclectic mixing of cultural elements and the erosion of consensus.
   a. Postmodernism
   b. Modernism
   c. Industrialization
   d. Preindustrialization
   e. Rationalization

13. Which of the following is NOT a source of globalization?
   a. expansion of international trade
   b. expansion of international investment
   c. migration and sustained contact among different ethnic groups
   d. growth of transnational organization
   e. All of the answers are sources of globalization.

14. Which is an aspect of postmodernism?
   a. decline of consensus around core values
   b. belief in the inevitability of progress
   c. respect for authority
   d. homogeneity
   e. All of these are correct.

15. Bureaucracies exemplify _____, because of their level of institutional efficiency towards clear goals.
    a. consumerism
    b. postmodernism
    c. modernism
    d. rationality
    e. rites of passage

16. The idea that experience, thought, and language interacts is know as the:
    a. Sapir-Whorf thesis.
    b. cultural lag.
    c. abstraction.
    d. rationalization.
    e. None of the above.

17. _____ is a number of people who interact, usually in a defined territory, and share a _____.
    a. Culture; society
    b. Material culture; nonmaterial culture
    c. Symbols; language
    d. Society; culture
    e. Language; symbols

18. Which is the best sociological definition of language?
    a. A system of symbols strung together to communicate thought
    b. A means of communication biologically wired into our brains
    c. The biologically rooted human potential to communicate
    d. The vocal sounds that come out of our mouths as we speak
    e. None of the above.

19. American Indian sovereignty movements are examples of:
    a. rites of passage.
    b. material culture.
    c. rights revolution.
    d. taming a subcultural revolt.
    e. None of the above.

20. While _____ culture is consumed mainly by upper classes, _____ culture is consumed by all classes.
    a. popular; mass
    b. mass; popular
    c. mass; high
    d. high; mass
    e. popular; high

21. The functions of genes are:
    a. well known with the advancements in scientific research.
    b. still unknown
    c. well known due to sociological contributions to biology.
    d. All of these are correct.
    e. None of the above.

22. Based on the Sapir-Whorf Thesis model displayed in Chapter 3, which process comes second?
    a. conceptualization
    b. experience
    c. verbalization
    d. All of these are correct.
    e. None of the above.

23. A sociological view of cow worship in India might suggest that this practice is:
    a. bizarre.
    b. irrational.
    c. consumeristic.
    d. taming of subcultural revolt.
    e. rational.

24. The idea of globalization first gained prominence in marketing strategies in the:
    a. 1950s
    b. 1960s
    c. 1970s
    d. 1980s
    e. 1990s

25. Of the 216 countries, more than _____ of them lists English as an official or majority language.
    a. one half
    b. one third
    c. 75%
    d. one fourth
    e. 90%

26. In 2004, English usage accounted for _____ of languages used on the internet.
    a. 25.5%
    b. 35.2%
    c. 75.5%
    d. 85.5%
    e. 15.5%

27. Which is <u>false</u> regarding *Werkglocken*?
    a. It exemplified the rational regulation of time.
    b. It marked the beginning of the workday, the timing of meals, and quitting time.
    c. It means "work clock" in German.
    d. It is an efficient example of culture that frees the human potential.
    e. All of the answers are true.

True False

1. Cooperation is the sum of sanctions in society.

    TRUE or FALSE

2. Due to technological innovation, industrialization lessened the fragmentation of culture.

    TRUE or FALSE

3. The fact that excessive consumption helps protect the environment is a function of consumerism.

    TRUE or FALSE

4. As the variety of jobs grew and the new political movements evolved, industrialization further fragmented culture.

    TRUE or FALSE

5. The need for educational curricula to reflect a more balanced story of America's inventors is an example of multiculturalism.

    TRUE or FALSE

6. Ethnocentrism is the belief that all cultures have equal value.

    TRUE or FALSE

7. Judging an ethnic group on the basis of its own cultural standards is an example of cultural relativism.

    TRUE or FALSE

8. Cultural diversity highlights the idea of culture as constraint.

    TRUE or FALSE

9. Norms are examples of material culture.

    TRUE or FALSE

10. Symbols are core elements of any culture.

    TRUE or FALSE

11. The United States continues to diversify culturally.

    TRUE or FALSE

12. A hallmark of postmodernism is the combining of cultural elements from different times.

    TRUE or FALSE

Short Answer

1. How does culture in general help you solve real-life problems? Identify three specific examples of functions of culture in your personal life. (p. 63-67)

    _____

    _____

2. List and define three main tools in human cultural survival kits. (p. 64-67)

    _____

    _____

3. Briefly explain the relationship between multiculturalism and cultural relativism. (p. 75-76)

    _____

    _____

4. List and describe two dysfunctions of consumerism. (p. 88-90).

    _____

    _____

5. Define Max Weber's notion of "rationalization". (p. 87-88)

    _____

    _____

Essay Questions

1. Explain the dysfunctional aspect of consumerism as obstacles of freedom in the case of Ozzy Osbourne. (p. 90-91)

2. Explain the impact of globalization on the ever-increasing cultural fragmentation of societies. (p. 78-80)

3. Explain the context of postmodern culture in terms of increasing individual freedom. Ironically, postmodern culture also places new constraints on contemporary culture. Contrast the idea of postmodern culture as a constraint versus freedom. (p. 81-84)

4. Explain how rationalization constrains culture by describing its relationship to time. (p. 84-88)

5. Describe the symbolic interactionist view of culture. Explain what is meant by the statement that culture is an *independent variable*. (p. 74)

Solutions

Practice Tests

Fill-In-The-Blank

1. social control
2. Consumerism
3. language
4. Society

5. multiculturalism
6. Culture
7. rites of passage
8. Ethnocentrism

9. cultural relativism
10. rights revolution

Multiple-Choice Questions

1. E, (p. 63-67)
2. D, (p. 64-66)
3. E, (p. 66)
4. B, (p. 67-69)
5. A, (p. 67)
6. A, (p. 67-69)
7. D, (p. 72)
8. C, (p. 67-69)
9. C, (p. 72)

10. E, (p. 75-76)
11. A, (p. 78-80)
12. A, (p. 81)
13. E, (p. 78-80)
14. A, (p. 81-84)
15. D, (p. 86-88)
16. A, (p. 70-71)
17. D, (p. 63-64)
18. A, (p. 69)

19. C, (p. 76-78)
20. D, (p. 64)
21. B, (p. 67-69)
22. A, (p. 71)
23. E, (p. 72)
24. C, (p. 78-80)
25. D, (p. 79-80)
26. B, (p. 76-77)
27. D, (p. 86-87)

True False

1. F, (p. 65)
2. F, (p. 78-80)
3. F, (p. 88-90)
4. T, (p. 71-80)

5. T, (p. 75-76)
6. F, (p. 72)
7. T, (p. 76)
8. F, (p. 75)

9. F, (p. 65)
10. T, (p. 65)
11. T, (p. 75-80)
12. T, (p. 81-84)

SOCIALIZATION

Student Learning Objectives

After reading Chapter 4, you should be able to:

1. Understand the importance of socialization on the development of our human potentials and self-identity.

2. Compare and contrast the theoretical perspectives of Freud, Cooley, Mead, Piaget, Kohlberg, Vygotsky, and Gilligan in terms of their contributions to our understanding of socialization.

3. Describe the significance of the family, schools, peer groups and the mass media as agents of socialization.

4. Explain the necessity of adult socialization.

5. Discuss the contradictory lessons among the agents of socialization that make childhood and adolescence more ambivalent and stressful today.

6. Explain the process of resocialization in the context of initiation rites and total institutions.

7. Describe the social factors that contribute to the fact that people's identities change faster, more often, and more completely, - and explain why the self has become more flexible.

8. Describe historical transformations of childhood and adolescence since the emergence of childhood as a distinct stage of socialization today.

9. Describe current dilemmas and problems for childhood and adolescent socialization today as a result of declining parental supervision, increasing adult responsibilities by youth, and declining participation in extracurricular activities.

Key Terms

socialization (95)

roles (96)

self (97)

id (98)

superego (98)

ego (98)

unconscious (98)

looking-glass self (99)

Detailed Chapter Outline

I. SOCIAL ISOLATION AND SOCIALIZATION

   A. Cases of socially isolated children suggest that the ability to learn culture and become human is only a potential. Socialization enables human potentials to be actualized.

   B. **Socialization** is the process by which people learn their culture (norms, values, and roles) and become aware of themselves as they interact with others.

   C. A **role** is the behavior expected of a person occupying a particular position in society.

   D. Socialization is also a life long process by which people develop a sense of self.

   E. Adolescence is a particularly crucial stage of the crystallization of self-identity, for "the central growth process in adolescence is to define the self through the clarification of experience and to establish self-esteem."

   F. In various institutional settings of socialization, referred to as "agents of socialization," we learn to control our impulses, think of ourselves as members of different groups, value certain ideals, and perform various roles.

II. THEORIES OF CHILDHOOD SOCIALIZATION

   A. Freud
      1. Main sociological contribution – the self emerges during early social interaction and early childhood experience exerts a lasting impact on personality development.
      2. Although infants initially are driven by elemental needs and gratifications, social interaction enables them to begin developing a self-image or sense of **self** – a set of ideas and attitudes about who they are as independent beings.
      3. Freud proposed the first social-scientific interpretation of self-emergence.
      4. Three parts of self:
         a. **Id** is the part of self that demands immediate gratification. Self-image emerges as soon as the id's demands are denied.

    b. **Superego** is the personal conscience, a moral sense of right and wrong and appropriate behavior.

    c. The **Ego**, in well-adjusted individuals, balances the conflicting needs of the pleasure-seeking id and the restraining superego.

5. The superego is a frustrating and painful process of repressed memories of denying the id, referred to as the **unconscious**.

6. Three criticisms of Freud:

    a. The connections between early childhood development and adult personality are more complex than Freud assumed.

    b. Many sociologists criticize Freud for gender bias in his analysis of male and female sexuality.

    c. Sociologists often criticize Freud for neglecting socialization after childhood.

B. Cooley's Symbolic Interactionism

1. American sociologist, Charles Horton Cooley, introduced the idea of the "**looking glass self**." Like our physical body reflected in a mirror, we see our social selves reflected in people's gestures and reactions to us.

    a. When we interact with others, they gesture and react to us.

    b. This allows us to imagine how we appear to them.

    c. We judge how others evaluate us.

    d. From these judgments, we develop a **self-concept** (a set of feelings and ideas about who we are).

C. Mead

1. George Herbert Mead expanded Charles Horton Cooley's idea of the **looking glass self**, which suggests that our feelings about who we are depend on how we see ourselves judged by others.

2. Like Freud, Mead:

    a. Noted the presence of a subjective and impulsive aspect of self from birth, called the **"I"**.

    b. Highlighted importance of social interaction.

    c. Argued that an objective repository of culturally approved standards emerges as part of the self, referred to as the **"me"**.

3. Unlike Freud, Mead:

    a. Focused on the need to see yourself objectively – the *"me"*.

    b. Emphasized the unique human capacity to *take the role of the other* (seeing yourself from the points of view of other people) as the source of the *"me,"* and hence the self (in contrast to Freud's focus on the denial of the id's impulses).

    c. Argued that the self emerges from people using symbols, gradually during social interaction (not traumatic as Freud's ideas).

4. Mead's Four Stages of Development: Role-Taking:

    a. Children learn to use language and other symbols by *imitating* **significant others**, important people in their lives.

    b. Children use their imaginations to pretend and *role-play* in childhood games like "house," "school," and "doctor."

    c. Children learn to play complex games requiring simultaneous role taking with several other people called the **generalized other**, which involves the application of cultural standards to the self.

D. Piaget
1. Jean Piaget's stages of the development of *cognitive* skills:
    a. *Sensorimotor* – Up to 2 years of life, children explore the world exclusively through their 5 senses.
    b. *Preoperational* – Children begin to think symbolically between 2 and 7 years of life, however are still unable to think in the abstract.
    c. *Concrete-Operational* – By about 7 to 11 years of life, children are able to recognize connections between cause and effect.
    d. *Formal Operational* – By about 12 years of life, children develop the ability to think more abstractly and critically.

E. Kohlberg
1. Lawrence Kohlberg's stages of the development of *moral reasoning* also develops through stages:
    a. *Preconventional* – Morality develops on the basis of whether something gratifies children's immediate needs.
    b. *Conventional* – Children understand right and wrong in terms of whether specific actions please their parents and teachers and are consistent with cultural norms.
    c. *Postconventional* – The capacity to think abstractly and critically about moral principles.

F. Vygotsky
1. Lev Vygotsky offered a more sociological approach to thinking, in that ways of thinking depended less on innate characteristics than on the nature of institutions in which people grow up, in other words the structure of society.

G. Gilligan
1. Carol Gilligan, likewise, emphasized the sociological foundations of moral development.
2. She attributed differences in the moral development of boys and girls to the different cultural standards parents and teachers pass on to them.

III. AGENTS OF SOCIALIZATION

A. Families
1. Freud and Mead understood that the family is the most important agent of **primary socialization**, which is the process of mastering the basic skills required to function in society during childhood.
2. Characteristics of the family as an agent of socialization:
    a. Small group
    b. Frequent face-to-face contact
    c. High motivation of parents to care for children, despite prevalence of abuse and neglect.
    d. Enduring influence on individuals over time.

3. The socialization function of the family was more pronounced in preindustrial societies, for many of its functions are performed by other institutions like schools and religion.

B. Schools
   1. Public schools emerged in response to the need for better trained and educated employees by American industry
   2. Responsible for **secondary socialization**, or socialization outside the family after childhood.
   3. In addition to the more obvious role of schools to instruct students in academic and vocational subjects, a **hidden curriculum** teaches students what will be expected of them in the larger society after graduation (to be "good citizens," punctuality, respect for authority, the importance of competition). Many students of poor and racial minority families reject school authorities because they are skeptical about the school's ability to open job opportunities.
   4. **Self-fulfilling prophecies**, expectations that help cause what it predicts, can occur in terms of students' beliefs that school does not lead to economic success; and in terms of teachers' expectations of students.
   5. W.I. Thomas and Dorothy Swaine Thomas had a similar idea known as the **Thomas theorem** – "Situations we define as real become real in their consequences."

C. Peer Groups
   1. **Peer groups** consist of individuals who are not necessarily friends but are about the same age and of similar status. **Status** refers to a recognized social position that an individual can occupy.
   2. Especially influential over lifestyle issues and the formation of an independent identity.
   3. Peer group values are often in conflict with values of the family, but families tend to be more influential and enduring.
   4. Peer groups nonetheless help integrate youth into the larger society.

D. The Mass Media
   1. The mass media include television, radio, movies, videos, CDs, audiotapes, newspapers, magazines, books, and the Internet.
   2. Used for stimulation and entertainment.

E. Gender Roles, the Mass Media, and the Feminist Approach to Socialization
   1. Allow adolescents to engage in **self-socialization**, the choosing of socialization influences from the wide variety of mass media offerings.
   2. People tend to choose media influences that are most widespread, fit cultural standards, and are made appealing by those who control mass media. In other words, not all media influences are created equal.
      a. For example, the social construction of **gender roles** involves widely shared expectations about how males and females are supposed to act.
      b. According to feminist sociologists, people *learn* gender roles, in part through the mass media.
      c. Boys and girls, however, do not passively accept gendered messages, but often interpret them in unique ways and resist them.
   3. Professional Socialization

a. When a person enters the labor force, secondary socialization enters a new phase that may be more or less stressful.

F. Resocialization and Total Institutions

1. **Resocialization** occurs when powerful socializing agents deliberately cause rapid change in people's values, roles, and self-conceptions, sometimes against their will.

   a. Example: staged **initiation rite** ceremonies signify the transition of the individual from one group to another and ensures his or her loyalty to the new group (i.e. fraternity, sorority, the marines, religious order).

   b. Three-stage ceremony:
      i. *Ritual rejection* involves separation from one's old status and identity.
      ii. *Ritual death* involves degradation, disorientation, and stress.
      iii. *Ritual rebirth* is the acceptance of the new group culture and status.

   c. **Total institutions** are settings where much resocialization occurs, and where people are isolated from the larger society and under the strict control and constant supervisions of a specialized staff.

   d. Example: Palo Alto experiment of a mock prison.

IV. SOCIALIZATION THROUGH THE LIFE COURSE

A. Adult socialization is necessary for four main reasons:

1. Adult roles are often discontinuous.
2. Some adults are largely invisible.
3. Some adults are unpredictable.
   a. To learn a predictable new role, we typically engage in **anticipatory socialization**, which involves beginning to take on the norms and behaviors of the role to which we aspire.
4. Adults change as we mature.

B. We see all four reasons in the role of a terminally ill person that is the simultaneous operation of role discontinuity, invisibility, maturation, and unpredictability.

V. THE FLEXIBLE SELF

A. Older sociological perspectives on adult socialization underestimate the flexibility of the self, for today people's identities change faster, more often, and more completely.

B. Factors that contribute to the growing flexibility of the self:

1. Globalization
2. Growing ability to change our bodies and hence our self-conception (i.e. bodybuilding, aerobic exercise, weight reduction, plastic surgery, sex-change operations, organ transplants, and whole-body transplants).
3. Complications of the Internet on the process of identity formation and **virtual communities**.

VI. THE DILEMMAS OF CHILDHOOD AND ADOLESCENT SOCIALIZATION

A. The Emergence of Childhood and Adolescence

1. In preindustrial societies:
   a. Children are perceived as small adults;

  b. Children are put to work as soon as they can contribute to the family;
  c. Achievement of adulthood is common by the age of 15 or 16.
2. In the late 1600s, the idea of childhood as a distinct stage of life emerged among well-to-do Europeans and North Americans, who thought that boys should be allowed to play games and receive an education that would allow them to develop emotionally, physically and intellectually. Girls continued to be viewed as "little women."
3. As we moved into the 20th century, social necessity and possibility influenced the emergence of the idea of childhood.
  a. Prolonged childhood was *necessary* in societies that required better educated adults to do complex work, because children needed a chance to prepare for adult life.
  b. Prolonged childhood was *possible* where improved hygiene and nutrition increased the human life span, - which is more common in wealthier and more complex societies.
B. Problems of Childhood and Adolescent Socialization Today
  1. The experience and meaning of childhood and adolescence seem to be changing radically today. Some analysts wonder if childhood and adolescence is disappearing.
  2. New developments are changing and creating problems for childhood and adolescence.
  a. Declining adult supervision and guidance.
  b. Increasing media influence.
  c. Declining extracurricular activities and increasing adult responsibilities.
  d. "The Vanishing Adolescent".

Study Activity: Applying The Sociological Compass

Think about your experiences as a child, and consider how your childhood was different from your parents' (or guardian figure's) childhood. Contrast your childhood with your parents' (or guardian figure's) childhood by listing the social constraints you faced, and the new freedoms you enjoyed. List some of the social sources for the constraints and freedoms of your childhood. List the contradictions and problems of these "new" constraints and freedoms.

Be mindful of how your experiences affect (or will affect) the way you socialize your children. If you do not plan on having children, think about how you might socialize children you might be in contact with in the future (i.e. nieces and nephews, children in your work setting, etc.).

Childhood Social Constraints

Parent's/Guardian's constraints:

1.

2.

3.

4.

Your constraints:

1.

2.

3.

4.

Childhood Social Freedoms

Parent's/Guardian's freedoms:

1.

2.

3.

4.

Your freedoms:

1.

2.

3.

4.

Social Sources of Your Constraints and Freedoms

1:

2.

3.

4.

Social Problems Associated with Your Constraints and Freedoms

1.

2.

3.

4.

How might your childhood experiences affect how you socialize your children, or the children you may be in contact with?

Infotrac College Edition Online Exercises

For the following exercises, log on to the online library of InfoTrac College Edition at http://www.infotrac-college.com/. Make note that InfoTrac has implemented a new registration system that will allow easier access to InfoTrac through the use of a personalized username and password. Once you've created your username and password you may proceed directly to the Log On page. To create an account, register your passcode packaged with your textbook, and create a username and password, by following the online prompt. After you are logged in, click on "Infotrac College Edition." You will arrive at a screen that enables you to search topics.

Keyword: **adolescence**. Identify and read at least four articles that discuss problems adolescents face today. Read the articles and compile a list of adolescent problems, and a list of good advice that might be useful to you personally.

Keyword: **hidden curriculum**. The education system is one of the agents of socialization. What role does the hidden curriculum play in the socialization process of humans? What role has the hidden curriculum played in your individual development?

Keyword: **cyber-language**. Cyber-language is an emerging language. Look for articles that deal with the impact of cyber-language in our society. What role do you think cyber-language will have on our socialization process?

Keyword: **sociobiology**. Look for articles that discuss the strengths and weaknesses of this argument. Read the articles and compile a list of the strengths and weaknesses of sociobiology.

After reading the evidence presented discuss if you personally think the sociological argument is valid.

Keyword: **cliques**. Look for articles that discuss the role of cliques in adolescent socialization.

What does your research conclude about the role of cliques in adolescent socialization? Explain how cliques would be viewed from the Marx perspective.

Keyword: **emotions**. Look for research on emotions from a global perspective. Find articles that deal with emotions from different cultural perspectives. For example, do some cultures not express emotion publicly?

Keyword: **symbols**. Look for articles that discuss how symbols are a part of the socialization process. Read the articles and compile a list of the ways symbols are part of the socialization process. Do symbols vary from culture to culture in the socialization process?

Internet Exercises

Read what the University of Minnesota's intellectual resources on issues important to children, youth and families are. Go to: http://www1.umn.edu/pres/cyf.html

This is a report prepared by the President's Initiative on Children, Youth and Families. It discusses important concerns in socialization. What key ideas do they suggest for sharing intellectual resources with children and families?

What impact has the National Vietnam Memorial had on American society? For a look at its impact link to http://www.saed.kent.edu/Architronic/v2n2/v2n2.05.html

and read the article titled *Ritual and Monument*. Make a list of 5 ways that this memorial has impacted American society.

The Boy Scouts of America and Girl Scouts of America are important agents of socialization for many boys and girls in American society. Go to: http://www.scouting.org/ and http://www.girlscouts.org/ for their views on socialization. Compare and contrast the purpose of each organization, their sponsors, and where they receive their funding.

Go to: http://www.nbcdi.org/ which is the web site of The National Black Child Development Institute (NBCDI). Their mission statement states they exist "to improve and protect the quality and life of African American children and their families". Look over the web site and discuss what you learn about socialization in a short paragraph.

Practice Tests

Fill-In-The-Blank

Fill in the blank with the appropriate term from the above list of "key terms."

1. _____ occurs when powerful socializing agents deliberately cause rapid change in people's values, roles, and self-conceptions, sometime against their will.

2. Socialization outside the family after childhood is referred to as _____ .

3. _____ is the process by which people learn their culture, and become aware of themselves as they interact with others.

4. According to Freud, the _____ is the personal conscience that provides a moral sense of right and wrong.

5. According to Mead, the presence of a subjective and impulsive aspect of self from birth is referred to as _____ .

6. _____ is a set of feelings and ideas about who we are.

7. Based on Mead's first stage of role taking, children learn by imitating _____ .

8. _____ refers to a recognized social position that an individual can occupy.

9. To learn a predictable new role, individuals engage in _____ , which involves the process of beginning to take on the norms and behaviors of the role to which one aspires.

10. _____ are settings where much resocialization occurs, whereby people are isolated from the larger society under strict control.

Multiple-Choice Questions

Select the response that best answers the question or completes the statement:

1. Cases of extremely socially isolated children described in the book, suggest that:
   a. we are born with our self-conceptions.
   b. the ability to learn culture and become human is only a potential.
   c. we are fully human at birth.
   d. the ability to use culture can develop in social isolation.
   e. All of these are correct.

2. Self-identity crystallizes:
   a. during early childhood.
   b. at birth.
   c. during old age.
   d. during adolescence.
   e. None of the above.

3. Institutional settings of socialization are referred to as:
   a. socialization encounters.
   b. interaction settings..
   c. back-stage socialization regions
   d. agents of socialization.
   e. None of the above.

4. _____ are expected behaviors of a person occupying a particular position in society.
   a. Roles
   b. Self-socialization
   c. Sanctions
   d. Self-conceptions
   e. Self-fulfilling prophecies

5. Which theorist proposed the first social-scientific interpretation of self-emergence?
   a. Freud
   b. Mead
   c. Piaget
   d. Kohlberg
   e. Vygotsky

6. Which is NOT one of Piaget's stages of the development of cognitive skills?
   a. sensorimotor stage
   b. preoperational stage
   c. preconventional stage
   d. formal operational stage
   e. All of the answers are Piaget's stages.

7. According to Mead, children learn to apply cultural standards when they:
   a. learn to role-take with the generalized other.
   b. imitate significant others.
   c. first use their imaginations to role-play.
   d. first acquire the use of language.
   e. All of these are correct.

8. The _____ is the part of the self that demands immediate gratification, according to Freud.
   a. id
   b. ego
   c. superego
   d. me
   e. None of the above

9. Which is a criticism of Freud?
   a. The connections between early childhood development and adult personality are more complex than Freud assumed.
   b. Freud is criticized for gender bias in his analysis of male and female sexuality.
   c. Freud is criticized for neglecting socialization after childhood.
   d. All of these are correct.
   e. None of the above.

10. Which concept is an expansion of Cooley's idea of the "looking-glass self?"
    a. superego
    b. unconscious
    c. cognitive development
    d. moral reasoning
    e. None of the above.

11. The ability of us to see ourselves objectively involves:
    a. the development of the self.
    b. the "me."
    c. taking the role of the other.
    d. All of these are correct.
    e. None of the above.

12. The capacity of children to think abstractly and critically about moral principles develops at which stage of moral reasoning?
    a. sensorimotor stage
    b. preoperational stage
    c. postconventional stage
    d. preconventional stage
    e. conventional stage

13. The _____ is the most important agent of primary socialization.
    a. family
    b. schools
    c. peer groups
    d. mass media
    e. All of these are correct.

14. Which is a characteristic of the family as an agent of socialization?
    a. small group
    b. frequent face-to-face interaction
    c. high motivation of parents to care for children
    d. enduring influence on individuals over time
    e. All of these are correct.

15. Which is an expectation that helps cause what it predicts?
    a. primary socialization
    b. secondary socialization
    c. tertiary socialization
    d. self-fulfilling prophecy
    e. self-socialization

16. _____ is a latent function of schools, because it teaches what will be expected of individuals in the larger society after graduation.
    a. The hidden curriculum
    b. The Thomas theorem
    c. Self-socialization
    d. None of the above.
    e. All of these are correct.

17. The Palo Alto experiment described in the book is an example of:
    a. virtual communities
    b. initiation rites
    c. the hidden curriculum
    d. total institutions
    e. None of the above

18. Which is NOT one of the stages of resocialization ceremonies?
    a. ritual rejection
    b. ritual death
    c. ritual rebirth
    d. ritual connection
    e. All of the answers are stages of resocialization ceremonies.

19. Which agent of socialization is most influential and long lasting to individuals?
    a. family
    b. schools
    c. peer groups
    d. mass media
    e. team sports

20. Which is a factor that contributes to the growing flexibility of the self?
    a. globalization
    b. growth of the domestic economy
    c. declining adult supervision
    d. All of these are correct.
    e. None of the above.

21. The research of Harry and Margaret Harlow on monkeys concluded that:
    a. emotional development requires affectionate cradling.
    b. baby monkeys raised with an artificial mother made of wire mesh without a soft cover were later unable to interact normally with other monkeys.
    c. when baby monkeys interacted with an artificial mother covered with a soft terry cloth, they revealed less emotional distress.
    d. All of these are correct.
    e. None of the above.

22. Sigmund Freud was the founder of:
    a. symbolic interactionism.
    b. psychoanalysis.
    c. the notion of the generalized other.
    d. Gestalt psychology.
    e. social behaviorism.

23. Carol Gilligan has made a major sociological contribution to our understanding of:
    a. childhood development.
    b. internal conversation.
    c. the interplay between the "I" and the "me."
    d. total institutions.
    e. extreme isolation.

24. Which occupational sector do adolescents prefer?
    a. Service including food, cleaning, and personal.
    b. Teachers, counselors, librarians, social workers, and religious workers.
    c. Lawyers and judges.
    d. Engineers, architects, natural and social scientists.
    e. Sales.

25. Gender segregation during schoolyard play:
    a. no longer exists in the United States.
    b. has been erased since the Women's Movement in the United States.
    c. can still be observed in U.S. schools.
    d. All of these are correct.
    e. None of the above.

26. According to the U.S. Department of Health and Human Services, which substance used by Americans age 12-17 showed the sharpest decline in percentage between 1990-1997 and 2003?
    a. Cocaine
    b. Marijuana
    c. Alcohol
    d. Cigarettes
    e. None of the above.

27. Which school age cohort spent the greatest number of hours watching television per day in 2002?
    a. Students with 0-11 years of schooling.
    b. Students with 12 years of schooling.
    c. Students with 13-14 years of schooling.
    d. Students with 15-16 years of schooling.
    e. Students with 17-20 years of schooling.

28. Which was the top waking activity for American women and men ages 18-24 from 1993 to 1995?
    a. television
    b. school
    c. visiting
    d. travel
    e. All of these are correct.

True False

1. Socialization is a process that ends at adulthood.

   TRUE or FALSE

2. According to Mead, the "me" is an objective repository of culturally approved standards that emerges as part of the self.

   TRUE or FALSE

3. According to Piaget, children are able to recognize connections between cause and effect at the sensorimotor stage.

   TRUE or FALSE

4. Primary socialization occurs outside the family after childhood.

   TRUE or FALSE

5. Self-socialization occurs when powerful socializing agents deliberately cause rapid change in people's values, roles, and self-conceptions, some against their will.

   TRUE or FALSE

6. Public schools emerged in response to the need for better-trained and educated employees by American industry.

   TRUE or FALSE

7. In the late 1600s, the idea of childhood as a distinct stage of life emerged among both the rich and poor, and for boys and girls.

   TRUE or FALSE

8. Improved hygiene and nutrition played a role in the emergence of the idea of childhood in the 20th century.

   TRUE or FALSE

9. Children were viewed as small adults during a significant part of human history.

   TRUE or FALSE

10. Although it was believed in the 1600s that boys should be allowed to play games and receive an education that would allow them to develop emotionally, physically and intellectually, girls continued to be treated as "little women."

    TRUE or FALSE

Short Answer

1. Provide an example of a self-fulfilling prophecy in schooling that illustrates the Thomas theorem. (p. 106)

   _____

   _____

2. How does mass media allow adolescents to engage in self-socialization? (p. 109-112)

   _____

   _____

3. List three factors that contribute to the growing flexibility of the self. (p. 115-117)

   _____

   _____

4. List and briefly describe the stages of Kohlberg's theory of the development of moral reasoning. (p. 102)

   _____

   _____

5. Explain the difference between primary and secondary socialization. (p. 103-107)

Essay Questions

1. Describe the central historical factors that led to the notion of childhood as a unique stage of socialization. (p. 117-121)

2. Some social scientists wonder if childhood and adolescence are disappearing today. Explain the rationale of this position by discussing the current sources of problems and dilemmas associated with childhood and adolescence. (p. 117-121)

3. Compare and contrast Freud and Mead's theories of the "self". How is Mead's theory more sociological? (p. 97-101)

4. Explain why feminist sociologists say that people learn gender roles. Describe five possible ways gender roles are learned through the mass media. (p. 109-110)

5. Explain what is meant by professional socialization. (p. 110-112)

Solutions

Practice Tests

Fill-In-The-Blank

1. Resocialization
2. secondary socialization
3. Socialization
4. superego
5. the "I"
6. Self-concept
7. significant others
8. Status
9. anticipatory socialization
10. Total institutions

Multiple-Choice Questions

1. B, (p. 95-96)
2. D, (p. 96-97)
3. D, (p. 103)
4. A, (p. 96)
5. A, (p. 97-98)
6. C, (p. 101)
7. A, (p. 99-101)
8. A, (p. 98)
9. D, (p. 98-99)
10. E, (p. 99)
11. D, (p. 99-101)
12. C, (p. 102)
13. A, (p. 103)
14. E, (p. 103-104)
15. D, (p. 106)
16. A, (p. 105)
17. D, (p. 112-113)
18. D, (p. 112-113)
19. A, (p. 103-104)
20. A, (p. 115-117)
21. D, (p. 96)
22. B, (p. 97)
23. A, (p. 103)
24. D, (p. 105)
25. C, (p. 107)
26. C, (p. 108)
27. B, (p. 110)
28. A, (p. 109)

True False

1. F, (p. 97)
2. T, (p. 99)
3. F, (p. 101)
4. F, (p. 103)
5. F, (p. 109)
6. T, (p. 104-105)
7. F, (p. 118)
8. T, (p. 118)
9. T, (p. 118)
10. T, (p. 118)

SOCIAL INTERACTION

Student Learning Objectives

After reading Chapter 5, you should be able to:

1. Explain the fundamental processes and concepts of social interaction.

2. Describe the historical context of emotions, and the role of emotions in social interaction in terms of emotion management and emotion labor.

3. Identify and explain the three modes of social interaction.

4. Describe the manipulative aspects of social interaction in terms of impression management.

5. Describe the role of verbal and nonverbal communication in the use and acquisition of language in a social and cultural context.

6. Explain the major points of interaction in terms of exchange/rational choice theory, symbolic interactionism, dramaturgical analysis, ethnomethodology, and conflict theory.

7. Discuss the ways of maintaining social interaction in the context of differential power.

8. Explain the relationship between micro-level interactions and higher-level structures.

Key Terms

social interaction (125)

status set (126)

ascribed status (126)

achieved status (126)

master status (126)

role set (126)

role conflict (127)

role strain (128)

emotion management (132)

emotion labor (132)

exchange theory (135)

rational choice theory (135)

dramaturgical analysis (137)

role distance (138)

ethnomethodology (139)

status cues (143)

Detailed Chapter Outline

I. WHAT IS SOCIAL INTERACTION?

   A. **Social interaction** involves people communicating face-to-face, acting and reacting in relation to each other, and is structured around statuses, roles, and norms.

   1. While *status* refers to recognized positions occupied by interacting people (i.e. flight attendants and passengers), a **status set** is the entire ensemble of statuses occupied by an individual (i.e. a flight attendant, a wife, and a mother at the same time. If a status is involuntary, it is an **ascribed status** (i.e. daughter). If it is involuntary, it is an **achieved status** (i.e. flight attendant). One's **master status** is the status that is most influential in shaping one's life over time.

   2. While people *occupy* statuses, they *perform* **roles**, which are sets of expected behaviors. A **role set** is a cluster of roles attached to a single status (i.e. in-flight safety expert and server).

   3. *Norms* are generally accepted ways of doing things that change over time. Some norms are *prescriptive*, whereby they suggest what a person is expected to do while performing a role. Other norms are *proscriptive*, whereby they suggest what a person is *not* to do.

   B. Case Study: Stewardesses and Their Clientele

   1. The changing role of stewardess

   C. The enforcement of norms.

   1. **Role conflict** occurs when two or more statuses held at the same time place contradictory role demands on a person.

   2. **Role strain** occurs when incompatible role demands are placed on a person in a single status.

   D. Change in status.

II. WHAT SHAPES SOCIAL INTERACTION?

   A. Norms, roles and statuses are the building blocks of all face-to-face communication.

   B. These building blocks structure face-to-face interaction.

III. THE SOCIOLOGY OF EMOTIONS

   A. Laughter and Humor

   1. Robert Provine's research on laughing in public places.

   a. Speakers laugh more often than listeners do (79.8 percent vs. 54.7 percent).

   b. Women laugh more than men do, especially when the speaker is a woman, and the listener is a man.

   2. Laugher is unevenly distributed across the status hierarchy.

   a. People with higher status get more laughs, and people with lower status laugh more.

   b. When social interaction takes place between status equals, they often direct their humor at perceived social inferiors.

    i. For example, "put down" jokes have the effect of excluding outsiders, making you feel superior, and reinforcing group norms and the status hierarchy itself.

  3. Disadvantaged people often laugh at the privileged majority, and also laugh at themselves.

  4. The more repressive a government, the more widespread are antigovernment jokes.

  5. Sometimes, humor has a political edge, and sometimes humor has no political content; yet all jokes are "little revolutions" as remarked by George Orwell.

  6. Sociologically, jokes enable us to see the structure of society that lies beneath our laughter, for they suddenly and momentarily let us see beyond the serious, taken-for-granted world.

 B. Emotion Management

  1. We *control* our emotions, for emotions don't just happen to us; we manage them.

  2. When we manage our emotions, we tend to follow cultural "scripts."

  3. Sociologist Arlie Hochschild is one of the leading figures in the study of what she coined **emotion management**, which involves people obeying "feeling rules" and responding appropriately to the situations they find themselves.

   a. Hochschild claims that "women, Protestants, and middle-class people cultivate the habit of suppressing their own feelings more than men, Catholics, and lower-class people do."

 C. Emotion Labor

  1. **Emotion labor** is emotion management that many people do as part of their job and for which they are paid, while emotion management is done by everyone in their personal life.

   a. Different occupations require different types of emotion labor (i.e. the flight attendant vs. the bill collector).

   b. All jobs that require emotion labor have in common the fact that "they allow the employer, through training and supervision, to exercise a degree of control over the emotional activities of employees."

   c. Emotion labor become a commodity, while the emotional life of workers is increasingly governed by organizations, and is therefore less and less spontaneous and authentic.

 D. Emotions in Historical Perspective

  1. Feeling rules take different forms under different social conditions, which vary historically.

  2. Three examples from social history of emotions.

   a. *Grief.*

   b. *Anger.*

   c. *Disgust.*

IV. MODES OF SOCIAL INTERACTION

 A. Interaction as Competition and Exchange

  1. Conversation and exchange:

   a. Conversation typically involves the *exchange* of attention, and can involve subtle competition for attention.

  2. Exchange and Rational Choice Theories

   a. **Exchange theory** proposes that social interaction involves trade in valued resources.

   b. **Rational choice theory** is a variant of exchange theory that focuses less on the resources and more on the way interacting individuals weigh the benefits and costs of interaction.

   c. Types of interaction that cannot be explained in terms of exchange and rational choice theories.

   i. Altruistic and heroic acts.

   ii. When people behave altruistically, they interact on the basis of *norms* that say they should act justly and help others in need, despite the substantial costs of doing so.

B. Interaction as Symbolic

  1. Dramaturgical Analysis: Role-Playing

   a. Haas and Shaffir's study of professional socialization.

   b. *Impression management* is the process of manipulating the way people present themselves so they appear in the best possible light.

   c. Three principles of *symbolic interactionism* according to Herbert Blumer:

   i. "Human beings act toward things on the basis of the meaning these things have for them."

   ii. "The meaning of a thing" emerges from the process of interaction.

   iii. "The use of meanings by the actors occurs through a process of interpretation."

   d. **Dramaturgical analysis** is a distinct approach to symbolic interactionism first developed by Erving Goffman, which views social interaction as a sort of play in which people present themselves so as to appear in the best possible light.

   i. We are constantly engaged in *role-playing*, as most clearly evident when we are "front stage."

   ii. When backstage people can relax, prepare and discuss the frontstage with fellow actors.

   e. If a role is stressful, we may engage in **role distancing**, which involves giving the impression that we are just "going through the motions" but actually lack serious commitment to a role.

  2. Ethnomethodology

   a. **Ethnomethodology** is the study of the methods ordinary people use during everyday interaction, often unconsciously, to make sense of what others do and say.

   b. Everyday interaction could not take place without *pre-existing* shared norms and understandings.

  3. Verbal and Nonverbal Communication

   a. Language is used in its *social and cultural context*. People are assisted in the task of learning the social and cultural contexts of language, and the meanings of words by *nonverbal* cues.

  4. Facial Expressions, Gestures, and Body Language

   a. Social interaction involves a complex mix of verbal and nonverbal messages including facial expressions, gestures, body language, and manipulation of "personal space" or "intimate zones."

   b. *Cosmopolitan* example "How to reduce otherwise evolved men to drooling, panting fools."

  5. Status Cues

   a. **Status cues** are a type of nonverbal communication that involves visual indicators of other people's social position.

      b. Example—Profiling strangers on the basis of skin color, age, gender, companions, clothing, jewelry, and objects to determine their "dangerousness."

      c. Status cues can degenerate into **stereotypes,** rigid views of how members of various groups act, regardless of whether individual group members really behave that way.

C. Power and Conflict Theories of Social Interaction

    1. Summary of four main points of interaction:

      a. Competitive exchange of valued resources is one of the most important forces that cement social interaction (exchange and rational choice theories).

      b. We mold values, norms, roles, and statuses to suit us as we interact with others (symbolic interactionism and dramaturgical analysis).

      c. Norms do not necessarily emerge spontaneously during interaction, for in general form they exist before any given interaction takes place (ethnomethodology).

      d. Nonverbal mechanisms of communication facilitate interaction.

    2. A Fifth Point: Power

      a. **Conflict theories of social interaction** emphasize that when people interact, their statuses are often arranged in a hierarchy, with those at the top enjoying more power.

      b. The degree of inequality affects the character of social interaction between the interacting parties.

      c. According to Weber, **power** is "the probability that one actor within a social relationship will be in a position to carry out his [or her] own will despite resistance." (male-female interaction).

    3. Types of Interaction

      a. **Domination** represents one extreme type of interaction, whereby nearly all power is concentrated in the hands of people of similar status while people of different status enjoy almost no power (i.e. guards vs. inmates, landowners vs. slaves in the antebellum South).

      b. **Cooperation** represents another extreme type of interaction whereby power is more or less equally distributed between people of different status, often based on feelings of trust (i.e. happy marriages where housework and child-care are shared).

      c. **Competition** a type of interaction between the two extremes whereby power is unequally distributed, but the degree of inequality is less that in systems of domination.

        i. Randall Collins' work on cooperative work environments.

V. MICRO, MESO, MACRO, AND GLOBAL STRUCTURES

A. Micro-level interaction often gives rise to higher-level structures.

B. Big structures set limits to the behavior of small structures.

C. It is within small structures, however, that people interpret, negotiate, and modify their immediate social settings, thus giving big structures their dynamism and life.

Study Activity: Applying the Sociological Compass

As you interact in your natural, everyday settings, such as at work, at school, shopping at the mall, etc., pay close attention to your interaction experiences with others, and take notes on your observations. You may recall from Chapter 2 that what you will engage in is an example of participant observation research. While observing, accomplish the following tasks:

Based on your observations and interpretations, make a list of examples of

Emotion labor:

1.

2.

3.

Impression management:

1.

2.

3.

Status cues:

1.

2.

3.

Additionally, during your everyday interactions, behave quite a bit nicer and more giving than usual. Perhaps, offer to help someone with a task, offer your place in line, offer your parking space on campus during rush hour, give a nice complement, and so forth. In other words, engage in altruistic behavior beyond the perceived norms. Behave very nicely and give without expecting anything in return, to the point of "disrupting" the situational norms, despite how people act around you. Make a list of reactions of people to your altruistic acts, as you keep in mind the principles of *ethnomethodology*. In other words, breach interaction by behaving overly altruistic, and list your observations of how others react to you.

1.

2.

3.

4.

Finally, make a list of explanations for: a) the reasons behind others' *frontstage* behaviors, and the reactions of people to your altruistic acts. Base your explanations on any relevant theories and concepts from Chapter 5.

Frontstage behaviors?

1.

2.

3.

Reactions to altruism?

1.

2.

3.

InfoTrac College Edition Online Exercises

For the following exercises, log on to the online library of InfoTrac College Edition at http://www.infotrac-college.com/. Make note that InfoTrac has implemented a new registration system that will allow easier access to InfoTrac through the use of a personalized username and password. Once you've created your username and password you may proceed directly to the Log On page. To create an account, register your passcode packaged with your textbook, and create a username and password, by following the online prompt. After you are logged in, click on "Infotrac College Edition." You will arrive at a screen that enables you to search topics.

Keyword: **dramaturgical analysis**. How is dramaturgical analysis used in research by sociologists? Compile a list of at least 5 ways that sociologists use this approach.

Keywords: **exchange theory** and **rational choice theory**. Identify and read at least four articles that provide research findings from an exchange or rational choice perspective. Read the articles and compile a list of evidence in support of the major principles of exchange and rational choice theories, and a list of criticisms.

Keyword: **community service**. Look for articles on how service learning is related to the community at large? What information do these articles bring to light about the interaction between the individual and the larger society?

Internet Exercises

For a choice example of cyberspace social interaction go to The MUD Connector at
http://www.mudconnect.com/

How is cyberspace social interaction different from face-to-face interaction?

Visit the Society of Social Interaction Web site at http://sun.soci.niu.edu/~sssi. This is a professional
organization for sociologists who study social interaction. Search their web site and compile a list of 10
things that you learned.

To become aware of the emotional and sociological implications of social interaction on the Internet go to:
http://www.levity.com/julian/bungle.html and read the article "A Rape in Cyberspace," by Julian Dibbell.
Make a list of the emotional and sociological implications of social interaction that Dibbell proposes.

Practice Tests

Fill-In-The-Blank

Fill in the blank with the appropriate term from the above list of "key terms."

1. A (An) _____ is the entire ensemble of statuses occupied by an individual.

2. One's _____ is the most influential status in shaping one's life over time.

3. _____ is a behavior people engage in as part of their job.

4. _____ proposes that social interaction involves trade in valued resources.

5. The process of manipulating the way people present themselves so they appear in the most favorable
   light is referred to as _____ .

6. _____ is a distinct approach to symbolic interactionism first developed by Erving Goffman,
   which views social interaction like a theatrical performance.

7. If a role is stressful, we may engage in _____ , which involves giving the impression that we are just "going through the motions" but actually lack serious commitment to the role.

8. _____ represents one extreme type of interaction, whereby nearly all power is concentrated in the hands of people of similar high-powered status.

9. _____ involves people communicating face-to-face, acting and reacting in relation to each other.

Multiple-Choice Questions

Select the response that best answers the question or completes the statement:

1. Interacting individuals:
   a. tend to conform to specific norms.
   b. act according to role expectations.
   c. assume a certain status.
   d. None of the above.
   e. All of these are correct.

2. Role conflict occurs when:
   a. two or more statuses held at the same time place contradictory role demands on an individual.
   b. incompatible role demands are placed on a person in a single status.
   c. individuals have difficulty with frontstage impression management.
   d. individuals are incapable of role distancing.
   e. All of these are correct.

3. Which is NOT a characteristic of interaction according to Goffman's dramaturgical analysis?
   a. We are constantly engaged in role-playing during frontstage interaction.
   b. People can relax in the backstage.
   c. We become aware of our public roles and our "true" selves.
   d. Interaction involves a trade in valued resources.
   e. We engage in role-distance during interaction.

4. _____ focuses less on resources and more on the way interacting individuals weigh the benefits and costs of interaction.
   a. Psychoanalytic theory
   b. Rational choice theory
   c. Symbolic interactionism
   d. Dramaturgical analysis
   e. Ethnomethodology

5. Which type of interaction can NOT be explained in terms of exchange or rational choice theory?
   a. altruism
   b. heroic acts
   c. a conversation between individuals competing for attention
   d. All of these are correct.
   e. a and b

6. Status cues can degenerate into _____ as in the example of racial profiling.
   a. role distance
   b. impression management
   c. backstage behavior
   d. stereotypes
   e. cooperation

7. _____ emphasize the fact that individuals often shape values, norms, roles and statuses to suit them as they interact.
   a. Exchange theorists
   b. Rational choice theorists
   c. Symbolic interactionists
   d. Functionalists
   e. Conflict theorists

8. A _____ is a type of nonverbal communication that involves visual indicators of other people's social position.
   a. role distance
   b. social category
   c. social role
   d. status cue
   e. backstage status

9. Which type of status do we assume involuntarily?
   a. Occupation status
   b. Achieved status
   c. Ascribed status
   d. Status conflict
   e. Status strain

10. Which term refers to a cluster of roles attached to a single status?
    a. Status set
    b. Status cluster
    c. Role achievement
    d. Role strain
    e. Role set

11. Who coined the term "emotion management"?
    a. Harold Garfinkel
    b. Erving Goffman
    c. Herbert Blumer
    d. Arlie Hochschild
    e. William Shaffir

12. Which is NOT one of the modes of interaction discussed in the Brym and Lie textbook?
    a. Interaction as internal meditation.
    b. Interaction as competition and exchange.
    c. Interaction as symbolic.
    d. Power and social interaction.
    e. All of the answers are modes of interaction discussed in the Brym and Lie text.

13. Which of the following statements is true of the role and status of the airline stewardess of the 1950s to the late 1970s?
    a. Stewardesses were not considered sex objects as they are today.
    b. Stewardesses of the 1950s to the late 1970s made relatively high incomes compared to today.
    c. The role of the stewardess of the 1950s to the late 1970s was certainly glamorous.
    d. All of these are correct.
    e. None of the above.

14. Robert Provine's research on conversations of people laughing in public places, found that, in general:
    a. speakers laugh more often than listeners do.
    b. listeners laugh more often than speakers do.
    c. speakers and listeners spend about equal time laughing.
    d. All of these are correct.
    e. None of the above.

15. According to the Brym and Lie textbook, which is true about "put down" jokes?
    a. It excludes outsiders.
    b. It makes one feel superior.
    c. It reinforces group norms and the status hierarchy itself.
    d. All of these are correct.
    e. None of the above.

16. The emotional life of workers seems to be:
    a. more and more spontaneous.
    b. increasingly governed by the organizations they work for.
    c. more and more authentic.
    d. All of these are correct.
    e. None of the above.

17. Which is NOT one of the examples from the social history of emotions discussed in the Brym and Lie text?
    a. Grief
    b. Anger
    c. Disgust
    d. All of these are correct.
    e. None of the above.

18. Who wrote that, "Every joke is a tiny revolution?"
    a. Erving Goffman
    b. Arlie Hochschild
    c. George Orwell
    d. Albert Einstein
    e. Herbert Blumer

19. The multiple difficulty of meeting the role expectations associated with being a mother, flight attendant, and wife is an example of:
    a. role conflict.
    b. role strain.
    c. master status.
    d. status set.
    e. role set.

20. The contradictory roles of stewardesses to be suggestive on the one hand, yet polite on the other hand, is an example of:
    a. role conflict.
    b. role strain.
    c. master status.
    d. status set.
    e. role set.

21. According to Robert Provine's research, the biggest discrepancy in laughing occurs when:
    a. the speaker is a woman and the listener is a man.
    b. the speaker is a man, and the listener is man.
    c. the speaker has greater status than the listener.
    d. All of these are correct.
    e. None of the above.

22. The examples of Chris Rock's on-stage jokes illustrated in the textbook indicate that:
    a. disadvantaged people often laugh at the privileged majority.
    b. disadvantaged people laugh at themselves.
    c. although disadvantaged people often laugh at the privileged majority, they do not always do so.
    d. All of these are correct.
    e. None of the above.

23. Manners in Europe during the Middle Ages were:
    a. quite acceptable by U.S. standards today.
    b. utterly disgusting based on the social norms of the Middle Ages in general.
    c. utterly disgusting by U.S. standards today.
    d. All of these are correct.
    e. None of the above.

24. According to Box 5.2 "Sociology at the Movies" in Chapter 5, Miss Congeniality is:
    a. an unsuccessful film because it fails to provide consumers with an entertaining escape from reality.
    b. an entertaining escape from reality's messiness.
    c. a poor guide to life as we actually live it.
    d. All of these are correct.
    e. Both 'B' and 'C'.

25. Box 5.3 Social Policy: What do You Think? "Allocating Time Fairly in Class Discussions", suggests that:
    a. in the wake of the feminist movements, female college students today tend to dominate class dialogue similar to the way males did in the past.
    b. systematically encouraging female college students to speak up in the classroom discussion is counterproductive to the learning environment.
    c. the realm of interpersonal interaction and conversation is an extremely difficult area in which to impose rules and policies.
    d. systematic allocation of equal time for female and male students to speak in classroom discussion would be relatively easy to institutionalize.
    e. All of these are correct.

True False

1. Interaction as exchange especially emphasizes the use of methods to make sense of what others do and say by ordinary people during everyday interaction.

    TRUE or FALSE

2. A job interview is an example of the frontstage.

    TRUE or FALSE

3. Nonverbal mechanisms of communication facilitate interaction.

    TRUE or FALSE

4. The assumption that the use of meanings by the actors occurs through the process of interaction is a major principle of symbolic interactionism according to Harold Garfinkel.

    TRUE or FALSE

5. Dramaturgical analysis is a distinct approach to symbolic interactionism first developed by Herbert Blumer.

TRUE or FALSE

6. Cooperation is a type of interaction whereby power is unequally distributed, but not to the degree as in a system of domination.

TRUE or FALSE

7. From the late 1950s to the late 1970s, both the role and pay of stewardesses were glamorous.

TRUE or FALSE

8. In social situations where people of different statuses interact, laughter is unevenly distributed across the status hierarchy.

TRUE or FALSE

9. People of higher status tend to laugh more, while people with lower status get more laughs.

TRUE or FALSE

10. Disadvantaged people often laugh at the privileged majority, but not always.

TRUE or FALSE

11. We virtually have no control over our emotions, which is a trademark of human beings as emotional beings.

TRUE or FALSE

12. According to Arlie Hochschild, Protestantism invites people to participate in an inner dialogue with God, whereas Catholicism offers sacraments and confession, which allow and encourage the expression of feeling.

TRUE or FALSE

13. Politicians should be well versed in the vocation of impression management because their success depend heavily on voter's opinions and perceptions of them.

TRUE or FALSE

14. Body language is a rare mode of communication that does not reflect gender roles.

TRUE or FALSE

15. In a relationship marked by cooperation, the imbalance of interpersonal power between superordinate persons and subordinate persons is greater than in relations of competition.

TRUE or FALSE

16. Domination, as a mode of interaction, yields the highest level of inequality and highest level of efficiency.

TRUE or FALSE

Short Answer

1. List and define three ways of maintaining social interaction in the context of power. (p. 144-147)

_____

_____

2. Describe examples of impression management, frontstage behavior, and back stage behavior. (p. 137-139)

_____

_____

3. What is the similarity and difference between the terms "emotion management" and "emotion labor"? (p. 131-133)

_____

_____

4. What is the difference between role conflict and role strain? Provide a brief example for each term. (p. 127-128)

_____

_____

5. Describe an example of role distancing. (p. 138)

_____

_____

Essay Questions

1. Compare and contrast exchange theory and dramaturgical analysis. (p. 135-139)

2. Compare and contrast symbolic interactionism and ethnomethodology in terms of their respective views on the role of social norms during social interaction. (p. 137-140)

3. Explain the cultural evolution of grief, anger, and disgust as examples of the social history of emotions. (p. 133-134)

4. Compare and contrast exchange theory and conflict theories of social interaction. (p. 135-147)

5. Based on readings and what you learned about the changing role of stewardesses in Chapter 5, examine and describe how the role of the nurse has changed in the last century. (p. 126-128)

Solutions

Practice Tests

Fill-In-The-Blank

1. status set
2. master status
3. Emotion labor

4. Exchange theory
5. impression management
6. Dramaturgical analysis

7. role distancing
8. Domination
9. Social interaction

Multiple-Choice Questions

1. E, (p. 124-129)
2. A, (p. 127)
3. D, (p. 137-139)
4. B, (p. 135)
5. E, (p. 135-137)
6. D, (p. 143-144)
7. C, (p. 124-129, 137-139)
8. D, (p. 143)
9. C, (p. 126)

10. E, (p. 126)
11. D, (p. 132)
12. A, (p. 134-147)
13. C, (p. 126-128)
14. A, (p. 129-131)
15. D, (p. 130)
16. B, (p. 132-133)
17. D, (p. 133-134)
18. C, (p. 129)

19. A, (p. 127)
20. B, (p. 128)
21. A, (p. 129-130)
22. D, (p. 130)
23. C, (p. 134)
24. E, (p. 138)
25. C, (p. 145)

True False

1. F, (p. 134-136)
2. T, (p. 137)
3. T, (p. 140-144)
4. F, (p. 139-140)
5. F, (p. 137)
6. F, (p. 146-147)

7. F, (p. 126-128)
8. T, (p. 129-131)
9. F, (p. 129-131)
10. T, (p. 130)
11. F, (p. 131-134)
12. T, (p. 132)

13. T, (p. 137)
14. F, (p. 141-143)
15. F, (p. 146-147)
16. F, (p. 146-147)

SOCIAL COLLECTIVITIES: GROUPS TO SOCIETIES

Student Learning Objectives

After reading Chapter 6, you should be able to:

1. Understand the influential role of networks, groups, and bureaucracies on individual behavior.

2. Describe the far-reaching connection between individuals through the dynamics of social networks.

3. Identify and discuss the building blocks of social networks.

4. Describe the range of social group experience.

5. Explain the power of conformity and "groupthink" in a social group context.

6. Discuss the group boundaries in terms of inclusion and exclusion.

7. Understand the human ability to interact with others in the imagination.

8. Describe the major sources of bureaucratic inefficiency from a sociological perspective.

9. Explain the importance of social networks to the operation of bureaucracies.

10. Discuss three types of leadership styles in terms of their effectiveness or lack thereof for bureaucracies.

11. Explain the influence of organizational environments on leadership characteristics and functions of bureaucratic organizations.

12. Discuss and explain how societies are collectivities of interacting people, and how these evolving collectivities change the relationship of humans to nature.

13. Explain the function of networks, groups, and organizations as foundations of constraint as well as sources of freedom.

Key Terms

bureaucracy (155)

social network (157)

dyad (160)

triad (160)

Detailed Chapter Outline

I. BEYOND INDIVIDUAL MOTIVES

   A. The Holocaust

      1. The Experience of Robert Brym's Father in the 1940s—Nazi Genocide

         a. How was it possible for many thousands of ordinary Germans to systematically murder millions of defenseless and innocent Jews, Roma ("Gypsies"), homosexuals, and mentally disabled people in death camps?

   B. How Social Groups Shape Our Actions

      1. Norms of solidarity demand conformity.

      2. Structures of authority tend to render people obedient.

      3. Bureaucracies are highly effective structures of authority.

         a. As Max Weber defined the term, **bureaucracy** is a large, efficient, impersonal organization composed of clearly defined positions arranged in a hierarchy.

II. NETWORKS

   A. It's a Small World

      1. The *Internet Movie Database* (2003).

         a. The example of Kevin Bacon.

      2. Jeffrey Travers and Stanley Milgram (1969) study.

   B. Network Analysis

      1. A **social network** is a bounded set of individuals linked by the exchange of material or emotional resources, from money to friendship. The patterns of exchange determine the boundaries of the network. Member exchange resources more frequently with each other

than with nonmembers. They also think of themselves as a network of members. Social networks may be formal or informal.

2. The "nodes" (units of analysis) in a social network can be individuals, groups, organizations, and countries; thus social network analysts have examined everything from intimate relationships to diplomatic relations.

3. In contrast to organizations, most networks lack names and offices. Networks lie beneath the more visible social collectivities.

C. Finding a Job

1. Mark Granovetter (1973) found that mere acquaintances are more likely to provide useful information about employment opportunities than friends and family members, because acquaintances are connected to diverse networks.

D. Urban Networks

1. Urban acquaintances tend to be few and functionally specific.

2. According to German sociologist Ferdinand Tonnies, a "community" is marked by intimate and emotionally intense social ties, whereas a "society" is marked by impersonal relationships held together by self-interest. Tonnies focuses on only sparse functionally specific ties, whereas network analysts found elaborate social networks, some functionally specific and some not.

E. From HIV/AIDS to the Common Cold

1. Network analysis helped to show that the characterization of HIV/AIDS as a "gay disease" was an oversimplification. It spread along friendship and acquaintanceship networks of people first exposed. In India and parts of Africa, it did not spread among gay men, but rather through a network of long-distance truck drivers and the prostitutes who catered to them.

2. Regarding health in general, People who are well integrated into cohesive social networks of family, extended kin, and friends are less likely to suffer heart attacks, complications during pregnancy, and so forth.

3. Research shows that you are less likely to come down with an infectious disease the more social contact you have.

F. The Building Blocks of Social Networks

1. The **dyad** is the most elementary network form, which is a social relationship between two nodes or social units.

2. A **triad** is a social relationship among three nodes.

3. As Georg Simmel showed early in the 20th century, the social dynamics of these two network forms are fundamentally different. When a third person enters the picture, relationships tend to be less intimate and intense.

III. GROUPS

A. Love and Group Loyalty

1. Group loyalty is often more powerful than romantic love.

B. Varieties of Group Experience

1. **Social groups** are composed of one or more networks of people who identify with one another and adhere to defined norms, roles, and statuses.

2. **Social categories** consist of people who share similar status but do not identify with one another.

3. In **primary groups**, norms, roles, and statuses are agreed upon but are not put into writing. Social interaction creates strong emotional ties, extends over a long period, and involves a wide range of activities.

4. **Secondary groups** are larger and more impersonal than primary groups, which creates weaker emotional ties, extends over a shorter period and involves a narrow range of activities.

C. Group Conformity

1. Conformity is an integral part of group life, and primary groups generate more pressure to conform than secondary groups.

2. Samuel Stouffer's (1945) classic study showed that primary group cohesion was the main factor motivating soldiers to engage in combat.

3. Social psychologist Solomon Asch's (1955) famous experiment demonstrates how group pressure creates conformity.

4. Research shows that several factors affect the likelihood of conformity.

   a. The likelihood of conformity increases as *group size* increases to three or four members.

   b. As *group cohesiveness* increases, so does the likelihood of conformity.

   c. *Social status* affects the likelihood of conformity. People with low status are less likely to dissent.

   d. The *appearance of unanimity* affects the likelihood of conformity. One dissenting voice increases the chance that others will dissent.

5. **Groupthink** is group pressure to conform despite individual misgivings.

   a. The dangers of groupthink are greatest in high-stress situations.

6. "Bystander apathy" − as the number of bystanders increases, the likelihood of any one bystander helping decreases because the greater the number of bystanders, the less responsibility any one individual feels.

D. Inclusion and Exclusion: In-groups and Out-groups

1. **In-group** members are people who belong to a group. **Out-group** members are people who are excluded from the in-group. Members of an in-group typically draw a boundary separating themselves from out-group members, and they try to keep out-group members from crossing the line.

2. Why do group boundaries crystallize?

   a. One theory is that group boundaries emerge when people compete for scarce resources.

   b. Another theory is that group boundaries emerge when people are motivated to protect their self-esteem.

   c. *The Robber's Cave Study* (1961), a classic experiment on prejudice, supports both theories.

3. Dominant groups construct group boundaries to further their goals.

E. Groups and Social Imagination

1. People interact with other group members in their imagination, through **reference groups**, which are composed of people against whom an individual evaluates his or her situation or conduct. Reference group members function as "role models."

a. Theodore Newcomb's (1843) study of students at Bennington College is a classic example of a reference group.

2. Most people highly value the opinions of in-group members.

3. We have to exercise our imagination because much social life in a complex society involves belonging to secondary groups without knowing or interacting with most group members.

4. **Formal organizations** are secondary groups designed to achieve explicit objectives.

IV. BUREAUCRACY

A. Bureaucratic Inefficiency

1. Contradiction of Weber's view of bureaucratic efficiency and the realities of bureaucratic inefficiency.

2. Four main sociological criticisms against bureaucracies:

a. **Dehumanization** occurs when bureaucracies treat clients as standard cases and personnel as cogs in a giant machine, which frustrates and lowers morale.

b. **Bureaucratic ritualism** is the preoccupation by bureaucrats with rules and regulations that make it difficult for the organization to fulfill its goals.

c. **Oligarchy** is the "rule of the few" that results from the tendency for power to become concentrated in the hands of a few people at the top of the organizational hierarchy.

d. **Bureaucratic inertia** occurs when bureaucracies become so large and rigid that they lose touch with reality and continue their policies even when their clients' needs change.

3. *Size* and *social structure* underlie bureaucratic inefficiency. Problems multiply with the increasing size of groups, and the increasing number and complexity of levels in bureaucratic structures.

B. Bureaucracy's Informal Side

1. A social network is important for the operation of bureaucracies. The patterns of exchange determine the boundaries of the network.

2. Examples:

a. 1930s Hawthorne plant of the Western Electric Company near Chicago.

b. Rosabeth Kanter's landmark 1970s study on the informal social relations in bureaucracies.

c. Socializing, business dinners.

C. Leadership

1. Leadership style affects bureaucratic performance.

2. Three types of leadership:

a. *Laissez-Faire* leadership—The least effective form of leadership that allows subordinates to work things out largely on their own, with almost no direction from above.

b. **Authoritarian leadership**—Authoritarian leaders demand strict compliance from subordinates. Effective in a crisis, and earn grudging respect for achieving goals, but rarely win popularity contests.

c. **Democratic leadership** offers more guidance than the *laissez-faire* type, but less control than the authoritarian type. Usually the most effective except for crisis situations, because democratic leaders attempt to include all group members in the decision-making process toward an integrated strategy.

D. Overcoming Bureaucratic Inefficiency

1. Smaller innovative firms are often more effective because they have flatter, more democratic and network-like organizational structures, and multiple lines of communication, which produces more satisfied workers, happier clients, and bigger profits.

E. Organizational Environments

1. An **organizational environment** is composed of a host of economic, political, and cultural factors that lie outside an organization and affect the way it works.
2. Japanese versus American business bureaucracies in the 1970s.
   a. *Japanese workers were in a position to demand efficiency in Japan.*
   b. *International competition encouraged bureaucratic efficiency in Japan.*
   c. *The availability of external suppliers allowed Japanese firms to remain lean.*

V. SOCIETIES

A. **Societies** are collectivities of interacting people who share a culture and a territory.
1. Networks, groups, and bureaucracies are embedded in societies.
2. Societies help shape human action.
3. The changing relationship between people and nature is the most basic determinant of how societies are structured, and how people's choices are constrained.
   a. There are six stages of human evolution characterized by a shift in the relationship between people and nature.

B. Foraging Societies

1. Until about 10,000 years ago, all people lived in **foraging societies**, in which people lived by searching for wild plants and hunting animals.
2. Inequality, the division of labor, productivity, and settlement size are very low.

C. Pastoral and Horticultural Societies

1. **Pastoral societies** are those in which people domesticate animals. This increases settlement size and permanence, the division of labor, productivity, and inequality above the levels of foraging societies.
2. **Horticultural societies** are those in which people domesticate plants and use simple hand tools to garden. This also increases settlement size and permanence, the division of labor, productivity, and inequality above the levels of foraging societies.

D. Agricultural Societies

1. **Agricultural societies** are those in which plows and animal power are used to substantially increase food supply and dependability.
2. This increases settlement size and permanence, the division of labor, productivity, and inequality above the levels of pastoral and horticultural societies.
3. Inequality between men and women during this era reached its historical high point.

E. Industrial Societies

1. **Industrial societies** use machines and fuel to greatly increase the supply and dependability of food and finished goods.

   2. Productivity, the division of labor, and average settlement size increased substantially compared with agricultural societies.
   3. While social inequality was substantial during early industrialism, it declined as the industrial system matured.

   F. Postindustrial societies
       1. **Postindustrial societies** are those in which most workers are employed in the service sector.
       2. Computers spur substantial increases in the division of labor and productivity.
       3. Gender inequality is reduced.
       4. Class inequality can increase.

   G. Postnatural Societies
       1. **Postnatural societies** are those in which genetic engineering enables people to create new life forms.
       2. Holds promise for improving productivity, feeding the poor, ridding the world of disease, etc.
       3. Social inequality could increase unless people democratically decide on acceptable risks of genetic engineering and the distribution of potential benefits.
       4. The invention of recombinant DNA marked the onset of postnatural society.
       5. **Recombinant DNA** involves removing a segment of DNA from a gene or splicing together segments of DNA from different living things, thus creating new life.

VI. FREEDOM AND CONSTRAINT IN SOCIAL LIFE

   A. Social collectivities are constraining.

   B. Paradoxically, to succeed in challenging social collectivities, people must sometimes form a new social collectivity.
       1. Seymour Martin Lipset, Martin A. Trow, and James S. Coleman's (1956) classic sociological study of the International Typographical Union.

Study Activity: Applying the Sociological Compass

On your next visit to the bank, the department of motor vehicles, admissions and records at your college, or your college major department, pay close attention to how employees interact and perform their roles. Take notes on your observations. In so doing, accomplish the following tasks:

Based on your observations, make a list of examples of

Dehumanization:

1.

2.

3.

Bureaucratic ritualism:

1.

2.

3.

Bureaucratic inertia:

1.

2.

3.

Finally, make a list of explanations for the reasons behind the bureaucratic behaviors you might have observed, based on what your learned from Chapter 6 of the Brym and Lie textbook.

1.

2.

3.

InfoTrac College Edition Online Exercises

For the following exercises, log on to the online library of InfoTrac College Edition at http://www.infotrac-college.com/. Make note that InfoTrac has implemented a new registration system that will allow easier access to InfoTrac through the use of a personalized username and password. Once you've created your username and password you may proceed directly to the Log On page. To create an account, register your passcode packaged with your textbook, and create a username and password, by following the online prompt. After you are logged in, click on "Infotrac College Edition." You will arrive at a screen that enables you to search topics.

Keyword: **tokenism**. In what ways is tokenism portrayed? Look for articles that give evidence that tokenism is a problem in the corporate world. Cite five articles that deal with this issue.

Keyword: **employment discrimination**. Identify and read at least three articles that deal with employment discriminations. Compile a list of patterns of employment discrimination that are similar in the three articles.

Keyword: **group service** and **community service**. Look for articles on how service learning is related to the community at large. What information do these articles bring to light about the interaction between groups and the larger society?

Keyword: **networks**. Identify four articles that discuss how networking helps to find employment. After reading the articles compile a list of things you can do now right now to apply the research findings of networking to your future job search. Why is it important for you to develop a "network" in your career area?

Internet Exercises

For a look at the Ray Mine Pilot Study (dealing with hazardous waste), put out by the Environmental Protection Agency go to: http://epawww.ciesin.org/raymine. Write a short paper discussing the role of organization in this operation.

Go to: http://www.dannen.com/decision/targets.html

This article discusses the decision to use the atomic bomb. Read the article and write a short paper on how Michel's' Iron Law of Oligarchy was at work in this situation.

Go to: http://www.geocities.com/mcdonaldization/

This site explores the meaning of McDonaldization in our society. Look over the site and then write a short opinion paper stating whether you agree or disagree with the McDonaldization of society in general.

Practice Tests

Fill-In-The-Blank

Fill in the blank with the appropriate term from the above list of "key terms."

1. As Max Weber defined the term, _____ is a large, efficient, impersonal organization composed of clearly defined positions arranged in a hierarchy.

2. The _____ is the most elementary network form.

3. _____ are composed of one or more networks of people who identify with one another and adhere to defined norms, roles, and statuses.

4. _____ are larger, more impersonal, and create weaker emotional ties.

5. People interact with other group members in their imagination, through _____ , which are composed of people against whom an individual evaluates his or her situation or conduct.

6. _____ are secondary groups designed to achieve explicit objectives.

7. _____ occurs when bureaucracies become so large and rigid that they lose touch with reality and continue their policies even when their client's needs change.

8. Although effective in a crisis, _____ earn grudging respect from subordinates, and rarely win popularity contests.

9. A (An) _____ is a bounded set of individuals linked by the exchange of material or emotional resources.

10. _____ is the preoccupation by bureaucrats with rules and regulations that make it difficult for the organization to fulfill its goals.

Multiple-Choice Questions

Select the response that best answers the question or completes the statement:

1. In _____, social interactions lead to strong, more enduring emotional ties.
   a. primary groups
   b. secondary groups
   c. tertiary groups
   d. social categories
   e. None of the above.

2. Which concept refers to "rule of the few" in bureaucracies?
   a. dehumanization
   b. bureaucratic inertia
   c. bureaucratic ritualism
   d. competition
   e. oligarchy

3. One evening after being in a traffic accident, Kiana was rushed to the nearest emergency room for moderate injuries. It took nearly two hours to be seen by a medical doctor because much time was spent by staff members following hospital procedure to determine if Kiana possessed medical insurance, and whether her insurance policy would indeed pay for care at the hospital in question. In short, the emergency room staff members lost site of the goals of the emergency room in favor of following regulations and procedures. Which terms best captures this scenario?
   a. bureaucratic inertia
   b. bureaucratic ritualism
   c. oligarchy
   d. monopolization
   e. None of the above.

4. Which type of leadership is most effective during times of crisis?
   a. *laissez-faire* leadership
   b. democratic leadership
   c. egalitarian leadership
   d. authoritarian leadership
   e. rational leadership

5. If a supervisor allows his or her subordinates to settle disputes and conflict without direction, he or she would be described as what type of leader?
   a. *laissez-faire* leadership
   b. democratic leadership
   c. egalitarian leadership
   d. authoritarian leadership
   e. rational leadership

6. The distinctions between Japanese and American corporations in the 1970s illustrate the impact of _____ on the way economic organizations operate.
   a. organizational deviance
   b. authoritarian leadership
   c. organizational environments
   d. social networks
   e. None of the above.

7. The function of business dinners and reservations to play golf with business associates illustrates the importance of_____ for the operation of economic bureaucracies.
   a. social encounters
   b. social networks
   c. competition
   d. domination
   e. ethnomethods

8. According to Mark Granovetter's research, which social network is more likely to provide useful information about employment opportunities?
   a. friends
   b. family members
   c. former supervisors
   d. former spouses
   e. acquaintances

9. Based on the research literature, which factor affects the likelihood of conformity?
   a. Group size
   b. Group cohesiveness
   c. Social status
   d. Appearance of unanimity
   e. All of these are correct.

10. The dangers of groupthink are greatest in:
    a. high-stress situations.
    b. low-stress situations.
    c. democratic situations.
    d. low anxiety situations.
    e. None of the above.

11. _____ members are people who belong to a group, whereas _____ members are people who are excluded from that group.
    a. Primary group; secondary group
    b. Secondary group; primary group
    c. In-group; out-group
    d. Out-group; in-group
    e. In-group; reference group

12. Which of the following is NOT a basis of criticism of bureaucracies?
    a. Dehumanization
    b. Efficiency
    c. Bureaucratic ritualism
    d. Oligarchy
    e. Bureaucratic inertia

13. The movie *Schindler's List* turns the history of Nazism into a play of morality of good against evil. Which of the following statements is true of this movie according to the textbook?
    a. It does not probe into the sociological roots of good and evil.
    b. As a rare case of a motion picture, it probes into the sociological roots of good and evil.
    c. It provides an authentic representation of many altruistic Nazis during the Holocaust.
    d. All of these are correct.
    e. None of the above.

14. Collectivities of interacting people who share a culture and a territory are called _____.
    a. cohorts
    b. peers
    c. communities
    d. societies
    e. cohabitating ethnicities

15. Which is one of the main sociological points of Box 6.1, Social Policy: What Do You Think? "Group Loyalty or Betrayal" in the textbook?
    a. Intense group loyalty usually produces an individual willing to betray the group due to the high pressure the group exerts over its members.
    b. The subordination of women tends to be lower in situations where there is high level of group loyalty.
    c. In social situations where there is much tolerance of male misconduct, compassion for the weak tends to be greater.
    d. An important question to consider is where we ought to draw the line between group loyalty and group betrayal.
    e. None of the above.

16. In the classic Milgram experiment on obedience to authority, which laboratory setting yielded the greatest percentage of subjects who administered "maximum shocks" to others?
    a. Same room; hand forced.
    b. Same room.
    c. Different rooms; see and hear.
    d. Different rooms; hand forced.
    e. Different rooms; see but not hear.

17. The 1968 movie, *The Graduate*, as mentioned in the textbook, illustrates the fact that:
    a. social networking does not necessarily lead to employment opportunities.
    b. parents can help their graduating children find jobs by plugging them into the right social networks.
    c. formal organizations are more effective than informal social networks at helping graduates find jobs.
    d. All of these are correct.
    e. None of the above.

18. Which of the following is NOT a characteristic of a triad?
    a. No "free riders" are possible.
    b. Intensity and intimacy are reduced.
    c. Coalitions are possible.
    d. A divide-and-conquer strategy is possible.
    e. It is possible to shift responsibility to the larger collectivity.

19. Which is NOT true of primary groups?
    a. Strong emotional ties.
    b. Extends over a long period of time.
    c. Narrow range of activities.
    d. Group members know each other well.
    e. The family is the most important primary group.

20. Which of the following is considered a social origin of evil according to Brym and Lie?
    a. Norms of solidarity demand conformity.
    b. Structures of authority tend to render people obedient.
    c. Bureaucracies are highly effective structures of authority.
    d. All of these are correct.
    e. None of the above.

21. Network analysis helped to show that the characterization of HIV/AIDS as "gay disease" was:
    a. true in the United States, although false in other countries.
    b. entirely true.
    c. an oversimplification.
    d. true in other countries, but false in the United States.
    e. None of the above.

22. _____ societies are those in which people live by searching for wild plants and hunting wild animals.
    a. baseline
    b. gleaning
    c. pre-agri
    d. primitive
    e. foraging

23. Which term captures the idea of people sharing similar status but do not identify with one another?
    a. Primary group
    b. Social category
    c. Social group
    d. Reference group
    e. None of the above.

24. Which invention helped create agricultural societies?
    a. aqueduct
    b. wheel
    c. silo
    d. sickle
    e. plow

25. Which of the following statements is true regarding the differences between Japanese and American business in the 1970s?
    a.   Japanese workers were in a position to demand efficiency in Japan.
    b.   International competition encouraged bureaucratic efficiency in Japan.
    c.   The availability of external suppliers allowed Japanese firms to remain lean.
    d.   All of these are correct.
    e.   None of the above.

True False

1.   Social groups and social categories are indistinguishable.

     TRUE or FALSE

2.   Because bureaucracies tend to take on a life of their own in the form of enduring social patterns, the size and social structure of bureaucracies have little to do with their efficiency.

     TRUE or FALSE

3.   Contemporary realities of bureaucratic inefficiency contradict Max Weber's view of bureaucratic efficiency.

     TRUE or FALSE

4.   In the business world, smaller innovative firms are often more effective in producing more satisfied workers, happier clients, and bigger profits than larger firms.

     TRUE or FALSE

5.   Nearly a quarter of all people who ever acted in a movie are separated from Kevin Bacon by just one or two links.

     TRUE or FALSE

6.   Inequality between women and men reached its historical high point during the era of agricultural societies.

     TRUE or FALSE

7.   In contrast to organizations, most networks lack names and offices.

     TRUE or FALSE

8.   A triad is fundamentally the same as a dyad, since only one person is added to the relations.

     TRUE or FALSE

9.   Romantic love is often more powerful than group loyalty.

     TRUE or FALSE

10. Although social networks are useful in finding jobs within a bureaucracy, social networks are not important for the operation of bureaucracies.

    TRUE or FALSE

11. An organizational environment is composed of economic, political, and cultural factors that lie outside an organization and affect the way it works.

    TRUE or FALSE

Short Answer

1. Chapter 6 opens with a narrative describing the experiences of Robert Brym's father with Nazi genocide in the 1940s. A perplexing question that struck Robert Brym's father was "How was it possible for many thousands of ordinary Germans − products of what he regarded as the most advanced civilization on earth − to systematically murder millions of defenseless and innocent Jews, Roma ("Gypsies"), homosexuals, and mentally disabled people in death camps?" Briefly answer this question based on what you learned about structures of bureaucratic authority. (p. 151-156, 161-172)

   _____

   _____

2. List and briefly describe the six stages of human evolution characterized by a shift in the relationship of people and nature. (p. 173-178)

   _____

   _____

3. Explain the relationship between social status and conformity. (p. 161-166)

   _____

   _____

4. List and define three types of leadership. (p. 171)

   _____

   _____

5. Illustrate the concepts of bureaucratic ritualism and democratic leadership by briefly describing the transformation of the character "Kanji Watanabe" in the 1952 film *Ikiru* described in Chapter 6. (p. 168)

   _____

   _____

Essay Questions

1. Explain the influence of organizational environments on bureaucratic structures by contrasting Japanese and American businesses in the 1970s. (p. 172-173).

2. Consider your occupation, or your future occupation (in other words engage in anticipatory socialization) for this question. Draw a bureaucratic structure of the organization in which you work (or hope to work). In so doing, identify each node and group. Next, explain the nature of secondary group relations, and possible types of leadership. (p. 157-158,160-162,166-173)

3. Explain four sociological criticisms against bureaucracies, and at least two strategies to overcome bureaucratic inefficiency. (p. 166-173)

4. Conduct a "mini" network analysis on your potential to find a job through networking with family, friends, and acquaintances. In your analysis, be sure to list the number of different possible job leads from family, friends, and acquaintances, respectively. Determine which social network (family, friends, or acquaintances) has the greatest potential of leading you to a job, and explain why. (p.157-159) Describe the transition between postindustrial and postnatural societies, and postulate whether the potential benefits from postnatural society will outweigh possible societal risks and perils, like increased social inequality. (177-178)

Solutions

Practice Tests

Fill-In-The-Blank

1. bureaucracy
2. dyad
3. Social groups
4. Secondary groups

5. reference groups
6. Formal organizations
7. Bureaucratic inertia
8. authoritarian leaders

9. social network
10. Bureaucratic ritualism

Multiple-Choice Questions

1. A, (p. 162)
2. E, (p. 167)
3. B, (p. 167)
4. D, (p. 171)
5. A, (p. 171)
6. C, (p. 172)
7. B, (p. 157)
8. E, (p. 158-159)
9. E, (p. 152-155, 162-164)

10. A, (p. 163-164)
11. C, (p. 164)
12. B, (p. 167-170)
13. A, (p. 152)
14. D, (p. 173)
15. D, (p. 154)
16. E, (p. 155)
17. B, (p. 158)
18. A, (p. 160-161)

19. C, (p. 162)
20. D, (p. 152-154)
21. C, (p. 160)
22. E, (p. 174)
23. B, (p. 161)
24. E, (p. 175)
25. D, (p. 172-173)

True False

1. F, (p. 161)
2. F, (p. 166-170)
3. T, (p. 166-170)
4. T, (p. 171-173)

5. T, (p. 157)
6. T, (p. 176)
7. T, (p. 157)
8. F, (p. 160-161)

9. F, (p. 161)
10. F, (p. 156-161)
11. T, (p. 172)

DEVIANCE AND CRIME

Student Learning Objectives

After reading Chapter 7, you should be able to:

1. Explain why definitions of crime and deviance vary across cultures, history, and social contexts.

2. Identify the general categories of deviance and crime.

3. Explain the social construction of crime and deviance in the context of differential social power.

4. Describe the major data sources of crime and delinquency, and their strengths and weaknesses.

5. Describe recent demographic characteristics (i.e. gender, age, and race) of crime and deviance.

6. Identify the major sociological theories of crime and deviance, their central features and key exemplars.

7. Compare and contrast motivational and constraint theories of crime and deviance.

8. Discuss the historical and contemporary trends in social control and punishment.

9. Identify and explain the major views on the function of prisons.

10. Describe the impact of moral panics on criminal justice policies and crime prevention.

11. Discuss the pros and cons of alternative forms of punishment.

Key Terms

deviance (184)

informal punishment (184)

stigmatization (184)

formal punishment (184)

social diversions (186)

social deviations (186)

conflict crimes (186)

consensus crimes (186)

white-collar crimes (188)

street crimes (188)

victimless crimes (189)

self-report surveys (189)

social control (192)

motivational theories (194)

Detailed Chapter Outline

I. THE SOCIAL DEFINITION OF DEVIANCE AND CRIME

A. Crime and Punishment in the United States

1. Considering television, cable, local, daytime, and late-night shows, one might conclude that the United States is a society obsessed with crime.

2. Punishment is also a big issue in the United States

a. The United States has more people behind bars than any other country on earth, and the number is increasing by 50,000-80,000 per year.

B. Definitions of deviance and crime change over time and vary across people.

1. Examples:

a. 1872 Susan B. Anthony and voting rights

b. Late 1950s and early 1960s Martin Luther King Jr. and civil rights

c. 1992 Los Angeles race riot

C. The Difference Between Deviance and Crime

1. **Deviance** involves breaking a norm.

2. **Informal punishment** is mild, and may involved raised eyebrows, gossip, ostracism, shaming, or **stigmatization** (negative evaluation because of a marker that distinguishes them from others).

3. **Formal punishment** results from people breaking *laws*, which are norms stipulated and enforced by government bodies.

4. Three dimensions of the various types of deviance (John Hagan):

a. *Severity of social response*.

b. *Perceived harmfulness*.

c. *Degree of public agreement*.

5. Four types of deviance and crime:

a. **Social diversions** are minor acts of deviance such as participating in fads or fashions like dyeing one's hair purple. Usually perceived as harmless.

b. **Social deviations** are more serious acts, large proportions of people agree they are deviant and somewhat harmful, and they are usually subject to institutional sanction.

c. **Conflict crimes** are deviant acts that the state defines as illegal, but the definition is controversial in the wider society.

d. **Consensus crimes** are widely recognized to be bad in and of themselves, and there is little controversy over their seriousness.

e. Conceptions of deviance and crime vary substantially over time and among societies.

D. Power and the Social Construction of Crime and Deviance

1. *Social constructionism* is the school of sociological thought that emphasizes that various social problems, including crime, are not inherent in certain actions themselves. *Powerful* groups are in positions to create norms and laws that suit their interests and stigmatize other people, while the *powerless* often struggle against stigmatization.

2. Crimes Against Women

a. Has the law been biased against women?

b. Many types of crimes against women were largely ignored in the United States and most parts of the world (i.e. rape, physical violence against women, sexual harassment).

c. The situation has improved today. Why?

i. Women's positions in the economy, the family, and other social institutions, has improved.

ii. Women's rights movement.

iii. Increased public awareness.

iv. Shift in the distribution of power.

3. White-Collar Crime

a. **White-collar crimes** refer to illegal acts "committed by a person of respectability and high social status in the course of his [or her] occupation" (i.e. embezzlement, false advertising, tax evasion, insider stock trading, fraud, unfair labor practices, copyright infringement, conspiracy to fix prices and restrain trade, and environmental degradation). Disproportionately committed by middle and upper classes.

b. **Street crimes**, by contrast, include arson, burglary, assault, and other illegal acts disproportionately committed by people from lower classes.

c. Many sociologists argue that white-collar crimes are *more costly* than street crimes.

d. White-collar crime is underreported.

e. White-collar crimes result in few prosecutions and fewer convictions. Why?

i. Much white-collar crime occurs in private and is therefore difficult to detect.

ii. Corporations can afford legal experts, public relations firms, and advertising agencies that advise their clients on how to bend laws, build up their corporate image in the public mind, and influence lawmakers to pass laws "without teeth."

f. Governments also commit serious crimes, and are also hard to punish.

E. Measuring Crime

1. A word about *official crime statistics*

a. Much crime is not reported to the police (i.e. common assaults and rape).

b. Authorities and the wider public decide which criminal acts to report and which to ignore.

c. Many crimes are not incorporated in major crime indexes published by the FBI, such as **victimless crimes** (violations of law in which no victim steps forward and is identified, such as white-collar crimes, prostitution and illegal drug use).

2. **Self-report surveys** are used to supplement official crime statistics in light of reporting problems. Respondents are asked to report their involvement in criminal activities, either as perpetrators or victims.

   a. The National Crime Victimization Survey is the main source of self-report victimization data in the United States.

   b. Survey data are influenced by people's willingness and ability to discuss criminal experiences; therefore, indirect measures of crime are sometimes used also.

3. Crime Rates

   a. Every hour in 2003, law enforcement agencies in the United States received verifiable reports on average of nearly 2 murders or nonnegligent manslaughter, 11 rapes, 47 robberies, 98 aggravated assaults, 144 motor vehicle thefts, 246 burglaries, and 802 larceny-thefts.

   b. 500% increase in the rate of violent crime, and a 150% increase in the rate of major property crimes, between 1960 and 1992.

   c. However the long crime wave that began its upswing in the early 1960s eased in the 1980s and fell in the 1990s.

   d. Why the decline? Four factors usually mentioned:

      i. Governments put more police in the streets and many communities established their own systems of surveillance and patrol.

      ii. America is aging and the number of young men in the population has declined.

      iii. 1990s economic boom.

      iv. Controversially, some researchers noted that the decline in crime started 19 years after abortion was legalized in the United States, thus proportionately fewer unwanted children in the population beginning in 1992.

   e. Note: The authors have not claimed that putting more people in prison and imposing tougher penalties for crime help account for lower crime rates. The authors explain why these actions do not generally result in lower crime rates later in the chapter when they discuss **social control** (methods of ensuring conformity) and punishment.

F. Criminal Profiles

   1. Gender

      a. 77% of all persons arrested in the United States in 2003 were men.

      b. Men account for 82% of arrests for violent crimes in 2003.

      c. Women are catching up.

      d. Men are still 3.3 times more likely than women to be arrested.

   2. Age

      a. Most crime is committed by people who have not reached middle age.

      b. The 15- to 19-year-old age cohort is the most crime prone.

   3. Race

      a. In 2000 whites accounted for 75.1% of the U.S. population, but only 70.6% of arrests in 2001.

      b. African Americans accounted for 27% of arrests but made up only 12.3% of the population.

      c. Why the disproportionately high arrest, conviction, and incarceration rates of African Americans?

      i. Bias in the way crime statistics are collected (i.e. absence of data on white-collar crime which are disproportionately committed by whites).

      ii. The low class position of blacks in American society.

      iii. Racial discrimination in the criminal justice system.

G. Explaining Deviance and Crime

  1. Explanations can be categorized into two basic types.

    a. **Motivational theories** identify the social factors that *drive* people to commit deviance and crime.

    b. **Constraint theories** identify the social factors that impose deviance and crime (or conventional behavior) on people.

  2. Learning the Deviant Role: The Case of Marijuana Users

    a. Three-stage learning process of becoming a regular marijuana user, based on Howard S. Becker's (1962) classic study of jazz musicians:

      i. *Learning to smoke the drug in a way that produces real life effects.*

      ii. *Learning to recognize the effects and connect them with the drug use.*

      iii. *Learning to enjoy the perceived sensations.*

  3. Motivational Theories

    a. Durkheim's Functional Approach

      i. Durkheim wrote that deviance is normal, necessary, and functional, and therefore it exists in all societies.

      ii. Deviance gives people the chance to define what is moral and what is not.

      iii. Our reactions to deviance clarify moral boundaries.

      iv. This clarification promotes social solidarity and encourages healthy social change.

    b. Strain Theory

      i. Robert Merton's **strain theory** argues that people may turn to deviance when they experience strain. Strain results when a culture teaches people the value of material success, and society fails to provide enough legitimate opportunities for everyone to succeed.

      ii. Five adaptations to strain:

      iii. *Conformity;*

      iv. *Innovation;*

      v. *Ritualism;*

      vi. *Retreatism;* and

      vii. *Rebellion.*

    c. Subcultural Theory

      i. **Subcultural theory** argues that gangs are a collective adaptation to social conditions. Distinct norms and values that reject the legitimate world crystallize in gangs.

      ii. Three features of criminal subcultures:

      iii. Delinquent youths turn to different types of crime, depending on the availability of different subcultures in their neighborhoods.

      iv. Members of criminal subcultures spin out a whole series of rationalizations for their criminal activities, called **techniques of neutralization**, or the creation of justifications and rationalizations that make deviance and crime seem normal, and enable criminals to clear their consciences and get on with the job.

    v.   Members of criminal subcultures are strict conformists to the norms of their own culture.

    vi. The main problem with strain theory is that it exaggerates the connection between class and crime. Research suggests that this connection must take into account the *severity* and *type* of crime.

  d. Learning Theory

    i.   Edwin Sutherland's **differential association theory** holds that people learn to value deviant or nondeviant lifestyles depending on whether their social environment leads them to associate more with deviants or nondeviants.

    ii. Differential association theory holds for people in all social class positions including white-collar crime.

4. Constraint Theories

  a. Labeling Theory: A Symbolic Interactionist Approach

    i.   **Labeling theory** argues that deviance results not so much from the actions of the deviant as from the response of others, who label the rule breaker a deviant.

    ii. Labeling may lead to immersion of the stigmatized juvenile into a deviant subculture and adopt "delinquent" as his [or her] **master status**, or one's overriding public identity.

    iii. Labeling may act as a self-fulfilling prophecy.

  b. Control Theory

    i.   **Control theory** proposes that the rewards of deviance and crime are ample; therefore, most of us would commit deviant and criminal acts if we could get away with it. The reason deviants and criminals break norms and laws are that social controls are insufficient to ensure their conformity.

    ii. Four types of social control according to Travis Hirschi and Michael Gottfredson: *attachments, opportunities, involvement;* and *beliefs.*

  c. The Conflict Theory of Deviance and Crime

    i.   **Conflict theory** maintains that the rich and powerful impose deviant and criminal labels on the less powerful members of society, especially those who challenge the existing social order. Conflict theory holds the view that the law applies differently to the rich and poor. The rich and powerful are also able to use their money and influence to escape punishment for their own deviant acts.

    ii. Examples: Thieves challenge private property, thus theft is a crime; "Bag ladies" and drug addicts drop out of society, thus are labeled deviant because of their refusal to engage in productive labor; Militant trade unionists represent threat to the social order, thus are defined as deviant or criminal; White-collar crimes are less severely punished than street crimes, regardless of the relative harm to society.

II. TRENDS IN CRIMINAL JUSTICE

  A. Social Control

    1. The *degree* and *form* of *social control* varies over time and from one society to the next.

2. Beginning in the late 19th century, sociologists argued that preindustrial societies are characterized by strict social control and high conformity, while industrial societies are characterized by less stringent social control and low conformity.

3. The more complex a society, the less likely many norms will be widely shared.

4. Some sociologists believe that social control has intensified over time. Many crucial aspects of life in the context of capitalist growth have become more regimented.

5. Electronic technology enhances the ability of authorities to exercise more effective social control (i.e. cameras, computers, satellites, intelligence services, and international telecommunications).

B. The Prison

1. Prisons are *agents of socialization*, and new inmates often become more serious offenders as they adapt to the culture of the most hardened, long-term prisoners.

2. Origins of Imprisonment
   a. In preindustrial societies, criminals were publicly humiliated, tortured, or put to death. Prisons emerged with industrialization, and were seen as less harsh and more "civilized."

3. Goals of Incarceration
   a. *Rehabilitation* involves the belief that prisoners can be taught how to become productive citizens, which was a predominant view in the United States in the 1960s and early 1970s.
   b. *Deterrence* involves a view that people will be less inclined to commit crimes if they know they are likely to get caught and serve long and unpleasant prison terms.
   c. *Revenge* involves a view that depriving criminals their freedom and forcing them to live in poor conditions is fair retribution for their illegal acts.
   d. *Incapacitation* involves the view that the chief function of the prison is simply to keep criminals out of society as long as possible to ensure they can do no harm.

C. Moral Panic

1. A **moral panic** occurs when wide sections of the public, including lawmakers and criminal justice officials, believe that some form of deviance or crime poses a profound threat to society's well being.

2. Examples of consequences in criminal justice:
   a. war on drugs
   b. "three strikes and you're out" law
   c. popularity of the death penalty

3. Crime prevention
   a. walls built around neighborhoods
   b. hiring of private security police to patrol
   c. security systems and steel bars
   d. handguns
   e. laws allowing concealed handguns in some states

4. The moral panic of the 1980s and 1990s occurred when crimes stabilized and dramatically declined.

5. Why the panic and who benefits?

a.   Mass media profits.
b.   Crime prevention and punishment industry profits.
c.   Some formerly depressed rural regions of the United States have become dependent on prison construction for economic well being of their citizens.
d.   The criminal justice system benefits because of increased spending on crime prevention, control and punishment.

D.   Alternative Forms of Punishment
1.   Capital Punishment
a.   The United States is one of the few industrial societies to retain capital punishment for the most serious offenders.
b.   The death penalty ranks high as a form of *revenge*, but is questionable as a *deterrent*. Why?
i.   Since murder is often committed in irrational rage, the murderer is unlikely to coolly consider the costs of his or her actions.
ii.   If rational calculation occurs, the perpetrator is likely to know that very few murders result in death sentences.
iii.   Capital punishment is hardly a matter of blind justice (racial bias), and mistakes are common (mistrial).
2.   Incarcerating Less Serious Offenders in Violent, "No Frills" Prisons
a.   Most of the increase in the prison population over the past 20 years is due to the conviction of nonviolent criminals, based on the hypothesis that imprisonment lowers the crime rate.
b.   Data shows a weak relationship between imprisonment and the crime rate.
c.   Prison teaches inmates to behave more violently.
d.   Rehabilitation and Reintegration
i.   Anecdotal evidence suggests that institutions designed to rehabilitate and reintegrate prisoners into society can work, especially for less serious offenders.
ii.   Cost-effective and workable alternative.

Study Activity: Applying the Sociological Compass

After reading the chapter, make a list of the motivational and constraint theories, and a list of key assumptions of each theory. Critically observe a motion picture from the "gangster" genre (*The Godfather* series, *Boys in the Hood*, *Menace to Society*, *Goodfellas*, and so forth). Apply relevant theories to the motion picture by accomplishing the following tasks. The tasks are divided by motivational and constraint theories. In applying the motivational theories, you are to describe the portrayal of the motion picture. In applying the constraint theories, you are to critique the portrayals.

Applying Motivational Theories

Make a list of causes and motivations of deviance and crime as portrayed in the motion picture in terms of strain theory, subcultural theory, and differential association.

Strain theory

1.

2.

3.

4.

Subcultural theory

1.

2.

3.

4.

Differential association

1.

2.

3.

4.

Applying Constraint Theories

In applying constraint theories, interpret the motion picture from a more critical point of view. Make a list of stereotypical portrayals. Then, in your interpretation, make a list of negative social consequences of the portrayals in terms of labeling. In other words, what are the negative consequences of the socially constructed stereotypes on the racial or ethnic group in question? Finally, make a list of reasons why the stereotypes are constructed in the manner in which they are from a conflict perspective.

Stereotypical portrayals

1.

2.

3.

4.

Labeling theory

1.

2.

3.

4.

<u>Conflict theory</u>

1.

2.

3.

4.

InfoTrac College Edition Online Exercises

For the following exercises, log on to the online library of InfoTrac College Edition at http://www.infotrac-college.com/. Make note that InfoTrac has implemented a new registration system that will allow easier access to InfoTrac through the use of a personalized username and password. Once you've created your username and password you may proceed directly to the Log On page. To create an account, register your passcode packaged with your textbook, and create a username and password, by following the online prompt. After you are logged in, click on "Infotrac College Edition." You will arrive at a screen that enables you to search topics.

Keyword: **Hazing**. Do a search for articles dealing with hazing. Is hazing considered a crime or an act of deviance? Assess the current trends in hazing.

Keyword: **Computer crime**. Look for articles on computer crime. What are the different types of computer crime? How can our society curtail computer crime?

Keyword: **crimes without victims**. Examine articles on victimless crime. What types of crimes are considered victimless crimes? Why?

Keyword: **deviant behavior**. Do a search for articles on deviant behavior. Select one area of interest and do an in-depth search in this area. Present your findings to the class.

Keyword: **Fraternity or Sorority Parties**. Look for articles that deal with binge drinking. What factors cause binge drinking?

Keyword: **Moral Panic**. Identify and read at least two articles that discuss a social issue involving a moral panic (i.e. war on drugs, child abuse in day-care centers, terrorism, three strikes initiative, get tough on crime initiatives, etc.). Read the articles and list the major findings of each article. What similarities and differences do you see in the different studies?

Internet Exercises

For statistics and information from The Drug Enforcement Administration go to http://www.usdoj.gov/dea/index.htm

Look up statistics in your state or area. What drugs are most likely to be used in your area or state?

The Federal Bureau of Investigation (FBI) website is found at http://www.fbi.gov/

Locate information about the Uniform Crime Reports. Locate information from your hometown. How often does auto theft occur in your hometown compared to the area where you are currently living?

The United States Government Bureau of Justice Statistics provides statistics, reports and other information about all areas of justice. Give an oral report on your findings. http://www.ojp.usdoj.gov/bjs/

For information on United States execution methods go to Amnesty International at http://www.amnesty-usa.org/abolish/methus.html

Write a short paper about your views on capital punishment execution methods.

Practice Tests

Fill-In-The-Blank

Fill in the blank with the appropriate term from the above list of "key terms."

1. _____ is defined as negative evaluation because of a marker that distinguishes groups and their members from others, which becomes the basis of labeling.

2. _____ are more serious acts that are usually subject to institutional sanctions.

3. Deviant acts that the state defines as illegal, which is however controversial in the wider society, are referred to as _____ .

4. In light of the reporting problems associated with official crime statistics, _____ are used to supplement official crime statistics.

5. _____ argues that gangs are a collective adaptation to social conditions.

6. Members of criminal subcultures often express a whole series of rationalizations for their criminal activities, which is referred to as _____ .

7. Labeling may lead to the immersion of a stigmatized juvenile into a deviant subculture, and therefore lead him or her to adopt "delinquent" as his or her _____ .

8. A (An) _____ occurs when wide sections of the public believe that some form of deviance or crime poses a profound threat to society's well being.

9. _____ results from breaking laws.

10. Arson, burglary, and assault are examples of _____ .

Multiple-Choice Questions

Select the response that best answers the question or completes the statement:

1. The experience of Martin Luther King Jr. at the hands of law enforcement in the late 1950s and early 1960s illustrates that:
   a. definitions of deviance and crime change over time.
   b. definitions of deviance and crime vary across social contexts.
   c. deviance and crime are socially constructed.
   d. All of these are correct.
   e. None of the above.

2. Raised eyebrows, shaming, and ostracism are examples of:
   a. deviance
   b. informal punishment
   c. formal punishment
   d. social diversions
   e. social deviations

3. Which is NOT a type or category of deviance and crime according to Brym and Lie?
   a. social diversions
   b. social deviations
   c. conflict crimes
   d. consensus crimes
   e. formal stigmatization

4. _____ is the school of sociological thought that emphasizes that various social problems, including crime, are not inherent in certain actions themselves, but are rather socially created.
   a. Functionalism
   b. Conflict criminology
   c. Social constructionism
   d. Social control theory
   e. Anomie theory

5. Women's situations involving their treatment at the hands of the criminal justice system has improved today. Which is a social source of this improvement?
   a. Women's position in the economy has improved.
   b. Women's rights movement.
   c. Increased public awareness.
   d. All of these are correct.
   e. None of the above.

6. _____ crimes are widely recognized to be bad in and of themselves, and there is little controversy over them.
   a. Conflict
   b. Consensus
   c. Dysfunctional
   d. Functional
   e. Constructive

7. White-collar crimes are disproportionately committed by:
   a. middle and upper classes
   b. lower classes
   c. the working class
   d. white ethnic groups
   e. None of the above.

8. Which is NOT a criticism of official crime statistics?
   a. Much white-collar crime is not reported to the police.
   b. Authorities and the wider public decide which criminal acts to report and which to ignore.
   c. Many crimes are not incorporated in major crime indexes such as victimless crimes.
   d. All of the answers are criticisms of official crime statistics.
   e. None of the above is a criticism of official crime statistics.

9. The long crime wave that began its upswing in the early 1960s eased in the 1980s and _____ in the 1990s.
   a. increased
   b. reached its climax
   c. fell
   d. remained consistent
   e. None of the above.

10. The _____ is the main source of self-report victimization data in the United States.
   a. Uniform Crime Report
   b. National Crime Victimization Survey
   c. U.S. Census
   d. Delinquency Index
   e. Index Crimes

11. Men accounted for _____ of arrests for violent crimes in 2003.
   a. 99%
   b. 77%
   c. 40%
   d. 29%
   e. 13%

12. Statistics indicate that African Americans have disproportionately high arrest, conviction, and incarceration rates. Which is an explanation of this pattern?
   a. Biases in the way crime statistics are collected.
   b. Absence of data on white-collar crime, which is disproportionately committed by whites.
   c. The disproportionate representation of African Americans in the lower classes.
   d. Racial discrimination in the criminal justice system.
   e. All of these are correct.

13. According to Durkheim, _____ gives people the opportunity to define what is moral and what is not.
   a. deviance
   b. control theory
   c. decadence
   d. anomie
   e. social control

14. Which is NOT an adaptation to strain according to Robert Merton?
   a. innovation
   b. ritualism
   c. retreatism
   d. rebellion
   e. stigmatization

15. According to Robert Merton's strain theory, drug dealing would be an example of:
    a. conformity
    b. innovation
    c. ritualism
    d. retreatism
    e. rebellion

16. According to Merton's typology of adaptations to anomie, an individual who accepts society's goals may be:
    a. an innovator or retreatist
    b. a conformist or ritualist
    c. a retreatist or ritualist
    d. an innovator or conformist
    e. a ritualist or innovator

17. Which theory proposes that deviance is tempting to everyone, thus social bonds with legitimate authority figures need to be established at an early age?
    a. differential association
    b. labeling theory
    c. strain theory
    d. control theory
    e. None of the above.

18. The theory of deviance that argues that deviant behavior is caused by the gap between culturally acceptable goals and the legitimate means of achieving them is:
    a. control theory
    b. labeling theory
    c. strain theory
    d. conflict theory
    e. differential association theory

19. Which is NOT a type of social control according to Travis Hirschi and Michael Gottfredson?
    a. attachments
    b. opportunities
    c. convictions
    d. involvements
    e. beliefs

20. Which criminological theory especially highlights the reasons why white-collar crimes are less severely punished than street crimes, regardless of the relative harm to society?
    a. strain theory
    b. control theory
    c. differential association theory
    d. conflict theory
    e. motivational theories

21. Which view of prisons believes that the chief function of the prison is simply to keep criminals out of society as long as possible to ensure they can do no harm?
    a. rehabilitation
    b. deterrence
    c. revenge
    d. incapacitation
    e. None of the above.

22. According to Table 7.1 in the textbook, in 2003 which age cohort was the most crime prone?
    a. 15-19
    b. 20-24
    c. 25-29
    d. 30-34
    e. 35-39

23. Until the early part of the 20th century, people considered cocaine a (an):
    a. medicine.
    b. hallucinogen.
    c. aphrodisiac.
    d. tonic.
    e. herbal supplement.

24. According to Figure 7.2 of the textbook, in 2000 the United States was _____ with respect to contact crime as reported in a definitive self-report survey of 17 countries.
    a. just above average
    b. just below average
    c. considerably above average
    d. considerably below average
    e. exactly average

25. Which type of violent crime remained relatively constant from 1978-2003, according to U.S. FBI statistics and Fig. 7.3?
    a. aggravated assault
    b. rape
    c. murder
    d. All of these are correct.
    e. Both 'B' and 'C.'

26. It is estimated that in 2002 there were _____ youth gangs in the United States with _____ members.
    a. 550; 7800
    b. 21,500; 731,500
    c. 3500; 48,500
    d. 103,500; 1,239,500
    e. 13,500; 284,000

27. Which is a major policy implication of Box 7.1 Social Policy: What Do You Think? "The War on Drugs" in the text?
    a. Drug policy needs to be discussed in a serious way in the United States.
    b. The "just say no" campaign has been highly effective.
    c. Spending most of the drug-control budget on trying to curb the supply of illegal drugs has been effective.
    d. All of these are correct.
    e. None of the above.

28. As depicted in *The Sopranos*, having family members and friends in the Mafia predisposes one to Mafia involvement. Which theory best captures this likelihood?
    a. conflict theory
    b. social control theory
    c. anomie theory
    d. labeling theory
    e. differential association theory

29. According to Fig. 7.6 in the textbook, from around 1995 to 2000, belief in capital punishment in the United States has:
    a. increased.
    b. declined.
    c. remained constant.
    d. All of these are correct.
    e. None of the above.

30. About what percentage of the world's countries allow capital punishment for at least some types of crimes?
    a. 12%
    b. 24%
    c. 38%
    d. 48%
    e. 75%

31. Which explanation of violence in America, does Michael Moore reject in his academy award-winning documentary, *Bowling for Columbine*?
    a. Americans may be violent because of the mass media.
    b. Americans may be violent because they have a violent history and culture.
    c. Americans have a high homicide rate, because guns are so readily available.
    d. All of these are correct.
    e. None of the above.

True False

1. In 1966, 77 percent of Americans believed that the main goal of prison was to rehabilitate prisoners; by 1994 only 16 percent held that position.

    TRUE or FALSE

2. The number of people behind bars in the United States has leveled off in recent years.

    TRUE or FALSE

3. Because women have long been perceived as in need of chivalrous protection, the law has often been biased in favor of protecting women.

    TRUE or FALSE

4. White-collar crimes are equally stigmatized as street crimes.

    TRUE or FALSE

5. Many sociologists argue that white-collar crimes are more costly than street crimes.

    TRUE or FALSE

6. Whites accounted for 70.6% of arrests in 2003.

    TRUE or FALSE

7. Members of criminal subcultures tend to be strict conformists to the norms of their own culture.

    TRUE or FALSE

8. According to conflict theory, thieves challenge private property, thus theft is a crime.

    TRUE or FALSE

9. The degree and form of social control tends to remain consistent over time.

    TRUE or FALSE

10. The more complex a society, the less likely many norms will be widely shared.

    TRUE or FALSE

11. The moral panic regarding the crime problem of the 1980s and 1990s occurred when crimes reached epidemic proportions.

    TRUE or FALSE

Short Answer

1. Briefly identify and explain three dimensions of the various types of deviance according to John Hagan. (p. 185-186)

    _____

    _____

2. Explain the premise of social constructionism in terms of the creation of deviance and crime in the context of social power. (p. 187-189)

_____

_____

3. List and define Hirschi and Gottfredson's four types social control. (p. 198-199)

_____

_____

4. Explain the major premise of labeling theory. (p. 198)

_____

_____

5. List and define four views of prison. (p. 202-203)

_____

_____

6. Define "moral panic" and provide an example of a moral panic. (p. 203-207)

_____

_____

Essay Questions

1. Compare and contrast a motivational and constraint theory of your choice. Be sure to provide a substantive example to illustrate the differences between the theories. (p. 195-200)

2. Explain why white-collar crimes are less stigmatized and result in fewer prosecutions in comparison to street crimes according to labeling theory and conflict theory. Be sure to integrate both theories. (p. 188-189,198,199-200)

3. Compare and contrast the strengths and weaknesses of official crime statistics and self-report surveys. (p. 189-193)

4. Explain the concept "techniques of neutralization" in terms of delinquent subcultures. Illustrate the concept by describing the process of delinquent gang formation. (p. 196-200)

5. Explain the process of becoming a marijuana user according to Howard S. Becker's research of jazz musicians and his theory of differential association. (p.193-196)

Solutions

Practice Tests

Fill-In-The-Blank

1. Stigmatization
2. Social deviations
3. conflict crimes
4. self-report surveys
5. Subcultural theory
6. techniques of neutralization
7. master status
8. moral panic
9. Formal punishment
10. street crimes

Multiple-Choice Questions

1. D, (p. 184)
2. B, (p. 184)
3. E, (p. 186)
4. C, (p. 187)
5. D, (p.187)
6. B, (p. 186)
7. A, (p. 188)
8. D, (p. 189)
9. C, (p. 189-192)
10. B, (p. 189)
11. B, (p. 192)
12. E, (p. 192-193)
13. A, (p. 195)
14. E, (p. 196)
15. B, (p. 196)
16. D, (p. 196)
17. D, (p. 198-199)
18. C, (p. 196)
19. C, (p. 198-199)
20. D, (p. 199-200)
21. D, (p. 203)
22. A, (p. 192)
23. A, (p. 185)
24. B, (p. 189)
25. E, (p. 190)
26. B, (p. 194)
27. A, (p. 194)
28. E, (p. 197)
29. B, (p. 204)
30. D, (p. 204)
31. D, (p. 205)

True False

1. T, (p. 203)
2. F, (p. 203)
3. F, (p. 187)
4. F, (p. 188-189)
5. T, (p. 188-189)
6. T, (p. 192)
7. T, (p. 196-197)
8. T, (p. 199-200)
9. F, (p. 201-202)
10. T, (p. 201)
11. F, (p. 203-207)

STRATIFICATION: UNITED STATES AND GLOBAL PERSPECTIVES

Student Learning Objectives

After reading Chapter 8, you should be able to:

1. Define stratification in terms of the gap between the rich and poor in the United States.

2. Describe the trends involving income and wealth inequality in the United States.

3. Identify and describe the categories of the upper class—that is the upper-upper class versus the lower-upper class, and "Old Money" versus "New Money."

4. Explain the impact of high-tech industries in the 1990s on the economy and social stratification in the United States.

5. Explain global inequality and discuss cross-national variations in internal stratification.

6. Describe the development of internal stratification in terms of societal development in general.

7. Discuss and contrast the contribution of Karl Marx and Max Weber to the sociological understanding of inequality and stratification.

8. Explain stratification and mobility according to the functionalist perspectives of Kingsley Davis and Wilbert Moore, and Peter Blau and Otis Dudley Duncan.

9. Explain the contemporary class theories of Erik Olin Wright and John Goldthorpe.

10. Contrast functionalist and contemporary class theories of stratification.

11. Describe the role of noneconomic dimensions of social class, including prestige, taste, and politics.

12. Explain the impact of government policy initiatives on poverty rates and perceptions of poverty.

13. Describe American perceptions of social class inequality in the United States based on cited research.

14. Dispel the myths of poverty and poor people based on sociological research.

Key Terms

social stratification (214)

vertical social mobility (219)

global inequality (222)

cross-national variations in
internal stratification (222)

The Gini index (224)

ascription (225)

achievement (225)

caste (225)

apartheid (225)

feudalism (227)

class consciousness (227)

class (227)

bourgeoisie (228)

proletariat (228)

petty bourgeoisie (228)

status groups (229)

parties (229)

the functional theory of stratification (230)

socioeconomic index of
occupation status (SEI) (233)

socioeconomic status (SES) (233)

intragenerational mobility (234)

intergenerational mobility (234)

structural mobility (234)

poverty rate (239)

Detailed Chapter Outline

I. SOCIAL STRATIFICATION

   A. Shipwrecks and Inequality
      1. Stories of shipwrecks and survivors provide space to make a point of social inequality—that allows the sweeping away of social convention and privilege.
      2. Examples:
         a. 1719 *Robinson Crusoe* portrays capitalism favorably, and the importance of hard work and inventiveness to get rich.
         b. Three messages in 1875 *Swept Away*:
            i.   It is possible to be rich without working hard because one can inherit wealth.
            ii. One can work hard without becoming rich.
            iii. Something about the structure of society causes inequality.
         c. 1997 *Titanic* shows, on the one hand that class differences are important. On the other hand, the movie suggests that class is insignificant in its optimistic tale of the American Dream.
      3. **Social stratification** is the way society is organized in layers or strata.

II. PATTERNS OF SOCIAL INEQUALITY

   A. Wealth
      1. *Wealth* is what you own.
         a. The gap between the rich and poor in the United States is tremendous.

      i.   For example, in 1995 George Soros, an American currency speculator, earned $1.5 billion, while the annual income of a minimum-wage worker was $8,840. Yet Soros is not the richest person in America ranking 24th.

      ii. In the mid-1990s, the richest 1% of American households owned nearly 39% of all national wealth, while the richest 10% owned almost 72%. The poorest 40% of American households owned a meager 0.2% of all national wealth. The bottom 20% has a negative net worth, which means they owed more than they owned.

  b. Wealth inequality has been increasing since the early 1980s. Sixty-two percent of the increase in national wealth in the 1990s went to the richest 1% of Americans, 99% of the increase went to the richest 20%. The United States has surpassed all other highly industrialized countries in wealth inequality.

  c. Only a modest correlation exists between wealth and income.

  2. Income

    a. *Income* is what you earn in a given period.

    b. Stratification is often divided into two categories of inequality:

      i.   *Income classes*- differ in lifestyles.

      ii. *Income strata*—equal-sized statistical categories.

    c. Two important facts about stratification:

      i.   Inequality has been rising in the United States for more than a quarter of a century. In 1974 the top fifth of households earned 9.8 times more than the bottom fifth. By 2003, the top fifth earned 14.6 times more than the bottom fifth. For the first time since data became available on the subject, the top 20% of households earned more than half of all of the national income, in 2003.

      ii. The middle 60% of income earners have been "squeezed" during the past 3 decades, with their share of national income falling from 52.5% to 46.9% of the total.

B. Income Classes

  1. Two categories of the upper class include the *upper-upper class* and *lower-upper class*.

  2. "Old Money"

    a. Comprises 1% of the U.S. population.

    b. Inherited wealth earned in older industries such as banking, insurance, and oil.

    c. Inhabited elite East Coast neighborhoods.

  3. "New Money"

    a. Substantial amount of "new money" entered the upper-upper class in the 1990s—booming high-tech industries.

    b. Concentrated in high-tech meccas in the West, such as San Francisco, San Mateo County, and Santa Clara County.

    c. Still overwhelmingly white and non-Hispanic.

C. How Green is the Valley?

  1. Although high-tech industries witnessed the longest economic boom in the 1990s it changed the pattern of social stratification in the United States, encouraged upward mobility, and increased

the median income. However, the changes were not always positive. It increased the cost of living and widened the gap between the rich and the poor.

2. High-technology industry helped to create a new division in the lower-upper class—the "Poor Rich."

3. High-technology industry squeezed the middle class and encouraged downward mobility
   a. This is referred to as **vertical social mobility**, or the movement up the stratification system ("upward mobility").

4. High-technology industry lowered the value of unskilled work, swelling the rank of the poor.

III. GLOBAL INEQUALITY

A. International Differences
   1. There is also a gap in income and wealth *between* countries, as there is a gap *within* countries. (i.e. United States, Canada, Japan, Australia, Germany, France, and Britain versus Angola, Africa, South Africa, South America, and Asia).
      a. **Global inequality** refers to the differences in the economic ranking of countries.
      b. **Cross-national variations in internal stratification** are differences between countries in their stratification systems.
      c. Inequality between rich and poor countries is staggering.
         i. Nearly a fifth of the world's population lack adequate shelter, and over a fifth lacks safe water.
         ii. About a third of the world's people are without electricity and over two-fifths lack adequate sanitation.
         iii. While the annual health expenditure in 2002 in the United States was $4887 per person, it was $70 for Angola.
         iv. Poor countries are more likely to experience mass suffering
      d. Much of the wealth of rich countries has been gained at the expense of poor countries, in terms of colonization, cheap labor, raw materials, economic dependence, and multinational companies.
   2. Measuring Internal Stratification
      a. The **Gini index** is the measure of income inequality. Its value ranges from zero (which means that every household earn exactly the same amount of money) to one (which means that all income is earned by a single household).
      b. The United States has the highest Gini index among highly developed, rich countries (0.368).

B. Economic Development and Internal Stratification
   1. What accounts for cross-national differences in internal stratification? *Political factors* and *socioeconomic development*.
   2. Socioeconomic development and internal stratification: over human history, as societies became richer and more complex, the level of social inequality first increased, then tapered off, then began to decline, and finally began to rise again.
      a. Foraging Societies
         i. First 90,000 years of human existence.

      ii.   Hunting and foraging.

     iii.   Shared food to ensure survival and subsistence.

     iv.   No rich and poor.

  b.   Horticultural and Pastoral Societies

      i.   About 12,000 years ago.

      ii.   First agricultural settlements.

     iii.   Based on horticulture (the use small hand tools to cultivate plants), and pastoralism (the domestication of animals).

     iv.   Production of wealth and surplus beyond subsistence.

      v.   Significant social stratification emerged.

  c.   Agrarian Societies

      i.   About 5,000 years ago.

      ii.   Plow agriculture.

     iii.   Surpluses grew.

     iv.   Religious beliefs justified growing inequality (i.e. "divine right")

      v.   Stratification was based on **ascription** (allocation of rank depends on the features a person is born with) than **achievement** (allocation of rank depends on one's accomplishments), - and there was little social mobility.

     vi.   Agrarian India was nearly purely ascriptive, and was divided into **castes** (occupation and marriage partners assigned on the basis of caste membership).

    vii.   **Apartheid** (caste system based on race) in South Africa is an example of caste systems in industrial times.

  d.   Industrial Societies

      i.   The tendency of industrialism to lower the level of stratification was not apparent in the first stages of industrial growth.

      ii.   Improvements in technology and social organization of manufacturing raised the living standards of the entire population

     iii.   Political pressure also played a role in reducing inequality (i.e. labor unions).

     iv.   Thus stratification declined as industrial societies developed.

  e.   Postindustrial Societies

      i.   In the United States and the United Kingdom, social inequality has been increasing for the last quarter century. The concentration of wealth in the hands of the top 1% is higher today than anytime in the past 100 years. The gap between the rich and the poor is greater today than is has been for at least 50 years. Other postindustrial societies, such as Germany, France, and Canada, social inequality has remained constant since the mid-1970s.

      ii.   Many high-tech jobs have been created at the top of the stratification system over the past few decades, which pay well. At the same time, new technologies make jobs routine, require little training, and pay poorly. Because the number of routine jobs is growing faster than the number of jobs at the top, the overall effect of technology today is to increase social inequality.

IV. THEORIES OF STRATIFICATION

A. Classical Perspectives
 1. Marx
   a. Marx observed economic changes in medieval Europe in the 1400s, 1600s, 1700s, during a time when society was arranged according to **feudalism**, a legal arrangement in preindustrial Europe that bound peasants to the land and obliged them to give their landlords a set part of the harvest. In exchange, landlords were required to protect peasants from marauders and open their storehouses and feed the peasants if crops failed.
   b. By the 1600s and 1700s, urban craftsman and merchants expanded production, and needed more workers to increase profits. Feudalism had to be destroyed so that peasants, who were legally bound to the land, could be turned into workers.
   c. The capitalist class grew richer and smaller, while the workers grew larger and more impoverished.
   d. Marx thought that workers would enhance their sense of **class consciousness**, or become more aware of belonging to the same exploited class, which would lead to the growth of labor unions and political parties to the point of creating a *communist system* based on shared wealth.
   e. Several points of Marx's theory:
     i. A person's **class** is determined by the *source* of his or her income, or one's "relationship to the means of production." The **bourgeoisie** are members of the capitalist class who own the means of production, including factories, tools, and land, and are in position to make profit. The **proletariat** are members of the working class who do physical labor, and do not own the means of production, —and earn wages.
     ii. There are more than two classes in any society. For example, the **petty bourgeoisie** is made up of small-scale capitalists who own the means of production but employ only a few workers or none at all. According to Marx, the petty bourgeoisie are bound to disappear as capitalism develops.
     iii. Some of Marx's predictions about the development of capitalism turned out to be wrong.
 2. A Critique of Marx
   a. Industrial societies did not polarize into two opposed classes engaged in bitter conflict. Instead, a large heterogeneous middle class of "white-collar" workers emerged, who have acted as a stabilizing force of capitalism.
   b. Marx did not expect investment in technology to make higher wages and toil fewer hours under less oppressive conditions for workers.
   c. Communism took root not where industry was most highly developed, but in semiindustrialized countries such as Russia in 1917 and China in 1947. And new forms of privilege emerged under communism, not a classless society of shared wealth (i.e. nomenklatura in communist Russia).
 3. Weber
   a. Weber foretold most of these developments Marx failed to see.
   b. Weber also saw classes as economic categories, but not in terms of a single criterion. For Weber, *class position* is determined by "market situation," including the possession of goods, opportunities for income, level of education, and degree of technical skill.

      c. Four main classes:
- i.   Large property owners.
- ii. Small property owners.
- iii. Propertyless, but relatively highly educated and well-paid employees.
- iv. Propertyless manual workers.

      d. Two types of groups other than class:
- i.   **Status groups** are distinguishable in the prestige and social honor they enjoy and in their life-style.
- ii. **Parties** are organizations that seek to impose their will on others, not just political groups.

B. An American Perspective: Functionalism
1. Functionalism
    a. Kingsley Davis and Wilbert Moore's **functional theory of stratification** argues that:
- i.   Some jobs are more important than others,
- ii. People have to make sacrifices to train for important jobs, and
- iii. Inequality is required to motivate people to undergo these sacrifices.

    b. Chief flaws of Davis and Moore's theory.
- i.   The question of which occupations are most important is not as clear-cut as Davis and Moore suggest. From a historical perspective, *none* of the jobs regarded by Davis and Moore as "important" would exist without the physical labor done by people in "unimportant" jobs through the ages.
- ii. Functional theory ignores the pool of talent lying undiscovered *because of* inequality.
- iii. Functional theory fails to examine how advantages are passed from generation to generation.

V. SOCIAL MOBILITY: THEORY AND RESEARCH

A. The opportunity to rise to the top is the crux of the American Dream, and to many observers, the main difference between the United States and Europe. The United States came to be viewed as a land where hard work and talent could easily overcome humble origins.
1. The rate of upward social mobility has been higher in America than Europe in the 19th century. However, there was little difference by the second half of the 20th century.
    a. The idea of America as a classless society has persisted.
2. Blau and Duncan: The Status Attainment Model
    a. Claimed stratification in America is based mainly on individual achievement and merit versus ascription and inheritance.
    b. Stratification system is not made up of distinct groups or classes, but rather a continuous hierarchy of occupations with hundreds of rungs.
    c. Saw little if any potential for class consciousness and action in the United States.
    d. Created a **socioeconomic index of occupation status (SEI)**, which combines for each occupation, average earnings and years of education of men employed full-time in the occupation. Other researchers then combined income, education, and occupational prestige data to construct an index of **socioeconomic status (SES)**.

    e. Their focus on the effects of family background and educational level on occupational achievement became known as the "status attainment model."

    f. Peter Blau and Otis Dudley Duncan's subsequent findings:

        i. Confirmed that the rate of social mobility for men in the United States is high and that most has been upward.

        ii. Since the early 1970s, substantial downward mobility has occurred.

        iii. Mobility for men within a single generation (**intragenerational mobility**) is modest, while mobility for men over more than one generation (**intergenerational mobility**) can be substantial.

        iv. Most social mobility is due to change in the occupational structure, which became known as **structural mobility**.

        v. There are only small differences in rates of social mobility among the highly industrialized countries.

  3. A Critique of Blau and Duncan

    a. Blau and Duncan sampled only men employed full-time, thus excluding women, part-time and unemployed workers.

    b. A disproportionately large number of unemployed and part-time workers are African American or Hispanic Americans.

    c. How can one make valid claims about the relative importance of ascription versus achievement when many of the most disadvantaged people in society are excluded from one's analysis?

    d. Group Barriers: Race and Gender

        i. Subsequent research confirmed that the process of status attainment is much the same for women and minorities as it is for white men (i.e. years of schooling have a greater influence than does father's occupation). However, when you compare people with the same level of education and similar family backgrounds, women and minority group members tend to attain lower status than white men.

B. The Revival of Class Analysis

  1. An adequate theory of class stratification must do two things:

    a. It must specify criteria that distinguish a *small number* of distinct classes.

    b. It must spark research that demonstrates *substantial gaps* between classes in economic opportunities, political outlooks, and cultural styles.

  2. Wright

    a. Update of Marx's class scheme. Wright's basic distinction is between property owners and nonowners. He also distinguishes between large, medium, and small owners.

    b. Three propertied classes—bourgeoisie, small employers, and petty bourgeoisie.

    c. Nine nonpropertied classes, with "expert managers" and proletariats at each extreme. Wage and salary earners differ in terms of:

        i. Different "skill and credential levels;" and

        ii. Different "organization assets" or decision-making authority.

  3. Goldthorpe

    a. Different classes are characterized by different "employment relations,"—employers, self-employed people, and employees (i.e. large and small employers, self-employed in agriculture and those outside agriculture, professionals, and those with labor contracts).

    b. Distinction of Goldthorpe's class scheme versus Wright's:

        i. The proletariat is made up of three classes of workers divided by skill and sector.

       ii. Groups large employers with senior managers, professionals, administrators, and officials, based on level of income and authority, political interests, lifestyle, etc.

VI. NONECONOMIC DIMENSIONS OF CLASS

A. Prestige and Power

    1. Elaborating on Weber, *status groups* distinguish their rank by means of material and symbolic culture in addition to prestige and honor, such as "taste" in fashion, food, music, literature, manners, travel, and so forth.

    2. Cultural objects that represent the best taste tend to be less accessible. Education and financial reasons account for accessibility.

    3. Example: Bach's *The Well-Tempered Clavier* versus Gershwin's *Rhapsody in Blue*.

    4. Clothing can act as a language that signals one's status to others.

B. Politics and the Plight of the Poor

    1. Politics has a profound impact on the distribution of opportunities and rewards in society (i.e. changing laws governing people's right to own property, entitlements to welfare benefits, and redistributing income through tax policies).

    2. Two main currents of American opinion on poverty, which correspond to the Democratic and Republican positions.

        a. Most Democrats want the government to play a role in helping to solve the problem of poverty.

        b. Most Republicans want to reduce government involvement with the poor so people can solve the problem themselves.

    3. Fluctuations in the **poverty rate** (the percentage of people living below the poverty threshold, which is three times the minimum food budget established by the United States Department of agriculture) reveals the impact of government policy on poverty.

    4. Government Policy and the Poverty Rate in the United States

        a. The 1930's: The Great Depression

            i. Introduction of Social Security, Unemployment Insurance, and Aid to Families with Dependent Children (AFDC) by President Franklin Roosevelt in response to the Great Depression in the mid-1930s.

        b. The 1960's: The War on Poverty

            i. President Lyndon Johnson's "War on Poverty" in the mid-1960s, in response to a new wave of social protest. Millions of southern blacks that migrated to the northern and western cities in the 1940s and 1950s were unable to find jobs.

        c. The 1980's: "War Against the Poor"

        i.   President Ronald Reagan from the 1980s cut government services and taxes that fund poor relief (i.e. AFDC budget, expenditures for employee training), based on the rationale that claims that relief worsens the problem of poverty.

    d. Poverty Myths

        i.   *Myth 1: The overwhelming majority of poor people are African- or Hispanic-American single mothers with children.*

       ii. *Myth 2: People are poor because they don't want to work.*

      iii. *Myth 3: Poor people are trapped in poverty.*

      iv. *Myth 4: Welfare encourages married women with children to divorce so they can collect welfare, and it encourages single women on welfare to have more children.*

       v. *Myth 5: Welfare is a strain on the federal budget.*

C. Perception of Class Inequality in the United States

    1. Surveys show that few Americans have trouble placing themselves in the class structure when asked to do so (i.e. General Social Survey).

    2. If Americans see the stratification system as divided by class, they also know that the gaps between classes are relatively large.

    3. Most Americans do not think big gaps between classes are needed to motivate people to work hard.

    4. Based on the 18-nation survey, most Americans agreed with the statement "inequality continues because it benefits the rich and powerful."

    5. Americans tend to be opposed to the government playing an active role in reducing inequality.

    6. Attitudes vary by class position.

Study Activity: Applying the Sociological Compass

After reading the chapter, study the classical, functionalist, and revived class theories of stratification—and re-read Box 8.3, You and the Social World "Perceptions of Class". The authors ask you a set of questions to consider, which are listed below. Answer the questions as suggested in the textbook. In addition, take it a step further and attempt to explain and support your answers based on the various theories and research evidence you learned in Chapter 8.

- "Do you think the gaps between the classes in American society are big, moderate, or small?" Why?

- "How strongly do you agree or disagree with the view that big gaps between classes are needed to motivate people to work hard and maintain national prosperity? What is the theoretical and research evidence to support your position?"

- "How strongly do you agree or disagree with the view that inequality persists because ordinary people don't join together to get rid of it?" Why?

InfoTrac College Edition Online Exercises

For the following exercises, log on to the online library of InfoTrac College Edition at http://www.infotrac-college.com/. Make note that InfoTrac has implemented a new registration system that will allow easier access to InfoTrac through the use of a personalized username and password. Once you've created your username and password you may proceed directly to the Log On page. To create an account, register your passcode packaged with your textbook, and create a username and password, by following the online prompt. After you are logged in, click on "Infotrac College Edition." You will arrive at a screen that enables you to search topics.

Keyword: **Social mobility** and **poverty**. Identify and read a few articles that discuss recent trends in mobility and poverty. Read the articles and make note of the major trends. Are the findings consistent with the research cited in Chapter 8? Make a list of supporting and refuting evidence.

Keyword: **American Dream**. What are the different ways that this term is used? How do the different ways relate to stratification?

Keyword: **global poverty**. Do a search under this topic. What are the key issues concerning global poverty? How do they relate to stratification in the United States?

Keyword: **black market**. What are the different ways that this term is used in the research literature? How do the different ways reflect stratification in our society?

Keyword: **runaways**. Look for ways that runaways earn their living. What are their main occupations? What are the economic issues facing runaways?

Internet Exercises

For information on the 400 richest people in America, visit the annual compilation of *Forbes* business magazine at: http://www.forbes.com/2002/09/13/rich400land.html

Look at the characteristics that the 400 richest people have. What commonalities and differences do you see? Compile a list of the differences and similarities.

Look up http://www.unicef.org/ to see how this United Nations Organization views global stratification. Compare it to the Karl Marx perspective of stratification.

Human trafficking has been a major news event in recent years. Children and women are most often sold into slavery. The Commission on Human Rights has issued a State Human Rights Report for 2001 in Thailand. Go to: http://www.hrw.org/wr2k1/asia/thailand.html and write a short paper about what they found in Thailand about this phenomenon.

The Bureau Labor Statistics provides current statistics for economic issues. Their Economy at a Glance page provides regional and state date. Go to: http://www.bls.gov/eag and find information about your state and region. Write a short paper discussing your findings.

To read about wage inequality go to: http://www.aei.org/cs/cs6931.htm

This takes you to a report called *Wage Inequality: International Comparisons of Its Sources* published by The American Enterprise Institute for Public Policy Research. Compare the United States to other countries on wage inequality. Discuss your findings in the class.

Practice Tests

Fill-In-The-Blank

Fill in the blank with the appropriate term from the above list of "key terms."

1. _____ are differences between countries in their stratification systems.

2. Fluctuations in the _____ reveals the impact of government policy on poverty.

3. Social mobility within a single generation is referred to as _____ .

4. _____ combines, for each occupation, average earnings and years of education in the occupation.

5. _____ is the measure of income inequality, and its value ranges from zero to one.

6. Stratification based on _____ depends on one's accomplishments.

7. Marx referred to small-scale capitalists who employed only a few workers if any, as _____

8. _____ are organizations that seek to impose their will on others.

9. Social mobility due to change in the occupational structure is called _____ .

10. _____ in South Africa is an example of a caste system in industrial times.

Multiple-Choice Questions

Select the response that best answers the question or completes the statement:

1. _____ is the way society is organized in layers or strata.
   a. Social status
   b. Social mobility
   c. Social stratification
   d. Economic inequality
   e. Income strata

2. _____ differ in their lifestyle, while _____ are equal-sized statistical categories.
   a. Income strata; income classes
   b. Income classes; income strata
   c. Wealthy groups; income classes
   d. Income strata; status groups
   e. Status groups; income classes

3. Which is an important fact about stratification in the United States?
   a. Inequality has been rising for the past quarter of a century.
   b. By 2001, the top fifth of households earned 14.3 times more than the bottom fifth.
   c. During the past 25 years, the middle 60% of income earners have been "squeezed," with their share of national income falling from 52.5% to 46.4% of the total population.
   d. All of these are correct.
   e. None of the above.

4. The emergence of "new money" in the United States in the 1990s is attributable to:
   a. inherited wealth.
   b. wealth earned in industries such as banking, insurance, and oil.
   c. booming high-tech industries.
   d. All of these are correct.
   e. None of the above.

5. Which is NOT a way that high-tech industries have widened the gap between the rich and the poor?
   a. High-technology industry helped to create a new division in the lower-upper class—the "poor rich."
   b. High-technology industry squeezed the middle class and encouraged downward mobility.
   c. High-technology industry lowered the value of unskilled work, swelling the rank of the poor.
   d. High-technology industry decreased the cost of living.
   e. All of the answers are ways the high-tech industry widened the gap between the rich and poor.

6. _____ is the movement up the stratification system.
   a. Vertical social mobility
   b. Horizontal social mobility
   c. Lateral social mobility
   d. Enhance social mobility
   e. None of the above.

7. Which term refers to the differences in the economic ranking of countries?
   a. cross-national variations in internal stratification
   b. Gini index
   c. intragenerational mobility
   d. global inequality
   e. None of the above.

8. Which is a fact regarding inequality between rich and poor countries?
   a. Nearly a fifth of the world's population lack adequate shelter, and over a fifth lacks safe water.
   b. About a third of the world's people are without electricity and over two-fifths lack adequate sanitation.
   c. While the annual health expenditure in the United States is $2,765 per person, it is $4 for Tanzania and Sierra Leone, and $3 for Vietnam.
   d. Poor countries are more likely to experience mass suffering.
   e. All of these are correct.

9. Which type of society is characterized by sharing for survival and subsistence with no rich and poor?
   a. foraging societies
   b. horticultural societies
   c. pastoral societies
   d. agrarian societies
   e. industrial societies

10. Evidence shows that inequality has been _____ in postindustrial societies like the United States.
    a. remaining constant
    b. tapering off
    c. increasing
    d. decreasing
    e. None of the above.

11. With industrialization, improvements in technology and social organization of manufacturing led to a (an) _____ in stratification in comparison to the agrarian period.
    a. increase
    b. dramatic climax
    c. decline
    d. All of these are correct.
    e. None of the above.

12. According to Marx, a crucial determinant of one's class position is one's:
    a. possession of cultural capital.
    b. occupation in powerful positions.
    c. political influence.
    d. relationship to the means of production.
    e. degree of upward mobility.

13. Which term did Marx refer to as the working class who do physical labor?
    a. bourgeoisie
    b. proletariat
    c. petty bourgeoisie
    d. parties
    e. status groups

14. Which is NOT a major criticism of Marx's theory of social class?
    a. Marx failed to recognize that investment technology would make high profits for capitalists possible.
    b. Industrial societies did not polarize into two opposed classes in the way Marx predicted.
    c. Marx did not expect investment in technology to make higher wages and less oppressive conditions for workers.
    d. Communism did not take root where industry was most highly developed.
    e. All of the answers are major criticisms.

15. Weber referred to _____ as distinguishable based on their prestige and social honor.
    a. class position
    b. expert managers
    c. parties
    d. status groups
    e. bourgeoisie

16. Which theorist(s) argued that inequality is necessary to motivate people to make sacrifices and work hard?
    a. Marx and Engels
    b. Weber
    c. W.E.B. Du Bois
    d. Davis and Moore
    e. Comte

17. Who created the "socioeconomic index of occupation status" (SEI)?
    a. Weber
    b. Blau and Duncan
    c. Davis and Moore
    d. Marx
    e. Engels

18. An adequate theory of class stratification must:
    a. specify criteria that distinguish a small number of distinct classes.
    b. spark research that demonstrates substantial gaps between classes.
    c. identify substantial gaps between classes in terms of economic opportunities, political outlooks, and cultural styles.
    d. All of these are correct.
    e. None of the above.

19. Taste in fashion, food and music are indicators of:
    a. class
    b. status groups
    c. political parties
    d. proletariats
    e. None of the above.

20. In the 1980s, President Ronald Reagan:
    a. enhanced social security, unemployment insurance, and Aid to Families with Dependent Children.
    b. continued the "War on Poverty" that Lyndon Johnson initiated in response to a contemporary wave of social protest from the working class.
    c. expanded government services and taxes to aid the poor.
    d. Cut government services and taxes to fund relief for the poor.
    e. None of the above.

21. The 1997 movie *Titanic* depicts a story of stratification and love, which:
    a. proves the sociological idea that love can overcome inequality.
    b. concludes that social stratification has no effect on love.
    c. suggests that social class overcomes love, and doesn't allow it to blossom.
    d. All of these are correct.
    e. None of the above.

22. Horatio Alger's books from the post-Civil War era to the end of the 19th century:
    a. depicted the brutal realities of social inequality in America.
    b. inspired Americans with tales of how courage, faith, honesty, hard work, and a little luck could help young people rise from rags to riches.
    c. inspired young readers to fight for social justice and social change toward an equal society.
    d. All of these are correct.
    e. None of the above.

23. George Soros was the _____ richest person in the United States in 2004.
    a. 2nd
    b. 10th
    c. 18th
    d. 24th
    e. 38th

24. In 2001, the annual household income of the lower-upper class was:
    a. $1 million.
    b. $100,000-$999,999
    c. $57,500-$99,999
    d. $37,500-$57,499
    e. $20,000-$37,499

25. As a real-life indicator of the concentration of wealth at the top of the social hierarchy in the United States, Bill Gates, the world's richest man, lives in a house that is valued at:
    a. more than $53 million.
    b. around $23 million.
    c. between $10 million and $17 million.
    d. around $8 million.
    e. None of the above.

True False

1. The debt of a significant proportion of Americans is greater than their assets.

   TRUE or FALSE

2. Income inequality has been increasing in the United States for three decades.

   TRUE or FALSE

3. The emergence of "new money" in the United States created proportionate representation of Hispanics in the upper-upper class.

   TRUE or FALSE

4. Within agrarian societies, stratification is based on ascription.

   TRUE or FALSE

5. Apartheid is a class system based on race.

   TRUE or FALSE

6. In contrast to Karl Marx, Max Weber's multidimensional approach to stratification negated the idea of classes as economic categories.

   TRUE or FALSE

7. Weber foretold most of the developments Marx failed to see.

   TRUE or FALSE

8. Weber's theory is considered a functionalist approach to stratification among most contemporary sociologists.

   TRUE or FALSE

9. Most Americans want to reduce government involvement with the poor so people can help themselves.

   TRUE or FALSE

10. Most Americans think big gaps between classes are needed to motivate people to work hard.

    TRUE or FALSE

11. The Gini index indicates that the United States has the highest level of income inequality, among wealthy postindustrial countries.

    TRUE or FALSE

Short Answer

1. Distinguish between intragenerational and intergenerational mobility. (p. 233-234)

   _____

   _____

2. What is apartheid? Be sure to explain how apartheid in South Africa is a contemporary form of a caste system based on race. (p. 225-226)

   _____

   _____

3. Identify three flaws of Davis and Moore's functional theory of stratification. You must first briefly articulate their major assumptions. (p. 233-235)

   _____

   _____

4. Briefly explain how Wright's theory updates Marx's class statement on social class. (p.235)

   _____

   _____

5. List three prevailing perceptions of class inequality among Americans based on major research findings. (p. 242-243)

   _____

   _____

6. Describe three ways the high-tech industry further widened the gap between the rich and poor in the United States in the 1990s. (p. 221-222)

   _____

   _____

Essay Questions

1. Describe the development and transformation of internal stratification in terms of the socioeconomic development of societies across foraging, horticultural and pastoral societies, agrarianism, and industrialization. (p. 224-227)

2. Compare and contrast any two of the following stratification theorists—Marx, Weber, Wright, and Goldthorpe. (p. 227-229, 236-237)

3. Contrast Davis and Moore's functional theory of stratification and Goldthorpe's theory of social class. (p. 229-232, 236-237)

4. Many sociologists claim that the United States took several steps backwards in the quest to alleviate problems of poverty, largely as a result of President Ronald Reagan's approach to the problem of poverty. Identify the five myths of Reagan's so-called "war against the poor"—and describe the research that dispelled these myths in the 1990s. (p. 240-245)

5. Compare and contrast how Marx and Weber used the word "class". Describe similarities and differences. What is your definition of the word? (p. 227-232)

Solutions

Practice Tests

Fill-In-The-Blank

1. Cross-national variations in internal stratification
2. poverty rate
3. intragenerational mobility
4. Socioeconomic index of occupation status (SEI)
5. The Gini index
6. achievement
7. petty bourgeoisie
8. Parties
9. structural mobility
10. Apartheid

Multiple-Choice Questions

1. C, (p. 214)
2. B, (p. 216)
3. D, (p. 213-222)
4. C, (p. 218)
5. D, (p. 221)
6. A, (p. 219)
7. D, (p. 222)
8. E, (p. 222-223)
9. A, (p. 224-225)
10. C, (p. 226)
11. C, (p. 226)
12. D, (p. 227)
13. B, (p. 228)
14. A, (p. 228)
15. D, (p. 229)
16. D, (p. 230-232)
17. B, (p. 233)
18. D, (p. 236)
19. B, (p. 229)
20. D, (p. 241)
21. E, (p. 214)
22. B, (p. 214)
23. D, (p. 215)
24. B, (p. 197)
25. A, (p. 218)

True False

1. T, (p. 216)
2. T, (p. 217)
3. F, (p. 217-218)
4. T, (p. 225)
5. T, (p. 225-226)
6. F, (p. 228-229)
7. T, (p. 228-229)
8. F, (p. 228-232)
9. T, (p. 244-245)
10. F, (p. 243)
11. T, (p. 224)

GLOBALIZATION, INEQUALITY, AND DEVELOPMENT

Student Learning Objectives

After reading Chapter 9, you should be able to:

1. Describe the historical evolution of globalization.

2. Discuss the advantages and disadvantages of globalization.

3. Explain the influence of globalization on the everyday lives of people.

4. Explain the importance of technology, politics, and economics on globalization.

5. Discuss the degree to which globalization is homogenizing the world on the one hand, and strengthening differences on the other hand.

6. Discuss rising antiglobalization and anti-Americanism.

7. Explain the concept of "development" in the context of global inequality.

8. Contrast modernization theory and dependency theory as explanations of global inequality, development, and underdevelopment.

9. Identify and explain the new means of exploitation of poor countries by rich countries, according to dependency theory.

10. Explain Immanuel Wallerstein's "world system" approach to capitalist development and global inequality.

11. Identify and describe the three types of countries according to the world system approach.

12. Describe and explain the differences between semiperipheral and peripheral countries.

13. Debate the advantages and shortcomings of neoliberal globalization.

14. Discuss the prospects for democratic globalization, and the contradictory support by the U.S. government of both antidemocratic and democratic regimes in the developing world.

Key Terms

imperialism (252)

global commodity chain (253)

transnational corporations (255)

McDonaldization (256)

glocalization (257)

regionalization (257)

colonialism (260)

modernization theory (263)

dependency theory (264)

core (266)

peripheral (266)

semiperipheral (266)

neoliberal globalization (268)

Detailed Chapter Outline

I. THE CREATION OF A GLOBAL VILLAGE

A. The world seems a much smaller place today than it did 25 years ago. Some people say that we have created a "global village." Is this uniformly beneficial, or does it have a downside too?

B. The Triumphs and Tragedies of Globalization
1. People throughout the world are now linked together as never before. Facts and indicators of globalization.
    a. Tourism.
    b. International organizations and agreements.
    c. International telecommunication.
    d. International trade and investment.
2. There are benefits of the rapid movement of capital, commodities, culture, and people across national boundaries.
3. Yet not everyone is happy. Inequality between rich and poor countries remains staggering, and in some respects is increasing. Many people oppose globalization because of its hurtful effect on local cultures and the natural environment. Antiglobalization activists suggest that globalization is a form of **imperialism**, the economic domination of one country by another. It contributes to the "homogenization" of the world, the cultural domination of less powerful by more powerful countries.

II. GLOBALIZATION

A. Globalization in Everyday Life
1. Everything influences everything else in a globalized world.
    a. Barbara Garson (2001) traced a small sum of money she invested in the Chase Manhatten Bank. Her money got caught up in the globalization of the world.
2. Most of us do not often appreciate that our actions have implications for people far away. When we buy a commodity we often tap into a **global commodity chain**, which is "a world-wide network of labor and production processes, whose end result is a finished commodity."

    a.   For example, athletic shoes like Nike as a commodity binds consumers and producers. *The new international division of labor* yielded high profits for Nike. When someone buys a pair of Nike athletic shoes they insert themselves in a global commodity chain.

   3. Sociology makes us aware of the complex web of social relations and interactions in which we are embedded. *The sociological imagination* allows us to link our biography with history and social structure.

B. The Sources of Globalization
   1. Technology
      a.   Technological progress has made it possible to move things and information over long distances quickly and inexpensively - for example, commercial jets and various means of communication.
   2. Politics
      a.   Politics is important in determining the level of globalization – for example, North Korea versus South Korea relations to the United States.
   3. Economics
      a.   Economics is an important source of globalization. Capitalist competition has been a major spur to international integration.
      b.   **Transnational corporations** – also called multinational or international corporations – are the most important agents of globalization in the world today. They are different from traditional corporations in five ways.
         i.   Traditional corporations rely on domestic labor and domestic production. Transnational corporations depend increasingly on foreign labor and foreign production.
         ii.   Traditional corporations extract natural resources or manufacture industrial goods. Transnational corporations increasingly emphasize skills and advances in design, technology, and management.
         iii. Traditional corporations sell to domestic markets. Transnational corporations depend increasingly on world markets.
         iv. Traditional corporations rely on established marketing and sales outlets. Transnational corporations depend increasingly on massive advertising campaigns.
         v.   Traditional corporations work with or under national governments. Transnational corporations are increasingly autonomous from national governments.
   4. Technological, political, and economic factors do not work independently in leading to globalization. Governments often promote economic competition to help transnational corporations win global markets.
      a.   For example, in the 1970s Phillip Morris and other tobacco companies pursued globalization as a way out of slumping domestic sales.

C. A World Like the United States?
   1. Much evidence supports the view that globalization homogenizes societies.
      a.   For instance, transnational organizations such as the World Bank and the International Monetary Fund have imposed economic guidelines for developing countries that are similar to those governing advanced industrial countries.

b. In the realm of politics, the United Nations (UN) engages in global governance whereas Western ideas of democracy, representative government, and human rights have become international ideals.

c. In the domain of culture, American icons circle the planet.

2. **McDonaldization**, as defined by George Ritzer, is a form of rationalization. Specifically, it refers to the spread of the principles of fast-food restaurants, such as efficiency, predictability, and calculability, to all spheres of life.

a. McDonald's has turned lunch preparation into a model of rationality.

b. McDonald's now does most of its business outside the United States.

c. McDonaldization has come to stand for the global spread of values associated with the United States and its business culture.

3. Some analysts find fault with the view that globalization is making the world a more homogeneous place based on American values. They argue that people always interpret globalizing forces in terms of local conditions and traditions − in fact, it sharpens some local differences. **Glocalization** is the simultaneous homogenization of some aspects of life and the strengthening of some local differences under the impact of globalization.

4. **Regionalization** is the division of the world into different and often competing economic, political, and cultural areas.

a. World trade is not evenly distributed around the planet or dominated by just one country.

b. Politically, regionalization can be seen in the growth of the European Union.

c. The expansion of Islam has become a unifying religious, political, and cultural force over Asia and Africa, and is a form of regional integration especially in its politically active form.

D. Globalization and Its Discontents: Antiglobalization and Anti-Americanism

1. In his book *Jihad vs. McWorld* published in 1992, Benjamin Barber book argued that globalization ("McWorld") was generating an antiglobalization reaction (*jihad*).

a. Traditionally, Muslims use the term, *jihad,* to mean perseverance in achieving a high moral standard.

b. The term can also suggest the idea of a holy war against those who harm Muslims. It represents an Islamic fundamentalist reaction to globalization. September 11, 2001 was the most devastating manifestation of fundamentalist Islamic *jihad.*

2. Another example involves Mexican peasants in the southern state of Chiapas, who staged an armed rebellion against the Mexican government in the 1990s.

3. As an example of the antiglobalization movement in the advanced industrial countries, when the WTO met in Seattle in December 1999, 40,000 union activists, environmentalists, supporters of worker and peasant movements in developing countries, and other opponents of transnational corporations protested, which caused property damage and threatened to disrupt the meetings.

4. The antiglobalization movement has many currents - some extremely violent, some nonviolent, some in rejection of the excesses of globalization, and other in rejection of globalization in its entirety.

E. The History of Globalization

1. The extent of globalization since about 1980 is unprecedented in world history. Sociologist Martin Albrow argues that the "global age" is only a few decades old.
2. Albrow and others exaggerate the extent of globalization.
3. Globalization is not as recent as Albrow suggests. Anthony Giddens argues that globalization is the result of industrialization and modernization which picked up pace in the late 19th century. In fact a strong case can be made that the world was highly globalized 100 or more years ago.
4. Some sociologists, such as Roland Robertson, note that globalization is as old as civilization itself and is in fact the *cause* of modernization rather than the other way around. Archeological remains show that long-distance trade began 5000 years ago.
5. Brym and Lie take an intermediate position and think of globalization as a roughly 500-year old phenomenon. They regard the establishment of colonies and the growth of capitalism as the main forces underlying globalization, and both **colonialism** (the control of developing societies by more developed, powerful societies) and capitalism began about 500 years ago, symbolized by Columbus's voyage to the Americas.

III. GLOBAL INEQUALITY

A. Levels of Global Inequality
   1. We find an even more dramatic gap between rich and poor at the global level, than the national level.
   2. The total worth of the world's 358 billionaires equals that of the world's 2.3 billion poorest people (45% of the world's population). The three richest people in the world own more than the combined GDP of the 48 least-developed countries. The richest 1% of the world's population earns as much income as the bottom 57%. The top 10% of U.S. income earners earn as much as the poorest 2 billion in the world.
   3. According to the UN, 800 million people in the world are malnourished and 4 billion people (2/3 of the world's population) are poor. A fifth of the developing world's population goes hungry every night. The citizens of the 20 or so rich, highly industrialized countries spend more on cosmetics, alcohol, ice cream or pet food, respectively, than it would take to provide basic education, water and sanitation, or basic health and nutrition for everyone in the world.
   4. Of the 1.3 billion people living on $1 a day or less around the world, 1 billion are women. Sixty-four percent of illiterate adults in the world are women. Racial and other minority groups often fare worse than their majority-group counterparts in the developing countries.
   5. Between 1975 and 2000, the annual income gap between the 20 or so richest countries and the rest of the world grew enormously.

IV. THEORIES OF DEVELOPMENT AND UNDERDEVELOPMENT

A. Modernization Theory: A Functionalist Approach
   1. According to **modernization theory**, global inequality and economic underdevelopment results from poor countries lacking Western attributes including Western values, business practices, level of investment capital, and stable governments. Thus, rich countries can help by transferring Western culture and capital to poor countries to invigorate their cultural, political and economic development.
   2. Dependency Theory: A Conflict Approach

a. Criticisms of modernization theory:
   i. For more than 500 years, the most powerful countries in the world deliberately impoverished the less powerful countries.
   ii. Focusing on internal characteristics blames the victim rather than the perpetrator.
b. **Dependency theory** views global inequality and economic underdevelopment as the result of exploitive relations between rich and poor countries.
c. Less global inequality existed in 1500 and even 1750 than today.
   i. However, beginning around 1500, the world's most power countries subdued and annexed or colonized most of the rest of the world. Around 1780, the Industrial Revolution began, which enabled the Western European countries, Russia, Japan, and the United States to accumulate enormous wealth, and extend globally. The colonies became sources of raw materials, cheap labor, investment opportunities, and markets for conquering nations. Colonizers prevented industrialization and locked the colonies into poverty.
d. Despite the political independence of nearly all colonies in the decades following World War II, exploitation by direct political control was replaced by new means of: substantial foreign investment, support for authoritarian governments, and mounting debt.
   i. *Substantial Foreign Investment.* Although they created some low-paying jobs, multinational corporations invested heavily in the poor countries to siphon off wealth in the form of raw materials and profits, which created more high-paying jobs in the rich countries where the raw materials were use to produce manufactured goods. Multinational corporations sold part of the manufactured goods back to the poor, unindustrialized countries for additional profit.
   ii. *Support for Authoritarian Governments.* Multinational corporations and rich countries continued to exploit poor countries in the postcolonial period by giving economic and military support to local authoritarian governments, which keep their populations subdued most of the time. Western governments have also sent in their own troops and military advisors, when authoritarian governments were unable to subdue their populations, which is called "gunboat diplomacy." The United States has been particularly active in using gunboat diplomacy in Central America, such as the case of Guatemala in the 1950s.
   iii. *Mounting Debt.* In order for the governments of poor countries to build transportation infrastructures, education systems, safe water systems, and basic health care, they had to borrow money from Western banks and governments. In 2002, foreign aid to the world's developing countries was only one-seventh the amount that developing countries paid to Western banks in loan interest.

B. Effects of Foreign Investment
   1. Almost all sociologists agree with dependency theory on one point. Since about 1500, Spain, Portugal, Holland, Britain, France, Italy, the United States, Japan, and Russia treated the world's poor with brutality to enrich themselves. Colonialism did have a devastating economic and human impact on poor countries. In the postcolonial era, the debt has crippled the development efforts of many poor countries.

2. Do foreign investment and liberalized trade policies have positive or negative effects today?
    a. Modernization theorists want more foreign investment in poor countries and freer trade to promote economic growth and general well being for everyone.
    b. Dependency theorists think foreign investment drains wealth out of poor countries. They want the poor countries to revolt against the rich countries, throw up barriers to free trade and investment, and find their own path to economic well being.
    c. A recent summary of research in this area cautiously reaches two conclusions.
        i. First, in the 1980s and 1990s but not in the 1960s and 1970s, openness to international trade and investment generally seems to have stimulated economic growth.
        ii. Second, in most well documented cases, openness to international trade and foreign investment also increased inequality.
    d. These findings suggest that it is a mistake to lump all periods of history and all countries together when considering the effects of openness on economic growth and inequality.

C. Core, Periphery, and Semiperiphery
    1. Immanuel Wallerstein argues that capitalist development has resulted in the creation of a "world system" composed of three tiers.
        a. **Core** capitalist countries, are rich such as the United States, Japan, and Germany that are the major sources of capital and technology.
        b. **Peripheral** countries are former colonies that are poor and are major sources of raw materials and cheap labor.
        c. **Semiperipheral** countries consist of former colonies that are making considerable headway in their attempts to become prosperous, such as South Korea, Taiwan, and Israel.
    2. The semiperipheral countries differ from the peripheral countries in four main ways: type of colonialism, geopolitical position, state policy, and social structure.
    3. Type of Colonialism
        a. In contrast to European colonizers of Africa, Latin America, and other parts of Asia, the Japanese built up the economies of their colonies, Taiwan and Korea, and established transportation networks and communication systems.
        b. The Japanese-built infrastructure served as a springboard to development for Korea and Taiwan.
    4. Geopolitical Position
        a. Countries with significant strategic importance to the United States receive more help.
        b. By the end of World War II, the United States began to feel its dominance threatened in the late 1940s by the Soviet Union and China, therefore poured unprecedented aid into South Korea and Taiwan in the 1960s fearing these countries might fall to the communists. Israel also received economic assistance for its strategic position in the Middle East.
    5. State Policy
        a. The Taiwanese and South Korean states developed on the Japanese-model: low worker's wages, restricted trade union growth, quasi-military discipline in the factories, high taxes on consumer goods, limitations on the import of foreign goods, and prevention of citizens to invest abroad.

b. This encourages their citizens to put their money in the bank, which created a large pool of capital that the state made available for industrial expansion.

c. From the 1960s on, South Korean and Taiwanese states gave subsidies, training grants, and tariff protection to export-based industries.

6. Social Structure

a. Social cohesion and solidarity in Taiwan and South Korea is strong and based partly on two main factors.

i. The sweeping land reforms in the late 1940s and early 1950s redistributed land to small farmers, eliminated the class of large owners, and thus eliminated a major source of conflict. In contrast, many countries in Latin America and Africa have not undergone land reform, whereby the United States has often intervened militarily to prevent land reform based on U.S. commercial interests in large plantations.

ii. Taiwan and South Korea are both ethnically homogeneous, thus limiting the ability of colonizers to instigate antagonism among the colonized. In contrast, British, French, and other West European colonizers kept tribal tensions alive to make it easier for imperial rule in Africa.

V. NEOLIBERAL VERSUS DEMOCRATIC GLOBALIZATION

A. Globalization and Neoliberalism

1. Resembling the modernization theorists of a generation ago, **neoliberal globalization** is a policy that promotes private control of industry, minimal government interference in the running of the economy, the removal of taxes, tariffs, and restrictive regulations that discourage the international buying and selling of goods and services, and the encouragement of foreign investment..

2. Many social scientists are skeptical of neoliberal prescriptions. Nobel Prize-winning economist and former Chief Economist of the World Bank, Joseph E. Stiglitz argues that the World Bank and other international economic organizations often impose outdated policies on the developing countries, putting them at a disadvantage *vis-à-vis* developed countries.

3. With the exception of Great Britain, neoliberalism was *never* a successful development strategy in the early stages of industrialization. Typically, it is only after industrial development is well under way and national industries can compete on the global market that countries begin to advocate neoliberal globalization to varying degrees.

a. Great Britain is the exception as it needed less government involvement to industrialize because it was the first industrializer and had no international competitors.

b. The United States, one of the most vocal advocates of neoliberalism today, invested a great deal of public money subsidizing industries and building infrastructure. Today, the United States still subsidizes large corporations in a variety of ways and maintains substantial tariffs on a whole range of foreign products.

B. Globalization Reform

1. On average, Americans believe we are spending 24 percent of our budget on aid to foreign countries. Few Americans believe we are doing too little to help the world's poor.

2. Perceptions are at odds with reality. The United States places at or near the bottom of the list of rich countries in terms of overall aid contributions.

C. Foreign Aid, Debt Cancellation, and Tariff Reduction
   1. If you think that we have a responsibility to compensate for centuries of injustice and that we are spending too little on foreign aid, you are in a minority. According to the GSS, foreign aid ranks a distant last on American's list of priorities. Most U.S. foreign aid goes for economic and military support to strategically important friends: Israel, Egypt, Jordan, and Colombia.
   2. Foreign aid as presently delivered is not an effective way of helping the developing world.
   3. Brym and Lie conclude that foreign aid is often beneficial. Poor countries need more of it. But strict government oversight is required to ensure that foreign aid is not wasted and that it is directed to truly helpful projects, such as improving irrigation and sanitation systems and helping people acquire better farming techniques.
   4. Many analysts argue that the world's rich countries and banks should simply write off the debt owed to them by the developing countries in recognition of historical injustices. This proposal of blunting the worst effects of neoliberal globalization may be growing in popularity among politicians in developed countries.
   5. A third reform proposed in recent years involves the elimination of tariffs by the rich countries. However, there is little room for optimism in this regard to date.

D. Democratic Globalization
   1. A final reform involves efforts to help spread democracy throughout the developing world. A large body of research shows that democracy lowers inequality and promotes economic growth, for several reasons.
      a. Democracies make it more difficult for elite groups to misuse their power and enhance their wealth and income at the expense of the less well to do.
      b. They increase political stability, thereby providing a better investment climate.
      c. Because they encourage broad political participation, democracies tend to enact policies that are more responsive to people's needs and benefit a wide range of people from all social classes.
   2. Although democracy has spread in recent years, by 2002 only 80 countries with 55% of the world's population were considered fully democratic by one measure. The United States has supported as many antidemocratic as democratic regimes in the developing world. The CIA refers to the unexpected and undesirable consequences of these actions as "blowback."
      a. For example, in the 1980s the U.S. government supported Saddam Hussein when it considered Iraq's enemy, Iran, the greater threat to U.S. security interests.
      b. The United States also funded Osama bin Laden when he was fighting the Soviet Union in Afghanistan.
      c. Only a decade later, these so-called allies turned into the United States' worst enemies.

Study Activity: Applying The Sociological Compass

After reading Chapter 9 of the textbook, re-read the sections "Theories of Development and Underdevelopment" and "Neoliberal Versus Democratic Globalization." As you re-read these sections, make a list of evidence cited in the textbook, in support of modernization theory, neoliberal globalization, dependency theory, and democratic globalization. Which theory implies the greatest potential of "freeing individuals" and which theory most implies that the world's poor people, who are the numerical majority of the world population, will likely experience greater constraints in the future. Which theory or theories seem most plausible based on the cited evidence? Explain why?

Modernization Theory:

Neoliberal Globalization:

Dependency Theory:

Democratic Globalization:

Infotrac College Edition Online Exercises

For the following exercises, log on to the online library of InfoTrac College Edition at http://www.infotrac-college.com/. Make note that InfoTrac has implemented a new registration system that will allow easier access to InfoTrac through the use of a personalized username and password. Once you've created your username and password you may proceed directly to the Log On page. To create an account, register your passcode packaged with your textbook, and create a username and password, by following the online prompt. After you are logged in, click on "Infotrac College Edition." You will arrive at a screen that enables you to search topics.

Keyword: **dependency theory**. Locate the article by Alex Dupuy called, "*Thoughts on globalization, Marxism, and the left*." Read the article and look for specific evidence to support and refute Marx's theory. How does the evidence, or lack of evidence, compare with the section in this chapter on Marx's theory?

Keywords: **Development Peru**. Look at all of the articles. There are not too many. What development issues are currently facing Peru? Make a list of all the issues pertaining to development.

How does your list compare to development issues discussed in the chapter?

Keywords: **Global village**. Find the article called, ***Toronto A global village,*** by Mary Vincent.

Read the article. How does evidence of a "global village" in the article compare to the section of a global village in this chapter? Do you agree or disagree with the idea of a global village?

Internet Exercises

The United Nations has many statisticians who have created indicators of human development for every country in the world. The indicators for the 2003 Human Development Indicators are available at: http://www.undp.org/hdr2003/indicator/index_indicators.html

Make a list of the indicators that you think are the most important. Why did you include each item on your list?

Go to: http://www.nologo.org/ for Naomi Klein's *No Logo: Taking Aim at the Brand Bullies*. This site is against the globalization movement. Look over the site and then list specific reasons why Klein is anti-globalization.

At this site: http://www.worldbank.org/ you will be able to find many reports and data on the state of the global economy and many other nations. This is the World Bank site. Look over a report of interest to you. Write a brief description stating its purpose and what information it contains.

Practice Tests

Fill-In-The-Blank

Fill in the blank with the appropriate term from the above list of "key terms."

1. Multinational corporations are also called _____ .

2. _____ is the simultaneous homogenization of some aspects of life and the strengthening of some local differences under the impact of globalization.

3. _____ is the economic domination of one country by another.

4. _____ views global inequality and economic underdevelopment as the result of exploitive relations between rich and poor countries.

5. A(An) _____ is a worldwide network of labor and production processes, whose end result is a finished commodity.

6. As defined by George Ritzer, _____ is a form of rationalization.

7. _____ is the division of the world into different and often competing economic, political, and cultural areas.

8. _____ countries consist of former colonies that are making considerable headway in their attempts to become prosperous.

9. _____ refers to the control of developing societies by more developed, powerful societies.

10. _____ is a variant of functionalism.

Multiple-Choice Questions

Select the response that best answers the question or completes the statement:

1. Antiglobalization activists suggest that globalization is a form of:
   a. modernization.
   b. imperialism
   c. development.
   d. All of these are correct.
   e. None of the above.

2. Transnational corporations are different from traditional corporations in what way?
   a. Traditional corporations extract natural resources or manufacture industrial goods. Transnational corporations emphasize skills and advances in design, technology, and management.
   b. Traditional corporations work with or under national governments. Transnational corporations are increasingly autonomous from national governments.
   c. Traditional corporation sell to domestic markets. Transnational corporations depend increasingly on world markets.
   d. Traditional corporations rely on domestic labor and domestic productions. Transnational corporations depend increasingly on foreign labor and foreign production.
   e. All of these are correct.

3. The fact that world trade is not evenly distributed around the planet or dominated by just one country is an example of:
   a. regionalization.
   b. glocalization.
   c. underdevelopment.
   d. development.
   e. None of the above.

4. Which theory suggests that global inequality is a result of poor countries lacking Western attributes such as Western values and business practices?
    a. dependency theory
    b. modernization theory
    c. world system theory
    d. All of these are correct.
    e. None of the above.

5. Which is NOT a new means of colonization today?
    a. substantial foreign investment.
    b. decolonization of the third world.
    c. support for authoritarian governments.
    d. mounting debt.
    e. All of the answers are new means of colonization.

6. Which tier of the "world system" is a major source of raw materials and cheap labor?
    a. core countries
    b. peripheral countries
    c. semiperipheral countries
    d. industrialized countries
    e. newly industrialized countries

7. Which is NOT one of the ways in which semiperipheral countries differ from peripheral countries, according to the textbook?
    a. type of colonialism
    b. geopolitical position
    c. state policy
    d. social structure
    e. number of traditional corporations

8. Neoliberal globalization resembles which theory of a generation ago?
    a. modernization theory
    b. dependency theory
    c. world system theory
    d. internal colony theory
    e. neo-Marxist theory

9. Neoliberal globalization promotes:
    a. public control of industry.
    b. moderate to strong government management in the running for the economy.
    c. the introduction of new taxes, tariffs, and regulations to check and balance the power of transnational corporations.
    d. the encouragement of foreign investment.
    e. exclusive national buying and selling of goods and services.

10. All of the following are indicators of globalization EXCEPT:
    a. tourism.
    b. international organizations and agreements.
    c. international telecommunications.
    d. international trade and investment.
    e. All of the answers are indicators of globalization.

11. In leading to globalization, technological, political and economic factors:
    a. do not work independently.
    b. operate independently in the context of global competition.
    c. are interrelated and functional exclusively on a domestic level.
    d. All of these are correct.
    e. None of the above.

12. McDonaldization:
    a. refers to the spread of the principles of fast-food restaurants.
    b. suggests that McDonald's has turned lunch preparation into a model of rationality.
    c. has come to stand for the global spread of values associated with the United States and its business culture.
    d. All of these are correct.
    e. None of the above.

13. Which is an example of antiglobalization and/or anti-Americanism?
    a. *jihad*
    b. the Mexican peasant rebellion in Chiapas in the 1990s.
    c. the protest in Seattle in December 1999 when the WTO met.
    d. All of these are correct.
    e. None of the above.

14. Which is a current of the antiglobalization movement?
    a. violent protest.
    b. rejection of the excesses of globalization.
    c. rejection of globalization in its entirety.
    d. All of these are correct.
    e. None of the above.

15. South Korea, Taiwan, and Israel are examples of which type of country?
    a. core countries
    b. peripheral countries
    c. semiperipheral countries
    d. underdeveloped countries
    e. None of the above.

16. According to Table 9.1 "Indicators of Globalization, Early 1980s - Circa 2003" in the textbook, which indicator showed the most change from 1980-81 to 1998-2003?
    a. foreign direct investment as percent of GDP.
    b. international tourist arrivals as a percent of the world population.
    c. Air freight and mail.
    d. number of international organizations.
    e. annual entries on globalization, *Sociological Abstracts*.

17. Michael Jordan:
    a. helps globalize Nike.
    b. is the icon of Nike, but primarily at a national level.
    c. is the primary reason why Nike exploits labor in third-world countries.
    d. All of these are correct.
    e. None of the above.

18. In the 1970s, the antismoking campaign:
    a. began to have an impact.
    b. led to slumping domestic sales for tobacco companies.
    c. sparked the pursuit of globalization by tobacco companies.
    d. All of these are correct.
    e. None of the above.

19. Figure 9.2 "The Size and Influence of the U.S. Economy, 2000" in the Brym and Lie textbook:
    a. shows how the GDP of various countries compares to that of each U.S. state.
    b. shows that the economy of each U.S. state is as big as that of a whole country.
    c. emphasizes just how large the U.S. economy is.
    d. All of these are correct.
    e. None of the above.

20. A half-hour's drive from the center of Manila, the capital of the Philippines, an estimated 70,000 Filipinos live:
    a. on a 55-acre mountain of rotting garbage.
    b. crowded on filthy canal barges.
    c. virtually homeless in vast, temporary villages.
    d. All of these are correct.
    e. None of the above.

21. The gross domestic product per capita across world regions from 1975-2000, indicates that:
    a. Latin American and Caribbean countries are fast approaching high-income countries in GDP.
    b. Countries in East Asia and the Pacific are fast approaching high-income countries.
    c. although the GDP in high-income countries is significantly higher than other world regions, the GDP in high-income countries has fallen since 1975.
    d. The GDP of high-income countries has increased since 1975, and remains much higher than other world regions.
    e. None of the above.

22. From 1820 to 1992, the share of world income going to the top 10% has:
    a. increased.
    b. declined.
    c. remained consistently proportionate to diverse populations.
    d. All of these are correct.
    e. None of the above.

23. Which country had the greatest percentage of people living for less than $1 a day in 1999?
    a. Sub-Saharan Africa
    b. East Asia (excluding China)
    c. South Asia
    d. Latin America
    e. East Europe and Central Asia

24. Which is true in the wake of the late 1800s British takeover of what became Rhodesia, and is now Zimbabwe?
    a. Every British trooper was offered about 9 square miles of native land.
    b. The Matabele and Mashona people were subdued.
    c. Native survivors were left without a livelihood.
    d. Forced labor was introduced by the British.
    e. All of these are correct.

25. Based on data on national priorities collected in 2002 by the National Opinion Research Center, which priority had the highest percentage of respondents answering "Too Little" in response to the question, "I'd like you to tell me whether you think we're spending too much money on it, too little money, or about the right amount?"
    a. Improving and protecting the nation's health.
    b. Dealing with drug addiction.
    c. Highways and bridges.
    d. The military, armaments, and defense.
    e. Foreign aid.

True False

1. When we purchase a product domestically, we seldom tap into the global commodity chain, because domestic commodities tend to be produced within a country independent of the global economy.

   TRUE or FALSE

2. The new international division of labor yielded high profits for producers like *Nike*.

   TRUE or FALSE

3. McDonald's now does most of its business outside the United States.

   TRUE or FALSE

4. According to Benjamin Barber's book Jihad vs. McWorld, jihad exclusively refers to the idea of holy war against those who harm Muslims.

   TRUE or FALSE

5. September 11, 2001 was the most devastating manifestation of fundamentalist Islamic *jihad*.

   TRUE or FALSE

6. Almost all manifestations of the antiglobalization movement are extremely violent, and occur in third-world countries.

   TRUE or FALSE

7. Technology speeds globalization.

   TRUE or FALSE

8. The GDP of California is equal to that of France.

   TRUE or FALSE

Short Answer

1. Briefly explain what it means to say that people have created a "global village". (p. 249-252)

   _____

   _____

2. Compare, contrast, and explain the connection between imperialism and colonialism. (p. 252, 260)

   _____

   _____

3. List five ways in which transnational corporations differ from traditional corporations. (p. 255)

   _____

   _____

4. List and explain four facts and indicators of globalization. (p. 250-251)

   _____

   _____

5. Explain glocalization, and provide one example of this phenomenon from the textbook. (p. 257)

   _____

   _____

6. List and explain the characteristics of each of the three types of countries on the world system, according to Immanuel Wallerstein. (p. 266)

_____

_____

Essay Questions

1. Explain the phenomenon of McDonaldization according to George Ritzer. In your answer, be sure to mention rationalization, efficiency, predictability, and calculability. (p. 256-258)

2. Contrast modernization theory and dependency theory as explanations of global inequality. In your answer, be sure to provide relevant historical and contemporary support for each theory. (p. 263-265)

3. Compare and contrast neoliberal globalization and democratic globalization in terms of their major assumptions, premises, and their prescriptions to alleviate global inequality. (p. 268-273)

4. Explain four ways in which semiperipheral countries differ from peripheral countries. (p. 266-268)

5. Describe how globalization has affected your life and your social conditions. Compare your life with that of someone from a place on the globe that has not been affected (yet) by globalization. (p. 249-273)

Solutions

Practice Tests

Fill-In-The-Blank

1. transnational corporations
2. Glocalization
3. Imperialism
4. Dependency theory
5. global commodity chain
6. McDonaldization
7. Regionalization
8. Semiperipheral
9. Colonialism
10. Modernization theory

Multiple-Choice Questions

1. B, (p. 252).
2. E, (p. 255)
3. A, (p. 257)
4. B, (p. 262)
5. B, (p. 264-265)
6. B, (p. 266)
7. E, (p. 266)
8. A, (p. 263-264, 268)
9. D, (p. 268)
10. E, (p. 250-251)
11. A, (p. 254-255)
12. D, (p. 256)
13. D, (p. 258-259)
14. D, (p. 258-259)
15. C, (p. 266)
16. E, (p. 251)
17. A, (p. 253)
18. D, (p. 255)
19. D, (p. 256)
20. A, (p. 261)
21. D, (p. 262)
22. A, (p. 263)
23. A, (p. 263)
24. E, (p. 264)
25. A, (p. 270)

True False

1. F, (p. 252-254)
2. T, (p. 253)
3. T, (p. 256)
4. F, (p. 258)
5. T, (p. 258)
6. F, (p. 258-259)
7. T, (p. 254)
8. T, (p. 256)

CHAPTER **10**

RACE AND ETHNICITY

Student Learning Objectives

After reading Chapter 10, you should be able to:

1. Engage the debate over the racial basis of athletic ability.

2. Identify social conditions involving prejudice and discrimination that lead certain racial groups into particular life chances.

3. Explain the notion of race and ethnicity as socially constructed ideas. How are these terms used to distinguish and categorize people based on perceived physical and cultural differences that have profound consequences for their lives?

4. Explain how racial and ethnic identities change over time and place in the context of insider self-conceptions and outsider labels.

5. Describe the historical experiences, demographic profiles and negotiations of racial and ethnic identity among the major Hispanic groups—Chicanos, Puerto Ricans, and Cubans.

6. Explain the notion of racial and ethnic identity as a choice versus an imposition.

7. Discuss how the identification with a racial or ethnic group can be economically, politically, and emotionally advantageous. What role does racial and ethnic identification play in ethnic community formation?

8. Identify and explain the major assumptions of ecological theory, the theory of internal colonialism, and split labor market theory.

9. Describe the historical experiences of Native Americans, Chicanos and African Americans in terms of internal colonialism—and the experiences of Irish Americans in terms of ecological theory.

10. Engage the "declining significance of race" debate, by discussing the pros and cons and the respective evidence for each position. In light of the evidence, what is the future outlook on racial and ethnic inequality in the United States?

11. Discuss recent trends and debate surrounding immigration.

12. Explain the idea of a vertical mosaic in the future of racial and ethnic relations. What can society do in terms of policy initiatives to decrease the verticality of the American ethnic mosaic?

Key Terms

prejudice (279)

discrimination (279)

race (282)

scapegoats (282)

ethnic group (282)

minority group (284)

ethnic enclave (287)

symbolic ethnicity (291)

racism (291)

institutional racism (291)

ecological theory (292)

segregation (292)

assimilation (292)

internal colonialism (293)

split labor markets (293)

expulsion (293)

genocide (293)

slavery (296)

transnational communities (301)

hate crimes (301)

affirmative action (302)

pluralism (306)

Detailed Chapter Outline

I. DEFINING RACE AND ETHNICITY

   A. The Great Brain Robbery
      1. Dr. Samuel George Morton, who died in 1851, claimed to show that social inequality in the United States and throughout the world had natural, biological roots, based on research on brain size of different races (indicated by skull size).
      2. Morton claimed that the people with the biggest brains were whites of European origin. Next were Asians. Then came Native Americans and African Americans.
      3. Morton's ideas were used to justify colonization and slavery.
      4. Three main issues that compromise Morton's research today.
         a. Even today archeologists cannot precisely determine race by skull shape.
         b. Morton's skulls formed a small, unrepresentative sample.
         c. Morton did not make sure that the sex composition of the white and black skulls was identical.

   B. Race, Biology, and Society
      1. Although biological arguments about racial differences became more sophisticated, their scientific basis is still shaky.
      2. Examples:
         a. "Blue blood" aristocrats in medieval Europe.

      b.   1920s United States IQ scores and biology.

   3. Is there evidence to support racial differences in singing ability, athletic prowess, and crime rates?

      a.   At first glance, there seems to be evidence of black athletic superiority in the NBA, NFL, and track and field.

      b.   Evidence falters in light of two points.

          i.   No gene connected to general athletic superiority has been identified.

         ii.   Athletes of African descent do not perform unusually well in many sports, such as swimming, hockey, cycling, tennis, gymnastics, and soccer.

C. Prejudice, Discrimination, and Sports

   1. For lack of other ways to improve their social and economic positions due to **prejudice** (an attitude that judges a person on his or her group's real or imagined characteristics) and **discrimination** (unfair treatment of people due to their group membership), people enter sports, entertainment, and crime in *disproportionate* numbers.

      a.   The idea that people of African descent are genetically superior to whites in athletic ability complements the idea that they are genetically inferior to whites in intellectual ability. Both ideas reinforce black-white inequality.

         i.   There are millions of professionals and probably fewer than 10,000 elite professional athletes in the United States.

         ii.   Promoting the Shaquille O'Neals of the world as role models for African-American youth, the idea of black athletic superiority combined with intellectual inferiority deflects the attention of African American youth from a much safer bet—that they can achieve upward mobility through academic excellence.

   2. The fact that one cannot neatly distinguish race based on genetic differences, is an additional problem with the argument that genes determine specific behaviors of different racial groups.

      a.   High level of genetic mixing.

D. The Social Construction of Race

   1. Many scholars believe we all belong to one human race.

   2. Although some biologists and social scientists therefore suggest we drop the term "race" from the vocabulary of science, most sociologists continue to use the term. Why?

      a.   *Perceptions* of race continue to have a profound effect on the lives of most people.

      b.   Race as a *sociological* concept is an invaluable analytical tool if we remember that it refers to socially significant physical differences, rather than biological differences.

      c.   Perceptions of racial differences are socially constructed and often arbitrary. **Race** is a social construct used to distinguish people in terms of one or more physical markers, usually with profound effect on their lives.

   3. Why Race Matters

      a.   It allows social inequality to be created and maintained.

      b.   Racial stereotypes embed themselves in literature, popular lore, journalism, and political debate to reinforce racial inequality.

      c.   Examples of systems of racial domination.

      i.   English colonization of Ireland.

     ii. American enslavement of people of Africa.

   iii. Use of Jews as **scapegoats** (a disadvantaged person or category of people who others blame for their own problems) by the Germans after World War I.

E. Ethnicity, Culture, and Social Structure

  1. While race is a category of people whose perceived *physical* markers are deemed socially significant, an **ethnic group** is composed of people whose perceived *cultural* markers are deemed socially significant (i.e. language, religion, customs, values, ancestors, etc.).

  2. Just as physical differences do not cause differences in the behavior of various races, ethnic values and other elements of ethnic culture have less of an effect on the way people behave than we commonly believe.

  3. Social structural differences underlie cultural differences.

  4. Examples:

     a. Class difference (rather than ethnic differences) between descendants of southern black slaves and West Indian immigrants with the African-American community.

     b. Class differences between West Indian immigrants in New York and London in the 1950s and 1960s.

     c. Impact of American immigration policy on social-structural conditions for Jews and Koreans on the one hand, versus descendants of Southern blacks.

  5. It is misleading to claim that race and ethnicity are quite different. Both race and ethnicity are rooted in social structure, not biology and culture. Groups socially defined as races may later be redefined as ethnic.

II. RACE AND ETHNIC RELATIONS

A. Labels and Identity

  1. **Minority group** refers to a group of people who are socially disadvantaged, though they may be in the numerical majority.

  2. Social contexts, and in particular the nature of one's relations with members of other racial and ethnic groups, shape and continuously reshape one's *racial and ethnic identity*. Change in social context changes your *racial and ethnic self-conception*.

  3. Examples:

     a. John Lie's migration experiences from South Korea to Japan, Japan to Hawaii, and from Hawaii to the American mainland.

     b. Italian immigrants in the United States around 1900.

B. The Formation of Racial and Ethnic Identities

  1. The development of racial and ethnic labels and identities, is a process of negotiation, as symbolic interactionists emphasize. Negotiations, between outsiders' imposition of a new label, and insiders' racial and ethnic self-conceptions and their response to the outsider labels (rejection, acceptance, or modification) results in the crystallization of a new, more or less stable ethnic identity.

C. Case Study: The Diversity of the "Hispanic American" Community

1. Although people scarcely used the terms "Hispanic American" and "Latino" 30 or 40 years ago, they are now common.
2. Who is Hispanic American?
   a. In 2002, 67% of Hispanic Americans were of Mexican origin, nearly 9% Puerto Rican origin, almost 4% of Cuban origin, and 14% were from El Salvador, the Dominican Republic, and other countries in Latin America.
3. What do members of these groups have in common?
   a. Spanish language.
   b. Based on a survey, most of them do not want to called "Hispanic American," and prefer instead to be referred to by their national origin.
   c. Many want to be referred to simply as "Americans."
4. The Formation of Ethnic Enclaves
5. Members of the three major Hispanic groups enjoy different cultural traditions, social class positions, and vote differently.
   a. Cubans
      i. More middle-class and professional because of the large wave of middle-class Cubans who fled Castro's revolution and arrived in the Miami area in the late 1950s and early 1960s.
      ii. They formed an **ethnic enclave**, a geographical concentration of ethnic group members who establish businesses that serve and employ mainly members of the ethnic group and reinvest profits in community businesses and organizations. This benefited subsequent waves of poor Cuban immigrants.
      iii. More easily integrated into American society as result of being procapitalist and anticommunist.
      iv. More likely to vote Republican.
   b. Chicanos and Puerto Rican Americans
      i. Immigrants from Mexico and Puerto Rico tend to be from the working class that has not completed high school.
      ii. Lower class origins, less upward mobility.
      iii. Weaker ethnic enclave formation.
      iv. More commonly vote Democrat.
6. What Unifies the Hispanic American Community?
7. Three reasons why the term "Hispanic American" is more widely used than it was 30 or 40 years ago, despite internal diversity:
   a. Many Hispanic Americans find it politically useful.
   b. The government finds the term useful for data collection and public policy purposes.
   c. Non-Hispanic Americans find it convenient, in light of the many ethnic divisions.

D. Ethnic and Racial Labels: Choice Versus Imposition
   1. There are wide variations over time and across societies in the degree to which people can choose their racial and ethnic identity freely.
   2. State Imposition of Ethnicity in the Soviet Union

   a. Until being formally dissolved in 1991, the Soviet Union was composed of 15
       republics—Russia, Ukraine, Kazakhstan, and so forth. The largest ethnic group in each
       republic was the so-called "titular" ethnic group: Russians in Russia, Ukranians in
       Ukraine, Kazakhs in Kazakhstan, and so forth.
   b. Over 100 minority ethnic groups lived in the republics.
   c. Administratively imposed ethnicity: leaders of the Soviet Union developed strategies
       to promote national unity to prevent the country from falling apart in light of its vast
       size and ethnic heterogeneity.
       i.   Weakening the boundaries between republics to create "a new historical community,
            the soviet people."
       ii. Creation of a system allowing power and privilege to be shared among ethnic groups.
            An "internal passport system," and ethnic quotas became important to determining the
            most fundamental aspects of life (i.e. residence, higher education, professional and
            administrative positions, and political posts). Granted advantages to titular ethnic
            groups within their republic.
3. Ethnic and Racial Choice in the United States
   a. In comparison to the Soviet Union before 1991, Americans are freer to choose their
       ethnic identity—and people with the most freedom to choose are white Americans
       whose ancestors came from Europe.
       i.   Example—The transformation of Irish identity from imposition to choice.
       ii. Involves what Herbert Gans defines as **symbolic ethnicity**, "a nostalgic allegiance to
            the culture of the immigrant generation, or that of the old country; a love for and a pride
            in a tradition that can be felt without having to be incorporated in everyday behavior."
4. Racism and Identity
   a. They make take pride in their cultural heritage, but their identity has been imposed on
       them daily by **racism**, the belief that a visible characteristic of a group, such as skin color,
       indicates group inferiority and justifies discrimination. **Institutional racism** is bias that is
       inherent in social institutions and is often not noticed by members of the majority group.
   b. Malcolm X noted that it doesn't matter to a racist whether an African American is
       a professor or a panhandler, a genius or a fool, a saint or a criminal. Where racism is
       common, racial identities are compulsory and at the forefront of one's self-identity.

III. THEORIES OF RACE AND ETHNIC RELATIONS

A. Ecological Theory
   1. Robert Park's **ecological theory** focuses on the struggle for territory that involves the process
      by which conflict between racial and ethnic groups emerges and is resolved.
   2. Five stages:
      a. *Invasion*;
      b. *Resistance*;
      c. *Competition*; and

d. *Accommodation and Cooperation*. Eventually, groups work out an understanding of what they should segregate, divide, and share. **Segregation** involves spatial and institutional separation of racial or ethnic groups.

e. *Assimilation*. **Assimilation** is the process by which a minority group blends into the majority population and eventually disappears as a distinct group.

3. Park's theory applies to some ethnic groups such as whites of European origin, but is too optimistic about the prospects for assimilation of African Americans, Native Americans, Hispanic Americans, and Asian Americans.

B. Internal Colonialism and the Split Labor Market

1. Robert Blauner's examination of **internal colonialism** (a condition that involves one race or ethnic group subjugating another in the same country. It prevents assimilation by segregating the subordinate group in terms of jobs, housing, and social contacts), emphasized the social-structural roots of race and ethnicity.

2. *Colonialism* involves people from one country invading another. *Internal colonialism* involves the same process but within the boundaries of a single country. Both conditions involve the following processes:

   a. Invaders change or destroy the native culture.

   b. Invaders gain virtually complete control over the native population.

   c. Invaders develop the racist belief that the natives are inherently inferior.

   d. Invaders confine the natives to work that is considered demeaning.

3. Edna Bonacich developed a second important theory that focuses on social-structural conditions hindering the assimilation of some groups. Racial identities are reinforced in **split labor markets**, where low-wage workers of one race and high-wage workers of another race compete for the same jobs. High-wage workers are likely to resent the presence of low-wage competitors, and conflict is bound to result. Consequently, racist attitudes develop or get reinforced.

4. Native Americans

   a. Expulsion and genocide best describe the treatment of Native Americans by European immigrants in the 19th century. **Expulsion** is the forcible removal of a population from a territory claimed by another population. **Genocide** is the intentional extermination of an entire population defined as a race or a people.

   b. Once the British won the battle for control of North America in the late 18th century, Indian-white relations focused on land issues. Forcing the Indians to become like the European settlers and forcing them off their land became the two dominant strategies that informed American policy for centuries.

   c. Indians and settlers signed treaties (Indian groups were often forced to sign), and defined the land belonging to each group. Treaty violations on the part of settlers were common.

   d. In 1830 the U.S. government passed the *Indian Removal Act*, which called for relocation of all Native Americans to land set aside for them west of the Mississippi. White European Americans fought a series of wars against various Native American tribes.

   e. The end of the war against Indians came in 1890 with the slaughter of hundreds of Sioux at *Wounded Knee* in South Dakota. The remaining small bands were placed on reservations under the rule of the *Bureau of Indian Affairs*.

f. What war could not accomplish, disease and the extermination of the buffalo did.

g. In the late 19th century, the U.S. government adopted a policy of forced assimilation, through the selling of some Indian land to non-Indians, and by taking and placing Native-American children in boarding schools to be "civilized."

h. In the 1930s and 1940s, the Roosevelt administration adopted a more liberal policy, which prohibited further break up of Indian lands and encouraged Native self-rule and cultural preservation. This was only a brief deviation from traditional policy.

i. In the 1950s, the U.S. government proposed to end the reservation system, deny the sovereign status of the tribes, cut off all government services, and stop protecting Indian lands held in trust for the tribes.

j. By the 1960s a full-fledged *Red Power* movement emerged.

k. In recent decades, Native Americans have used the legal system to fight for political self-determination and the protection of their remaining lands.

l. Many Native Americans still suffer the consequences of internal colonialism.

m. Urban Indians are less impoverished than those who live on the reservations, but still fall below the national average in terms of income, education, occupation, employment, health care, and housing.

5. Chicanos

a. In light the U.S. desire for land, the U.S. war with Mexico in 1848 led to the United States winning Arizona, California, New Mexico, Utah, and part of Colorado and Texas (formerly Northern Mexico).

b. Due mainly to discrimination, Chicanos tend to be occupationally and residentially segregated from white European-American neighborhoods, -which has prevented upward mobility and assimilation.

    i. Until the 1970s, most Chicanos lived in *barrios*.

    ii. Still mainly agricultural and unskilled laborers.

    iii. Ironically, some Mexicans are called "illegal migrants" in places where their ancestors were expelled, such as California and Texas.

    iv. Since the 1960s, many Chicanos have participated in a movement to renew their culture and protect and advance their rights.

6. African Americans

a. Millions of Africans were brought to America by force and enslaved. **Slavery** is the ownership and control of people.

b. The cotton and tobacco economy of the American South depended completely on African slave labor.

c. Slavery and later Jim Crow laws kept blacks segregated, from voting, attending white schools, and in general from participating equally in American social institutions (i.e. 1896 *Plessy v. Ferguson*).

d. In the 19th and early 20th centuries, the opportunity to integrate the black population quickly and completely into the American mainstream was squandered in favor of policy makers choosing to encourage white European immigration.

e. Jobs in northern and western industries that went to African Americans who migrated from the South since the 1910s, enabled them to compete with European immigrants. Here we see the operation of a *split labor market*, which fueled deep resentment, animosity, and anti-Black riots on the part of working class whites.

f. Black migration northward and westward signaled the permanence and vitality of the new black communities, especially by the mid-1960s.

g. Although some gains among African Americans (i.e. residential segregation in poor neighborhoods decreased, some intermarriage with the white community occurred, and many children of migrant blacks finished high school, establishment of ethnic enterprises, and employment in civil service) new and old social-structural barriers limited prosperity and assimilation.

7. Chinese Americans

a. In 1882 Congress passed an act prohibiting the immigration of "lunatics, idiots, and Chinese" into the United States for 10 years. The act was extended for another decade in 1892, made permanent in 1907, and repealed in 1943, to grant a quota of 105 Chinese immigrants per year. The California gold rush of the 1840s and the construction of the transcontinental railroad in the 1860s had drawn in tens of thousands of Chinese immigrants, whereas the majority of them were young men who worked as unskilled laborers. They were the objects of one of the most hostile anti-ethnic movements in American history.

b. The film and tourist industries did much to reinforce fear of the "yellow peril" in the first half of the 20th century.

c. A *split labor market* caused anti-Chinese prejudice to boil over into periodic race riots and exclusionary laws. Where competition for jobs between Chinese and white workers was intense, trouble brewed and ethnic and racial identities were reinforced. Recent research shows that anti-Chinese activity was especially high where white workers were geographically concentrated and successful in creating anti-Chinese organizations.

d. Chinese Americans have experienced considerable upward mobility in the past half-century. However, a social structural factor − split labor markets − did much to prevent such mobility and assimilation until the middle of the 20th century.

IV. SOME ADVANTAGES OF ETHNICITY

A. Three main factors enhance the value of ethnic group members for some white European Americans who have lived in the country for many generations.

1. Ethnic group membership can have economic advantages.

2. Ethnic group membership can be politically useful.

3. *Ethnic group membership tends to persist because of the emotional support it provides*, especially for those who experience high levels of prejudice, discrimination involving expulsion and attempted genocide.

B. Transnational Communities

1. Inexpensive international travel and communication has enabled some ethnic groups to become **transnational communities** whose boundaries extend between countries, which has

further enabled ethnic group members to maintain strong ties to their motherland beyond the second generation.

V. THE FUTURE OF ETHNICITY

A. The Declining Significance of Race?
 1. The growth of racist organizations and hate crimes are increasing problems in the United States.
  a. More than 500 racist and Neo-Nazi organizations have sprung up in the United States.
  b. The FBI recorded 7489 **hate crimes** in 2003, which are criminal acts motivated by a person's race, religion, or ethnicity. African Americans are the most frequent victims, who were the object of 35% of all offenses.
 2. While white supremacists form only a tiny fraction of the American population, many more Americans engage in *subtle* forms of racism. African Americans continue to suffer high levels of covert and overt racial prejudice and discrimination.
  a. Based on interviews with a sample of middle class blacks in 16 cities, Joe Feagin and Melvin Sykes (1994) found:
   i. Respondents often have trouble hailing a cab.
   ii. If they arrive in a store before a white customer, a clerk commonly serves them afterwards.
   iii. When they shop, store security often follows them around to make sure they don't shoplift.
   iv. Police often stop middle-class African American men in their cars without apparent reason.
  b. Blacks are less likely than whites of similar means to receive mortgages and other loans.
 3. William Julius Wilson's position that race is declining in significance as a force shaping the lives of African Americans.
  a. Argues that the civil rights movement helped to establish legal equality between blacks and whites (i.e. 1954 *Brown v. Board of Education*, 1964 Civil Rights Act, 1965 Voting Rights Act, 1968 Civil Rights Act)
  b. Reforms allowed a large black middle class to emerge. Today a third of the African-American population is middle class.
  c. Many facts support Wilson's view—increase in median family income and higher education, decrease in the number of blacks living below the poverty line, increase in white tolerance of blacks,
  d. The third of blacks living below the poverty line are not much different than other Americans in similar economic circumstances according to Wilson.
  e. Calls for "color-blind" public policies intended to improve the class position of the poor; and opposes race-specific policies, such as **affirmative action** (a policy that gives preference to minority group members if equally qualified people are available for a position).
 4. Wilson's critiques claim that racism remains a big barrier to black progress.
  a. Racial differences in wages and housing patterns.
  b. Many African Americans continue to live in inner-city ghettos, where they experience high rates of poverty, crime, divorce, teenage pregnancy, and unemployment.

5. Wilson argues that ghettos exist not because of racism but for three economic and class reasons.
   a. Since the 1970s, older manufacturing industries have closed down in cities where the black working class was concentrated.
   b. Many middle- and working—class African-Americans with good jobs moved out of the inner city, which deprived young people of successful role models.
   c. The exodus of successful blacks eroded the inner city's tax base at the same time conservative federal and state governments cut budgets for public services.
6. However, Wilson's position on ghettoization does not explain the persistently high level of residential segregation among many middle-class blacks who left the ghettos for suburbs since the 1960s.
   a. A "segregation index" measures the extent of the problem.
   b. Douglas Massey and Nancy Denton show black segregation in housing persists due to racism (i.e. white flight, discrimination by real estate agents and mortgage lenders to protect real estate values).
7. Hispanic and Asian American experiences indicate less segregation in housing, and faster desegregation than African Americans.
8. Puerto Ricans are nearly as segregated in housing as blacks, and are improving as slowly.
9. Affirmative Action
   a. Survey research shows that European Americans often express resentment against real and perceived advantages enjoyed by African Americans.
   b. Supporters of affirmative action feel that African Americans should get preference if equally qualified for a job or college admission, to correct historical injustices.
   c. Many whites object and decry "reverse discrimination."
   d. Although most blacks favor affirmative action, some opponents argue that affirmative action demeans their accomplishments and contributes to belief in black inferiority.
   e. Today's political climate is generally unfavorable to affirmative action.

B. Immigration and the Renewal of Racial and Ethnic Groups
   1. A steady flow of immigrants gives new life to racial and ethnic groups. We can expect many vibrant racial and ethnic communities to invigorate American life for a long time.
   2. Of the broad U.S. Census Bureau's racial and ethnic categories, the fastest growing is "Asian American." Numbering nearly 12 million in the 2000 census, most in this category arrived after the mid-1960s, when legislators eliminated racial selection criteria and instead designed an immigration law emphasizing the importance of choosing newcomers who can contribute to the U.S. economy. Many Asian immigrants are middle-class professionals and business people.
   3. Some people hold up Asian Americans as models. This argument has two main problems:
      a. First, some substantial Asian-American groups, including several million Vietnamese, Cambodians, Hmong, and Laotians, do not fit the "Asian model." Members of these groups came to the United States as political refugees, and are disproportionately poor and unskilled, with a poverty rate higher than that of African Americans.
      b. Second, most economically successful Asian Americans were selected as immigrants precisely because they possessed educational credentials, skills, or capital. They arrived

with advantages. Saying that unskilled Chicanos or the black descendants of slaves should follow their example ignores the social-structural brakes on mobility and assimilation.

C. A Vertical Mosaic
   1. Today, U.S. society is based on segregation, assimilation, and **pluralism**, the retention of racial and ethnic culture combined with equal access to basic social resources.
   2. The United States has come a long way over the last 200 years in terms of social tolerance, especially in comparison to other countries.
   3. The growth o tolerance is taking place in the context of increasing ethnic and racial diversity.
   4. However, if current trends continue, the racial and ethnic mosaic will be vertical with some groups disproportionately clustered at the bottom.
   5. Policy initiatives could decrease the verticality of the American ethnic mosaic, thus speeding up the movement from segregation to pluralism and assimilation for the country's most disadvantaged groups.

Study Activity: Applying The Sociological Compass

After read Chapter 10 of the textbook, review the terms "race", "ethnic group", racial and ethnic "labels", and "identities". Re-read the section, "Ethnic and Racial Choice in the United States". Then, answer the following questions.

1. What is your race? Why do you self-identify racially in this manner?

2. What is your ethnicity? Why do you self-identify ethnically in this manner?

3. What is your ancestry?

4. How do you prefer to identify yourself?

5. Is your decision to self-identify a matter of choice, imposition, or both?

6. What is it about your choice of terms that indicates who you are as a person? Make a list of reasons why you chose to identify yourself as indicated in questions one, two, and four.

7. How do you think others (outside of your racial or ethnic group) perceive the labels you chose? Do their perceptions have any influence on your choices of self-identification? Make a list of reasons as to why or why not.

8. Are you choices of racial and ethnic identification consistent with your ancestry?

Infotrac College Edition Online Exercises

For the following exercises, log on to the online library of InfoTrac College Edition at http://www.infotrac-college.com/. Make note that InfoTrac has implemented a new registration system that will allow easier access to InfoTrac through the use of a personalized username and password. Once you've created your username and password you may proceed directly to the Log On page. To create an account, register your passcode packaged with your textbook, and create a username and password, by following the online prompt. After you are logged in, click on "Infotrac College Edition." You will arrive at a screen that enables you to search topics.

Keyword: **affirmative action**. Look for specific articles that address affirmative action for minority groups in the United States. What are the key arguments "for" and "against" affirmative action? Write a short paper addressing your findings and your own opinion on affirmative action.

Keyword: **racial profiling**. Racial profiling remains a "hot topic" in the area of race and ethnicity. What types of subjects does your racial profiling search produce? Do any articles discuss what measures can be taken to reduce racial profiling? Do you think racial profiling is justified now that the United States is at war with "terrorism?"

Keyword: **human rights violations**. Look for articles that address contemporary human rights violations. What other terms can be used interchangeably with human rights violations. Where are most human rights violations taking place?

Internet Exercises

Go to: http://www.census.gov/population/www/cen2000/briefs.html. Click on "Overview of Race and Hispanic Origin". You will access a brief report that defines new terms and categories for racial and ethnic identification, and explains the reasons for the changes in race data. Make a list of the new terms and categories. Make a list of reasons for the changes. Based on the report, combined with what you learned in Chapter 10, do you think the new census enables greater choice or places more constraint on our freedom to racially and ethnically self-identify? Why?

The Federal Bureau of Investigation (FBI) publishes different types of statistics. Go to: http://www.fbi.gov/ucr/hatecrime.pdf to find the UCR Hate Crimes Date Collection Guidelines. Read the short report and write a short paper about how race and ethnicity are viewed in the guidelines. Do you agree or disagree with these guidelines? Why or why not?

For information on southern women and women of color, The Center for Research on Women at the University of Memphis has the following website available: http://cas.memphis.edu/isc/crow

Write a summary paper examining what type of research projects have been conducted in recent years.

Read Thomas Sowell's essay on Race, Culture and Equality at http://www-hoover.stanford.edu/publications/he/23/23a.html

Do you agree or disagree with Sowell's ideas?

Practice Tests

Fill-In-The-Blank

Fill in the blank with the appropriate term from the above list of "key terms."

1. _____ is a social construct used to distinguish people in terms of one or more physical markers, usually with profound effects on their lives.

2. A(An) _____ is a group of people who are socially disadvantaged although they may be in the numerical majority.

3. _____ is bias that is inherent in social institutions and is often not noticed by members of the majority group.

4. In _____ , low-wage workers of one race and high-wage workers of another race compete for the same jobs.

5. _____ is the ownership and control of people.

6. Communities whose boundaries extend between countries are referred to as _____ .

7. _____ is the retention of racial and ethnic culture combined with equal access to basic social resources.

8. _____ is unfair treatment of people due to their group membership.

9. A(An) _____ is a disadvantaged person or category of people whom others blame for their own problems.

10. A(An) _____ is a spatial concentration of ethnic group members who establish businesses that serve and employ mainly members of the ethnic group and reinvest profits in community business and organizations.

11. _____ involves the spatial and institutional separation of racial or ethnic groups.

12. _____ involves one race or ethnic groups subjugating another in the same country.

Multiple-Choice Questions

Select the response that best answers the question or completes the statement:

1. In the 19th century, brain size was falsely held to be one of the main indicators of intellectual capacity that varied by race. Dr. Samuel George Morton was especially known for research in this area. Today, we know that the main issues compromise his findings. Which is NOT among the evidence that compromises Morton's research?
    a. Archeologists cannot precisely determine race by skull shape.
    b. Morton's skulls formed a small, unrepresentative sample.
    c. Morton did not ensure that the sex composition of black and white skulls was identical.
    d. Subsequent analysis of Morton's evidence actually suggested that skull size of Native Americans indicated higher intelligence.
    e. All of the answers are pieces of evidence that compromise Morton's research.

2. _____ is an attitude that judges a person on his or her group's real or imagined characteristics.
    a. prejudice
    b. discrimination
    c. scapegoating
    d. racism
    e. expulsion

3. According to most sociologists, minority group refers to a group of people:
    a. who are numerically smaller in population size.
    b. who tend to hold negative racial and ethnic self-conceptions
    c. who have no choice in their racial and ethnic identities.
    d. who are socially disadvantaged.
    e. None of the above.

4. John Lie's migration experience from South Korea to Japan, Japan to Hawaii, and from Hawaii to the U.S. mainland illustrates:
   a. that racial and ethnic labels change from country to country.
   b. that racial and ethnic identities are reshaped from one social context to another.
   c. that continuous negotiations take place between one's racial and ethnic self-conceptions, and outsider labels.
   d. All of these are correct.
   e. None of the above.

5. Which example illustrates the sociological assumption that social structural differences underlie cultural differences?
   a. Social class differences between descendants of southern black slaves and West Indian immigrants within the African American community.
   b. The differential impact of American immigration policy on Jews and descendants of southern blacks.
   c. Social class differences between West Indian immigrants in New York and London in the 1950s and 1960s.
   d. All of these are correct.
   e. None of the above.

6. Which Hispanic group is more likely to be middle-class and professional?
   a. Chicanos
   b. Puerto Ricans
   c. Cubans
   d. All of these are correct.
   e. None of the above.

7. Which is NOT a reason why the term "Hispanic American" is more widely used that it was 30 to 40 years ago?
   a. Many Hispanic Americans find it politically useful.
   b. The term captures the internal diversity of Hispanic Americans.
   c. The government finds the term useful for data collection and public policy purposes.
   d. Non-Hispanic Americans find the term convenient.
   e. All of the answers are reasons.

8. _____ is a nostalgic allegiance to the culture and traditions of the immigrant generation, or that of the old country.
   a. assimilation
   b. accommodation
   c. multiculturalism
   d. pluralism
   e. symbolic ethnicity

9. Which is NOT a stage of Robert Park's ecological theory?
   a. expulsion
   b. invasion
   c. resistance
   d. competition
   e. accommodation

10. _____ is the process by which a minority group blends into the majority population and eventually disappears as a distinct group.
   a. pluralism
   b. symbolic ethnicity formation
   c. assimilation
   d. expulsion
   e. colonialism

11. Which concept best describes the treatment of Native Americans by European immigrants in the 19th century?
   a. assimilation
   b. expulsion
   c. accommodation
   d. competition
   e. pluralism

12. Which concept best describes contemporary experiences of Irish Americans?
   a. assimilation
   b. expulsion
   c. colonialism
   d. genocide
   e. competition

13. Which government policy or historical event had an especially detrimental effect on the eviction of Native Americans from their lands.
   a. 1830 Indian Removal Act
   b. 1964 Civil Rights Act
   c. 1954 Brown v. Board of Education
   d. 1848 Treaty of Guadalupe Hidalgo
   e. None of the above.

14. Racist organizations are _____ today.
   a. declining
   b. virtually nonexistent
   c. growing
   d. made up of a large proportion of the American population
   e. None of the above.

15. The research of Joe Feagin and Melvin Sykes indicates that:
    a. African Americans continue to be most victimized by hate crime.
    b. the civil rights movement helped to establish legal equality between blacks and whites.
    c. reforms allowed the black middle-class to emerge.
    d. African Americans continue to suffer high levels of covert and overt racial prejudice and discrimination.
    e. All of these are correct.

16. As a solution to racial inequality in America, William Julius Wilson calls for:
    a. continued implementation of affirmative action to enhance African American representation in higher education and professional employment.
    b. an initiative to aid middle-class blacks to further their patterns of upward mobility.
    c. For "color-blind" public policies intended to improve the class position of the poor, which would likely help poor African Americans as well.
    d. All of these are correct.
    e. None of the above.

17. Which is NOT a reason why ghettos exist according to Douglas Massey and Nancy Denton?
    a. The exodus of white residents in formerly predominantly white neighborhoods where blacks move in.
    b. Discrimination against blacks by real estate agents to protect property values.
    c. Racism
    d. Many middle-class blacks with good jobs moved out of the inner city, which deprived African American youth of successful role models.
    e. All of the answers are reasons according to Massey and Denton.

18. Which is NOT true about affirmative action today?
    a. European Americans often express resentment against real and perceived advantages enjoyed by African Americans.
    b. Many whites decry "reverse discrimination."
    c. Some black opponents of affirmative action argue that it demeans their accomplishments.
    d. Today's climate is generally favorable to affirmative action in light of a more progressive-minded population.
    e. Most blacks favor affirmative action.

19. Among the major Hispanic groups in the United States, which group especially illustrates the formation of an ethnic enclave?
    a. Chicanos
    b. Puerto Ricans
    c. Cubans
    d. Costa Ricans
    e. Mexican nationals

20. Which statement is correct regarding the cultural emphasis on African-American sports heroes?
    a. Athletic heroes like Shaquile O'Neal are held up as role models for African American youth, since their chance of "making it" in professional sports is greater than their chance of going to college, due to discrimination.
    b. The cultural emphasis on African-American sports heroes dispels long-standing stereotypes of African Americans in favor of realistic images that highlight the superiority.
    c. The cultural emphasis on African-American sports heroes reinforces incorrect racial stereotypes about black athletic prowess and intellectual inferiority.
    d. All of these are correct.
    e. None of the above.

21. According to Figure 10.1 "The Vicious Circle of Racism," which step is the process of perceptions of behavioral differences creating racial stereotypes that get embedded in culture?
    a. first step
    b. second step
    c. third step
    d. fourth step
    e. fifth step

22. The United States is becoming a more ethnically and racially diverse society due to:
    a. a comparatively high immigration rate.
    b. a comparatively high birth rate among non-Hispanic whites.
    c. the level of out-migration of European Americans.
    d. All of these are correct.
    e. None of the above.

23. Which racial or ethnic group is projected to grow the fastest in terms of composition by the year 2050?
    a. White
    b. Hispanic
    c. Black
    d. Pacific Islander
    e. Native American

24. Which region of the United States had the highest percentage of Puerto Ricans in 2002?
    a. Northwest
    b. Midwest
    c. South
    d. West
    e. Pacific

25. Who suggested that "it doesn't matter to a racist whether an African American is a professional or a panhandler, a genius or a fool, a saint or a criminal. Where racism is common, racial identities are compulsory and at the forefront of one's self-identity?"
    a. Martin Luther King Jr.
    b. William Julius Wilson
    c. Robert Blauner
    d. Robert Park
    e. Malcolm X

True False

1. Researchers who were eager to prove correlations between brain size, intellectual capacity, and race in the 19th century are now widely dismissed as practitioners of racist and sexist quasi-science.

   TRUE or FALSE

2. Promoting the Shaquille O'Neal's and Kobe Bryant's as role models for African-American youth deflects the attention of African American youth from more probable avenues of upward mobility through academic excellence.

   TRUE or FALSE

3. Race as a biological concept is a useful analytical tool to explain inequality.

   TRUE or FALSE

4. The United States has long been a society where nearly all racial and ethnic groups have exercised their freedom to choose their racial and ethnic identities.

   TRUE or FALSE

5. The experience of Puerto Ricans is best captured by ecological theory.

   TRUE or FALSE

6. African Americans lack the freedom to enjoy symbolic ethnicity.

   TRUE or FALSE

7. Because the United States has come a long way in social tolerance over the past 200 years, assimilation is for the most part equally accessible for all racial and ethnic groups.

   TRUE or FALSE

8. Asian Americans are the most frequent victims of hate crimes today.

   TRUE or FALSE

9. Although colonized minority groups have often expressed cultural forms of resistance, they rarely exercised political resistance in light of high levels of oppression.

   TRUE or FALSE

10. Puerto Ricans and Chicanos have similar experiences as minority groups in the United States.

    TRUE or FALSE

11. Like most Americans, Collin Powell fits neatly into an ethnic category.

    TRUE or FALSE

Short Answer

1. If race is merely a social construct and not a useful biological term, explain why race continues to matter in the United States today? (p. 280-284)

   _____

   _____

2. Explain the difference between an ethnic enclave and residential segregation. (p. 287, 292, 302-303)

   _____

   _____

3. Briefly explain the similarities and differences between groups of the "Hispanic American Community" in terms of immigration experience, economic status, and community formation. (p. 285-288)

   _____

   _____

4. Briefly describe some advantages of ethnicity. (p. 298-300)

   _____

   _____

5. Define and describe the relationship between prejudice and discrimination. (p. 279-280)

   _____

   _____

Essay Questions

1. Compare and contrast ecological theory and the theory of internal colonialism as explanations of racial and ethnic relations. In doing so, be sure to mention the major assumptions, conditions, and critiques of each theory. (p. 292–298).

2. Explain the pros and cons for the "declining significance of race" debate. Take a stance on the issue by providing research, statistical, and/or historical support for your position cited in the textbook. (p. 301-306)

3. Explain the contradictory relationship between the notion of black athletic superiority and the idea of black intellectual inferiority, in their perpetuation of racial inequality. (p. 277-284)

4. Compare and contrast the terms "race" and "ethnic group". Why are these two terms often used interchangeably, and is it misleading to claim that race and ethnicity are quite different? Why? (p. 281-284)

5. Describe the relationship between the ethnic groups of the Hutus and Tutsis in Rwanda. What were the factors contributing to the increasing conflict between them that led up to the genocide, as portrayed in the film *Hotel Rwanda*? (p. 308-309)

Solutions

Practice Tests

Fill-In-The-Blank

1.	Race	5.	Slavery	9.	scapegoat
2.	minority group	6.	transnational communities	10.	ethnic enclave
3.	Institutional racism	7.	Pluralism	11.	Segregation
4.	split labor markets	8.	Discrimination	12.	Internal colonialism

Multiple-Choice Questions

1. D, (p. 277-278)	10. C, (p. 292)	19. C, (p. 287)
2. A, (p. 279)	11. B, (p. 293)	20. C, (p. 280)
3. D, (p. 284)	12. A, (p. 292)	21. C, (p. 282)
4. D, (p. 284)	13. A, (p. 293-295)	22. A, (p. 286)
5. D, (p. 281-284)	14. C, (p. 301)	23. B, (p. 286)
6. C, (p. 287)	15. D, (p. 301)	24. A, (p. 286)
7. B, (p. 287-288)	16. C, (p. 301-304)	25. E, (p. 291)
8. E, (p. 291)	17. D, (p. 304)	
9. A, (p. 292)	18. D, (p. 304-305)	

True False

1. T, (p. 277-278)	5. F, (p. 285-288, 292)	9. F, (p. 292-298)
2. T, (p. 280-281)	6. T, (p. 291)	10. T, (p. 287)
3. F, (p. 281)	7. F, (p. 292-298)	11. F, (p. 281)
4. F, (p 284-291)	8. F, (p. 301)	

SEXUALITY AND GENDER

Student Learning Objectives

After reading Chapter 11, you should be able to:

1. Explain the differences and connections between sex and gender.

2. Contrast essentialist and social constructionist theories of gender differences.

3. Explain the role of gender socialization and the mass media in constructing our ideas of masculinity and femininity.

4. Discuss the patterns of male-female interaction in terms of gender socialization, especially in the family, schools, and the workplace.

5. Discuss social reactions to people who resist traditional gender roles (transgendered, transsexuals, homosexuals, and bisexuals).

6. Describe the origins and creation of gender inequality in terms of societal development.

7. Identify and describe the main factors that account for the earnings gap between men and women in the United States.

8. Explain the social forces and sources of rape and sexual harassment of women by men.

9. Describe current trends regarding equality between men and women in a comparative perspective.

10. Discuss the future prospects and obstacles for nontraditional socialization of children, affirmative action, child care reform, and comparable worth laws, as avenues toward gender equality.

11. Describe the historical emergence of the women's movement.

12. Identify and explain the major currents of the modernist feminist movement.

Key Terms

intersexed (313)

sex (314)

gender (314)

gender identity (314)

Detailed Chapter Outline

I. SEX VERSUS GENDER

  A. Is It a Boy or a Girl?

    1. The "twins" case raises several issues and questions about sex and gender.

      a. Whether hormonal and surgical measures combined with socialization can alter the sex and gender identity of **intersexed** individuals or **hermaphrodites** (people born with ambiguous genitals due to a hormone imbalance in the womb), and other ambiguous children, as Dr. John Money suggested and attempted.

      b. That being male or female partly involves biology, but also certain "masculine" and "feminine" feelings, attitudes, and behaviors. Unlike sex, gender is not determined just by biology.

      c. Biology is not destiny, but that the *social learning* of gender begins early in life.

  B. Gender Identity and Gender Role

    1. **Sex** depends on whether you were born with distinct male or female genitals and a genetic program that released either male or female hormones to stimulate the development of your reproductive system. **Gender** is your sense of being male or female and your playing masculine and feminine roles in ways defined as appropriate by your culture and society.

    2. **Gender identity** is your identification with, or sense of belonging to a particular sex—biologically, psychologically, and socially.

    3. **Gender role** is the set of behaviors associated with widely shared expectations about how males and females are supposed to act.

  C. The Social Learning of Gender

    1. The first half of this chapter:

      a. Helps you better understand what makes us male or female, by outlining competing theories of gender differences.

      b. Examines how people learn gender roles during socialization in the family and at school.

      c. Discusses how members of society enforce **heterosexuality**, the preference for members of the opposite sex as sexual partners.

2. The second half of Chapter 11 examines gender inequality.

II. THEORIES OF GENDER

A. Essentialism

1. **Essentialism** is a school of thought that sees gender differences as a reflection of biological differences between women and men. In contrast, **social constructionism** is a school of thought that sees gender differences as a reflection of the different social positions occupied by women and men.

2. All humans are assumed to instinctively attempt to ensure their genes get passed generationally, however men and women develop different strategies. "Universal features of our evolved selves" contribute to the survival of the human species.

   a. Because women produce a small number of eggs, they have a bigger investment in ensuring the survival of their offspring. It is thus in the best interest of a woman to locate the best mate to intermix her genes.

   b. Men can develop hundreds of millions of sperm every 24–48 hours, thus a man increases the chance of his and only his genes will get passed on. Because men compete with other men for sexual access to women, men evolve competitive, aggressive and violent dispositions.

B. Functionalism and Essentialism

1. Functionalists reinforce the essentialist viewpoint when they claim that traditional gender roles help to integrate society.

2. Each successive generation learns to perform the complementary roles of women at home and men in the paid labor force by means of *gender role socialization.*

3. Larger society also promotes *gender role conformity.*

4. In the functionalist view learning the essential features of femininity and masculinity integrates society and allows it to function properly.

C. A Critique of Essentialism from the Conflict and Feminist Perspectives.

1. Four main criticisms:

   a. *Essentialists ignore the historical and cultural variability of gender and sexuality.*

      i. Notions of the male "good provider," and domestic skills for females, decrease in societies with low levels of gender inequality.

      ii. Social situations involving competition and threat stimulate production of testosterone in women, causing more aggressive behavior.

      iii. Gender differences are declining rapidly.

   b. *Essentialism tends to generalize from the average, ignoring variations within gender groups.*

   c. *No evidence directly supports the essentialists' major claims.*

   d. *Essentialists' explanations for gender differences ignore the role of power.*

2. Conflict theorists, going back to Engels, have located the root of male domination in class inequality.

   a. As industrial capitalism developed, male domination increased because it made men wealthier and more powerful while it relegated women to subordinate, domestic roles.

3. Feminist theorists doubt that male domination is so closely linked to the development of industrial capitalism.
   a. They note that gender inequality is greater in agrarian than in industrial capitalist societies.
   b. Male domination is evident in socialist or communist societies.
   c. Male domination is rooted in patriarchal authority relations, family structures, and patterns of socializations and culture in societies.

D. Social Constructionism and Symbolic Interactionism
   1. **Social constructionism** is the view that *apparently* natural or innate features of life, such as gender, are sustained by *social* processes.
      a. Conflict and feminist theories are types of social constructionism.
      b. Symbolic interactionism is also a type of social constructionism.
   2. Gender Socialization
      a. Toys
         i. When they play with Barbie, girls learn the ideal woman and body image.
         ii. G I Joe teaches boys stereotypical male roles.
      b. Research shows that, from birth, parents, especially fathers, treat infant boys and girls differently.
         i. Girls are: perceived as delicate, cute, weak and beautiful; encouraged in cooperative and role-playing games; and praised for compliance.
         ii. Boys are: treated as strong, alert, and well coordinated; encouraged in boisterous and competitive play; and praised for assertiveness.
      c. Teachers and other authority figures impose their ideas of appropriate gender behavior on children.
   3. Gender Segregation and Interaction
      a. Barrie Thorne's research indicates that children are actively engaged in the process of constructing gender roles, and the content of children's gendered activities is not fixed.
      b. By adolescence, **gender ideologies** (sets of interrelated ideas about what constitutes appropriate masculine and feminine roles and behavior) are well formed.

E. The Mass Media and Body Image
   1. Children, adolescents, and adults continue to negotiate gender roles as they interact with the mass media.
   2. Women are more frequently seen cleaning house, caring for children, modeling clothes, and acting as objects of male desire, while men are more frequently seen in aggressive, action-oriented, and authoritative roles. This reinforces the normality of traditional gender roles.
   3. As body image became more important for one's self-definition during the 20th century, the ideal body image became thinner, especially for women.
      a. Why did body image become more important for self-definition during the 20th century? Why was slimness stressed?
         i. As Americans became better educated, they grew increasingly aware of the health problems associated with being overweight, which led to the desire to slim down.

ii. Extremely thin body shapes in modeling promote a virtually unattainable image (not health) for good business (i.e. diet and frozen low-cal, diet and self-help, fitness, cosmetic surgery, and advertising industries).

b. Survey data show how widespread dissatisfaction with our bodies is and how important a role the mass media plays in generating our discomfort. North Americans' anxiety about their bodies increased substantially between 1972 and 1997.

c. Body ideals are influenced by gender.

i. Women are more concerned about their stomachs.

ii. By 1997 men were more concerned about their chests than women were about their breasts.

d. Body dissatisfaction motivates many women to diet, and even prompts some people to take dangerous and life threatening measures to lose weight (i.e. anorexia and bulimia).

F. Male-Female Interaction

1. Early gender role socialization is the basis of children's social interactions as adults. Examples

a. Boys learn competition, conflict, self-sufficiency, hierarchical relationships, taking center-stage, and boasting, by playing sports.

b. Girls learn to maintain cordial relationships, avoid conflict, and resolve differences through negotiation, give advice, and not be bossy, by playing with dolls.

c. Misunderstandings between men and women are thus common.

d. Gender-specific interaction styles have implications for who gets heard and who gets credit at work, which sometimes leads to complaints about a **glass ceiling**, a social barrier that makes it difficult for female managers to rise to the top level of management. Example: Deborah Tannen's research.

G. Homosexuality

1. A minority of people resists traditional gender roles.

a. **Transgendered** people are individuals who want to alter their gender by changing their appearance or resorting to medical intervention.

b. Some transgendered people are **transsexuals**, who believe they were born with the "wrong" body, and therefore identify with and want to live fully as members of the "opposite" sex. They often undergo a sex change operation.

c. **Homosexuals** are people who prefer sexual partners of the same sex.

d. **Bisexuals** prefer sexual partners of both sexes.

2. Homosexuality has existed in every society, was identified as a distinct category in the 1860s, and has become less stigmatized over the past century.

3. Sociologists are more interested in the way homosexuality is socially constructed, that is the ways it is expressed and repressed.

4. In the 1940s, American sexologist Alfred Kinsey and others concluded that homosexual practices were so widespread that homosexuality could hardly be considered an illness affecting a tiny minority.

5. Since the middle of the 20th century, gays and lesbians have gone public with their lives and built large communities and subcultures.

6. Opposition to people who do not conform to conventional gender roles remains strong at all stages of the life cycle. Some people are prepared to back up their beliefs with force and violence.

7. Due to widespread animosity toward homosexuals, many people who have wanted or who had sex with members of the same sex do not consider themselves gay, lesbian or bisexual.

8. Antigay violence
   a. Recent research suggests some antigay crimes may result from repressed homosexual urges on the part of the aggressor.
   b. Aggressors may be "homophobic" or afraid of homosexuals because of their inability to cope with their own homosexual impulses.
   c. Some assaults occurred to prove toughness, heterosexuality, alleviate boredom and have fun.
   d. Some antigay crimes may result from repressed homosexual urges on the part of the aggressor. From this point of view, the aggressors are **homophobic**, or afraid of homosexuals.
   e. Antigay violence is not just a question of abnormal psychology, but a broad cultural problem, and seems to be growing in America. Example: the 1998 murder of Matthew Shepard.

III. GENDER INEQUALITY

A. The Origins of Gender Inequality
   1. Long-Distance Warfare and Conquest
      a. Anthropological and archeological evidence suggest that women were about equal in status in nomadic hunting-gathering societies (90% of human history), due to the fact that women produced up to 80% of the band's food.
      b. Between 4,300 and 4,200 B.C.E., Old Europe was invaded by successive waves of warring people from the Asiatic and European northeast (the Kurgans) and the deserts to the south (the Semites), whose societies were based on hierarchical structures of male dominance.
      c. God became a male who willed that men should rule women. Traditional Judaism, Christianity, and Islam embody ideas of male dominance and they all derive from the tribes who conquered Old Europe in the fifth millennium B.C.E.
      d. Law reinforced women's sexual, economic, and political subjugation to men.
   2. Plow Agriculture
      a. Since men were on average stronger, men owned land, and women were restricted in their activities by pregnancy, nursing, and childbirth, plow agriculture made men more powerful socially.
   3. The Separation of Public and Private Spheres
      a. During industrialization, men became wage and salary workers and their work moved out of the household, and into the factory and the office.
      b. Most women remained in the domestic or private sphere.
      c. The idea of a natural division of labor developed.

B. The Earnings Gap Today

1. In the first quarter of 2001, women over the age of 15 working full-time in the paid labor force earned only 76% of what men earned. Four main factors account for the earnings gap.
   a. *Gender discrimination.*
      i. 1985 Microsoft example
      ii. **Gender discrimination** is the rewarding of women and men differently for the same work.
      iii. Antidiscrimination laws have helped increase the **female-male earnings ratio** (women's earnings as a percentage of men's earnings), which has increased 17.3% between 1960 and 2000.
   b. *Heavy domestic responsibilities reduce women's earnings.*
      i. Women do more housework, child rearing and elderly care than men, which substantially decreases the time women spend training, and doing paid work.
      ii. Women who work full-time in the paid labor force continue to shoulder a disproportionate share of domestic responsibilities.
   c. *Women tend to be concentrated in low-wage occupations and industries.*
      i. High school and college courses tend to limit women to jobs in low-wage occupations.
   d. *Work done by women is commonly considered less valuable than work done by men because it is viewed as involving fewer skills.*

C. Male Aggression Against Women
   1. Why do men commit more frequent (and more harmful) acts of aggression against women than women commit against men? Greater physical power is more likely to be used to commit acts of aggression when norms justify male domination and men have much more *social* power than women.
   2. Rape
      a. Although some rapists suffer from psychological disorders, and others misinterpret signals in what they regard as sexually ambiguous situations, such cases account for only a small proportion of total rapists.
      b. Rape is sometimes not about sexual gratification, but all rape involves domination and humiliation as principal motives.
      c. Social situations increase the rate of rape.
         i. In war, conquering male soldiers often use rape to humiliate the vanquished.
         ii. Spousal abuse is common among police officers, in which aggressiveness is an important part of their work.
         iii. The relationship between male dominance and rape is evident in research on college fraternities.
         iv. There are proportionately more rapists among men who participate in athletics than among nonathletes, although the overwhelming majority of athletes are not rapists.
         v. The incidence of rape is highest in situations where early socialization experiences predispose men to want to control women, where norms justify the domination of women, and where a big power imbalance between men and women exists.
   3. Sexual Harassment
      a. Two types of sexual harassment:

        i. **Quid pro quo harassment** takes place when sexual threats or bribery are made a condition of employment decisions.

        ii. **Hostile environment harassment** involves sexual jokes, comments, and touching that interferes with work or creates an unfriendly work setting.

   b. Research suggests that:

        i. relatively powerless women are most likely to be sexually harassed.

        ii. sexual harassment is most common in work settings that exhibit high levels of gender inequality and a culture of male domination of women.

        iii. women who are young, unmarried, and employed in nonprofessional occupations are the most likely targets.

   c. Many aspects of our culture legitimize male dominance.

   d. Gender inequality is the foundation of aggression against women.

D. Toward 2050

   1. Equality between men and women in many countries grew in the 20th century. Why?

      a. Women started having fewer children, because they became more costly to raise and less economically useful.

      b. The growth of the service sector increased the demand for women in the paid labor force.

      c. The legalization and availability of contraception enabled women to control their bodies.

      d. The women's movement.

   2. The "Gender Empowerment Measure" (GEM) indicates that:

      a. Scandinavian countries were the most gender-egalitarian.

      b. The United States ranked 14th in the world.

      c. There is more gender equality in rich than in poor countries.

      d. Gender equality is a function of government policy.

      e. American women still have a long way to go before they achieve equality with men.

   3. Socializing children at home and in school to understand that women and men are equally adept at all jobs is important to motivate girls to excel in nontraditional fields. **Affirmative action**, which involves hiring more qualified women to diversify organizations, is important in helping to compensate for past discrimination in hiring.

E. Child Care

   1. High quality, government -subsidized, affordable childcare is widely available in most Western European countries, but not in the United States. Consequently, many American women with small children are unable to work outside the home or able to work only on a part-time basis.

   2. Childcare options and the quality of childcare vary by social class for women who have small children and work outside the home.

   3. A third of all day-care facilities in the United States do not meet children's basic health and safety needs, due to lack of government regulation and low wages for child care workers.

   4. Governments throughout the country recognize the crisis in childcare (i.e. 1997 recruitment of welfare mothers to start at-home, for profit day-care facilities).

   5. Many companies, schools, and religious organizations in the United States provide high-quality day care.

6. Comparable Worth
   a. Researchers attempted to establish gender-neutral standards to judge the dollar value of work to identify pay inequities between men and women to promote equal pay for jobs of **comparable worth**, even if men and women did different jobs.
   b. A number of U.S. states adopted laws requiring equal pay for work of comparable worth. Minnesota leads the country. Obstacles:
      i. The Laws do not apply to most employers.
      ii. Comparable worth assessments have been challenged in the courts.
      iii. No federal legislation on comparable worth is in development.

IV. THE WOMEN'S MOVEMENT

A. Progress has always depended in part on the strength of the organized women's movement, not just on the sympathy of government and business leaders.

B. In the 1840s, the chief demand of the "first wave" of the women's movement was the right to vote (paralleled the oppression of black slaves).

C. In the mid-1960s, the "second wave" was inspired by the successes of the civil rights movement, and advocated equal rights with men in education and employment, the elimination of sexual violence, and women's control over reproduction (i.e. Equal Rights Amendment ERA).

D. Three main streams of the modern feminist movement:
   1. *Liberal feminism* is the most popular current of feminism, and advocates nonsexist methods of socialization and education, more sharing of domestic tasks between men and women, and extending to women equal educational, employment, and political rights.
   2. *Socialist feminists* regard women's relationship to the economy (namely capitalism) as the main source of women's disadvantages—and advocate the elimination of private property and the creation of economic equality to end oppression of all women.
   3. *Radical feminists* regard patriarchy as the larger source of gender inequality, and conclude that the very idea of gender in the context of patriarchy needs to be changed to bring an end to male domination.

E. *Antiracist* and *postmodernist* feminists have extended the relevance of feminism to previously marginalized groups.

F. Some feminist ideas have gained widespread acceptance in American society over the last three decades, partly due to the political and intellectual vigor of the women's movement. (i.e. 1998 General Social Survey).

G. Traditional patterns of gender socialization weigh heavily on men—"But socialization is not destiny."

Study Activity: Applying The Sociological Compass

After reading Chapter 11, review essentialism and social constructionism. Consider this scenario. You are at your neighborhood park playing with your five year-old son and seven year-old daughter (or your five-year old brother, nephew, or male cousin and your seven year-old sister, niece, or female cousin). Your "son" falls off the swing and scrapes his knee. What is a likely masculine response to this incident? How might you react to the situation? Five minutes later, your "daughter" falls off the monkey bars and bruises her elbow. In both cases, injuries are minor. What is a likely feminine response? Are there any differences in the way you might react to your son's versus daughter's injury? Based on your personal experiences, make a list of masculine reactions to the injury of male five-year old boy, and a list of feminine reactions to the injury of the seven-year old girl. Then, make a list of your first impressions when pondering the questions of how you would respond. Finally, consider whether essentialist or social constructionist assumptions about gender differences guide the responses to each situation by writing the letter 'E' for essentialist or 'SC' for social constructionist next to each response. Briefly explain why you consider the responses as examples of essentialism and social constructionism, respectively.

Traditional Masculine Reactions

1.

2.

3.

Traditional Feminine Reactions

1.

2.

3.

4.

Your Reactions

1.

2.

3.

4.

Essentialist and Social Constructionist Explanations:

Infotrac College Edition Online Exercises

For the following exercises, log on to the online library of InfoTrac College Edition at http://www.infotrac-college.com/. Make note that InfoTrac has implemented a new registration system that will allow easier access to InfoTrac through the use of a personalized username and password. Once you've created your username and password you may proceed directly to the Log On page. To create an account, register your passcode packaged with your textbook, and create a username and password, by following the online prompt. After you are logged in, click on "Infotrac College Edition." You will arrive at a screen that enables you to search topics.

Keyword: **glass ceiling**. Identify and read the latest article/periodical that discusses women and the glass ceiling. Read the article, identify its major contribution, and make a list of major findings.

Keyword: **fatherhood**. Is the role of fatherhood changing? Is so, how, and in what ways?

Keyword: **women's movement**. Examine the women's movement in third world countries. Are the issues and concerns the same as those involved in the women's movement in westernized countries?

Keyword: **masculinity**. Examine the different ways that masculinity is defined. Is there a masculine men's movement? Is it credible? What evidence can you provide for your responses?

Keyword: **date rape**. The relationship between date rape and drugs has increasingly become an issue of concern. Research the recourse victims have to this gender-based problem. Make a list of your findings.

Internet Exercises

Look up the Web site: http://www.fordham.edu/halsall/mod/Senecafalls.html to see a Declaration of Sentiments written in 1848 by Elizabeth Cady Stanton. How have things changed for women in the last 155 years? Write a short paper discussing your findings.

The media is always portraying images of the *ideal* body type. For more information on this subject link to: http://www.intergrouprelations.uiuc.edu/dim2000/page3.html

This is an article written on body image by DeNeishia Manual. What do you think is the ideal body type? Compare your ideal body type with Manual's.

To learn about the men's movement, link to the National Organization for Men's homepage at: http://www.tnom.com/

You can also link to the National Organization for Women's homepage at http://www.now.org/

Compare and contrast the purposes of the National Organization for Men with the National Organization for Women.

The National Council for Research on Women has put out a Report on Girls and IQ in Science.

Go to http://www.ncrw.org/ to investigate their research report. What does it say about the relationship between gender and science IQ? Be sure to click on Research for Action.

Practice Tests

Fill-In-The-Blank

Fill in the blank with the appropriate term from the above list of "key terms."

1. _____ refers to the equal dollar value of different jobs, which is established in gender-neutral terms by comparing jobs in terms of the education and experience needed to do them.

2. Your _____ depends on whether you were born with distinct male or female genitals and a genetic program that released either male or female hormones to stimulate the development of your reproductive system.

3. _____ is one's identification with, or sense of belonging to, a particular sex.

4. _____ are sets of ideas about what constitutes appropriate masculine and feminine roles and behavior.

5. The _____ is a social barrier that makes it difficult for women to rise to the top level of management.

6. The _____ is women's earnings expressed as a percentage of men's earnings.

7. _____ involves sexual jokes, comments, and touching that interfere with work or creates an unfriendly work setting.

8. A (An) _____ is the set of behaviors associated with widely shared expectations about how males or females are supposed to act.

9. _____ is a school of thought that sees gender differences as a reflection of biological differences between women and men.

10. _____ are people who prefer sexual partners of both sexes.

Multiple-Choice Questions

Select the response that best answers the question or completes the statement:

1. _____ is the set of behaviors associated with widely shared expectations about how males and females are supposed to act.
   a. Sex
   b. Sex type
   c. Gender role
   d. Gender prescription
   e. None of the above.

2. Which school of thought sees gender differences as a reflection of biological differences between men and women?
   a. essentialism
   b. social constructionism
   c. functionalism
   d. feminism
   e. None of the above.

3. _____ believed that men gained substantial power over women when preliterate societies were first able to produce more than their members needed for their own subsistence.
   a. Durkheim
   b. Weber
   c. Freud
   d. Engels
   e. Mills

4. _____ and _____ suggest that "universal features of our evolved selves" contribute to the survival of the human species.
   a. Essentialists; social constructionists
   b. Sociobiologists; evolutionary psychologists
   c. Evolutionary psychologists; social constructionists
   d. Sociobiologists; feminists
   e. None of the above.

5. Which is a critique of essentialism?
   a. Essentialists ignore the historical and cultural variability of gender and sexuality.
   b. Essentialism tends to generalize from the average, ignoring variations within gender groups.
   c. Essentialists' explanations for gender differences ignore the role of power.
   d. All of these are correct.
   e. None of the above.

6. Barry Thorne's research indicates that:
   a. children are actively engaged in the process of constructing gender roles.
   b. because men compete with other men for sexual access to women, men evolve competitive, aggressive and violent dispositions.
   c. the very idea of gender needs to be changed to bring an end to male domination.
   d. All of these are correct.
   e. None of the above.

7. Based on social constructionism, children, adolescence, and adults continue to negotiate gender roles as they _____ the mass media.
   a. ignore
   b. reject
   c. interact with
   d. All of these are correct.
   e. None of the above.

8. As body image became more important for one's self-definition during the 20th century, the ideal body for women became:
   a. heavier
   b. thinner
   c. more muscular
   d. All of these are correct.
   e. None of the above.

9. Research on gender socialization shows that:
   a. girls are perceived as delicate and weak.
   b. girls are perceived as cute and beautiful.
   c. girls are encouraged in cooperative and role-playing games.
   d. All of these are correct.
   e. None of the above.

10. Which is NOT a reason why body image became more important for self-definition according to the textbook?
    a. As Americans became better educated, they grew increasingly aware of health problems associated with being overweight.
    b. Extremely thin body shapes in modeling began to promote virtually unattainable images.
    c. The women's movement fueled the celebration of modern day Victorian images of women.
    d. Images constructed by the modeling industry promoted good business in terms of other complementary industries, such as the self-help, fitness, and cosmetic surgery industries.
    e. All of the answers are reasons why body image became more important for self-definition.

11. _____ believe they were born with the "wrong" body, and therefore identify with and want to live fully as members of the "opposite" sex.
    a. Homosexuals
    b. Bisexuals
    c. Transsexuals
    d. All of these are correct.
    e. None of the above.

12. Homosexuality:
    a. has existed in every society.
    b. was identified as a distinct category in the 1860s.
    c. has become less stigmatized over the past century.
    d. All of these are correct.
    e. None of the above.

13. Regarding homosexuality, sociologists are interested in:
    a. the way homosexuality is socially constructed.
    b. the ways homosexuality is expressed.
    c. the ways homosexuality is repressed.
    d. All of these are correct.
    e. None of the above.

14. In which societal type do women enjoy the highest levels of equality relative to men?
    a. nomadic hunting-gathering societies
    b. horticultural societies
    c. agrarian societies
    d. industrial societies
    e. All of these are correct.

15. Which is a main factor for the earning gap between men and women?
    a. gender discrimination
    b. Heavy domestic responsibilities reduce women's earnings.
    c. Women tend to be concentrated in low-wage occupations and industries.
    d. Work done by women is commonly considered less valuable than work done by men.
    e. All of these are correct.

16. In 2001, women over the age of 15 working full-time in the paid labor force earned what percentage of men's earnings?
    a. 96%
    b. 88%
    c. 76%
    d. 47%
    e. 29%

17. Which is UNTRUE about rape?
    a. There is a relationship between male dominance and rape according to research on college fraternities.
    b. In war, conquering male soldiers often use rape to humiliate the vanquished.
    c. The majority of rapes are committed by strangers.
    d. All rape involves domination and humiliation.
    e. Rape is sometimes not about sexual gratification.

18. Sexual jokes, comments, and touching that creates an unfriendly work setting are examples of:
    a. quid pro quo harassment
    b. sexist labor exploitation
    c. hostile environment harassment
    d. sexist ideologies
    e. All of these are correct.

19. The "Gender Empowerment Measure" (GEM) indicates that:
    a. The United States ranked 14th in gender-egalitarianism in the world.
    b. Scandinavian countries were the most gender-egalitarian.
    c. There is more gender equality in rich than in poor countries.
    d. American women still have a long way to go to achieve equality with men.
    e. All of these are correct.

20. Which strand of feminism advocates nonsexist methods of socialization and education, more sharing of domestic tasks between men and women, and extending to women equal educational, employment, and political rights?
    a. liberal feminism
    b. socialist feminism
    c. radical feminism
    d. antiracist feminism
    e. postmodernist feminism

21. By conventional North American standards, the ceremonial dress of male Wodaabe nomads in Niger may appear:
    a. masculine
    b. feminine
    c. transgendered
    d. hyper masculine
    e. bisexual

22. A movement to market more gender-neutral toys emerged in the:
    a. 1940s and 1950s
    b. 1960s and 1970s
    c. 1980s and 1990s
    d. 2000s
    e. None of the above.

23. Between 2001-2005, the percent of university students who severely assaulted a dating partner in the past year in the United States was:
    a. 27.5%
    b. 13.8%
    c. 1.7%
    d. 6.5%
    e. 38.4%

24. The low-calorie and diet food industry promotes an ideal body type that:
    a. is attainable with proper nutrition and exercise.
    b. could generate widespread body satisfaction among otherwise overweight women.
    c. enhances self-esteem for both men and women.
    d. All of these are correct.
    e. None of the above.

25. Which part of the body were men and women most dissatisfied with in 1997?
    a. stomachs
    b. thighs
    c. buttocks
    d. All of these are correct.
    e. None of the above.

26. Which country recognized full and equal marriage rights for homosexual couples in April 2001?
    a. Mexico
    b. The United States
    c. The Netherlands
    d. Guatemala
    e. Canada

27. Which statement is an accurate description of the movie *Boys Don't Cry*, as discussed in Box 11.2 of the textbook?
    a. Although her only transgression was that she wanted to be a man, the character "Teena Brandon" was murdered.
    b. Although wanting a sex change, "Teena Brandon" took the inexpensive route by stuffing a sock down the front of her jeans to pose as a man.
    c. "Brandon" develops a romantic relationship with a woman named "Lana Tisdel," who despite later discovering that "Brandon" is female, continues to love her anyhow.
    d. All of these are correct.
    e. None of the above.

True False

1. Gender depends on whether you were born with distinct male or female genitals.

   TRUE or FALSE

2. Social constructionism is a main alternative to essentialism.

   TRUE or FALSE

3. Research shows that toys have an influence on gender socialization.

   TRUE or FALSE

4. With the changes influenced by the women's movement, the mass media portrays far less gender stereotypes.

   TRUE or FALSE

5. Body dissatisfaction motivates many women to diet, and even prompts some people to take dangerous and life-threatening measures to lose weight.

   TRUE or FALSE

6. The female-male earnings ratio indicates that women's earnings as a percentage to men's earnings have decreased slightly since 1960.

   TRUE or FALSE

7. Because traditional patterns of gender socialization weigh so heavily on men in the context of a male-dominant world, changing socialization patterns is virtually impossible.

   TRUE or FALSE

8. The chief demand of the "first wave" of the women's movement was the right to vote.

   TRUE or FALSE

9. Federal legislation on comparable worth is on the books in a number of U.S. states.

   TRUE or FALSE

Short Answer

1. Briefly explain why women were about equal in status to men in nomadic hunting-gathering societies? (p. 330-331)

   _____

   _____

2. Define and briefly explain the connection between the terms transgendered and transsexual. (p. 326)

   _____

   _____

3. Explain the idea of the "glass ceiling" for women in managerial professions. (p. 326)

   _____

   _____

4. What is the difference between sex and gender as categories of distinction between men and women? How are the two terms related? (p. 314-315)

   _____

   _____

5. List four reasons why equality between men and women grew in the 20th century in many countries. (p. 332-333)

   _____

   _____

Essay Questions

1. Describe the development of the women's movement in the United States from its "first wave" to its "third wave." In doing so, explain the three major streams of the modern feminist movement, and the additional contribution of antiracist and postmodernist feminists. (p. 339-342)

2. Compare and contrast the major assumptions of essentialist and social constructionist perspectives on gender differences. In doing so, identify the major critiques of essentialism in terms of one of the essentialist theories discussed in the textbook. (p. 315-326)

3. Explain the origins and development of gender inequality from the past to the present. In your answer, be sure to discuss long-distance warfare, plow agriculture, and public and private spheres, and make appropriate connections to the earnings gap today. (p. 330-339)

4. Describe how parents may gender socialize their sons and daughters differently. How do children interpret, negotiate, and/or resist conventional gender roles? (p. 315-322)

5. Social constructionism is the view that *apparently* natural or innate features of life, such as gender, are sustained by *social* processes. Compare and contrast the different types: feminist, conflict, and symbolic interactionist theories of social constructionism. Where do they agree? Where do they differ? (p. 315-326)

Solutions

Practice Tests

Fill-In-The-Blank

1. Comparable worth
2. sex
3. Gender identity
4. Gender ideologies

5. glass ceiling
6. female-male earnings ratio
7. Hostile environment sexual harassment

8. gender role
9. Essentialism
10. Bisexuals

Multiple-Choice Questions

1. C, (p. 314)
2. A, (p. 315)
3. D, (p. 318)
4. B, (p. 316)
5. D, (p. 316-318)
6. A, (p. 320-322)
7. C, (p. 322)
8. B, (p. 323)
9. D, (p. 318-322)

10. C, (p. 322-325)
11. C, (p. 326)
12. D, (p. 326-329)
13. D, (p. 326-329)
14. A, (p. 330)
15. E, (p. 332-333)
16. C, (p. 332)
17. C, (p. 334-335)
18. C, (p. 335)

19. E, (p. 336-338)
20. A, (p. 340)
21. B, (p. 317)
22. B, (p. 319)
23. B, (p. 334)
24. E, (p. 323)
25. A, (p. 324)
26. C, (p. 327)
27. D, (p. 330)

True False

1. F, (p. 314)
2. T, (p. 315)
3. T, (p. 318-320)

4. F, (p. 322-325)
5. T, (p. 322-325)
6. F, (p. 332)

7. F, (p. 318-322)
8. T, (p. 339)
9. F, (p. 339)

SOCIOLOGY OF THE BODY: DISABILITY, AGING, AND DEATH

Student Learning Objectives

After reading Chapter 12, you should be able to:

1. Understand the social and cultural contexts of the body in terms of height, weight, and social status.

2. Discuss social, economic, and technological bases of enhancing one's body image to conform to prevailing norms in the context of urbanization and industrialization.

3. Explain the social construction of disability across different times and places, and the normality of disability.

4. Discuss ablism in terms of active prejudice, discrimination, and unintended neglect, which results in the social disadvantage of disabled people.

5. Explain age as yet another dimension of social stratification and inequality, and specific social problems of the elderly.

6. Discuss the increased power and wealth of the elderly in recent decades.

7. Discuss attitudes toward death and dying, as a reflection of the nature of society and culture.

Key Terms

impaired (352)

disabled (352)

rehabilitation (352)

ablism (353)

life course (356)

rite of passage (356)

life expectancy (357)

age cohort (357)

age roles (357)

generation (357)

age stratification (358)

gerontocracy (359)

ageism (363)

euthanasia (366)

Detailed Chapter Outline

I. BOB DOLE'S BODY

   A. Age divides two categories of people who can identify Bob Dole. People who think of Bob Dole mainly as a political figure tend to be 30 years old or more. People who think of Bob Dole chiefly as an advertising icon tend to be under 30.

   B. Bob Dole's body has different meanings for the two categories of people. If you are 30 or older, you are probably aware that Dole has a disabled right arm. If you are younger than 30, you probably remember Dole as a poster boy for Viagra and Pepsi.

   C. The example of Bob Dole is not trivial, for in it lies embedded a sociological principle. In addition to being a biological wonder, the human body is also a sociological wonder and a subject of growing interest in the discipline. Its parts, its disabilities, its aging, and its death mean different things and have different consequences for different cultures, historical periods, and categories of people.

   D. The relationship between the body and society is especially clear in the case of disability.

II. SOCIETY AND THE HUMAN BODY

   A. The Body and Social Status
   1. Height
      a. Believing that physical stature reflects social stature, students in an experiment correlated social status with height.
      b. Do tall people really tend to enjoy high social status? Why are some people tall in the first place? Genes are the most important determinant of an individual's height. However different *populations* are approximately the same genetically. A complex series of *social* factors determine the average height of any population.
      c. The impact of family income on stature.
      d. Within countries, there is also a correlation between stature and class position. Members of upper classes are on average taller than middle-class members, who are in turn taller than working-class members.
      e. The *consequences* of stature are important. On average, tall people live longer than others, earn more money, and reach the top of their profession more quickly. In only three of the U.S. presidential elections held over the past century did the shorter candidates win.
      f. Part of the reason short people tend to be less successful in some ways than tall people is that they experience subtle discrimination based on height.
   2. Weight
      a. Body weight influences status because of the cultural expectations we associate with it.
         i. One study found that in comparison to women who are not overweight, overweight women tend to complete four fewer months of school, are 20% less likely to be married, make $7000 (household income), and are 10% more likely to live in poverty.
         ii. The consequences for being overweight are less serious for men, but still, overweight men are 11% less likely to be married.
      b. The negative effects of being overweight are evident even for women matched in terms of social and economic background, which suggests the need to revise the conventional view

that poverty encourages obesity. While it is true that poor women have less access to eat healthier, exercise, and lose weight, its is also true that obesity in and of itself tend to make women poorer.

    c. In preindustrial societies, people generally favored well-rounded physiques because they signified wealth and prestige.

    d. In our society, being well rounded usually signifies undesirability.

    e. The percentage of overweight people is big and growing quickly in the United States.

B. Sociology of the Body

    1. The Body and Society

        a. Our bodies are not just biologically defined, but also socially defined.

        b. In the United States and other highly industrialized countries, people tend to think they have rights over their bodies. For example, the feminist movement asserts the right of every woman to control her own reproductive functions. Yet people have not always endorsed this view, as in the case of slave's bodies as property of their masters, and women's bodies as property of their husbands.

        c. Despite the widespread view that people have rights over their bodies, most people are influenced by norms of body practice.

        d. For social, economic, and technological reasons, the enhancement of one's body image to conform to prevailing norms became especially important in urban, industrial societies.

            i. *Socially*, urbanized societies present people with many more opportunities to meet and interact with strangers. Manipulating body image helps grease the wheels of social interaction by making it clear to strangers who you are.

           ii. *Economically*, industrialized societies enable people to afford body enhancement.

          iii. *Technologically*, we have created many new techniques for transfiguring the body – for example, dental hygiene, dentistry, orthodontics, and plastic surgery.

III. DISABILITY

A. The Social Construction of Disability

    1. Historical cases of the treatment of left-handed people. Almost universally, people have considered left-handedness a handicap.

    2. We don't think of left-handed people, who make up roughly 10% of the population, as **impaired** or deficient in physical or mental capacity. Nor do we think of them as **disabled** or incapable of performing with a range of "normal" human activity. The fact that so many people once thought otherwise suggests that definitions of disability vary socially, historically, and even in any one time and place when people disagree.

B. Rehabilitation and Elimination

    1. Modern Western approaches to disability emerged in the 19th century. All scientists and reformers of the time viewed disability as a biological reality, and some sought **rehabilitation**.

    2. Other scientists and reformers sought to eliminate disability altogether by killing the disabled or sterilizing them.

3. One of the most ugly chapters in our history involves the federally funded, forced sterilization of native North American women from the 1920s to the 1970s, in which the "disability" of these women was being native North Americans.

C. Ablism
   1. Because society is structured around norms of the able-bodied, disabled people are disadvantaged, especially if they are elderly, women, lower class, or racial or ethnic minorities.
   2. People routinely stigmatize and stereotype the disabled. The resulting prejudice and discrimination against the disabled is called **ablism**.
   3. In addition to active prejudice and discrimination, ablism also involves unintended *neglect* of the conditions of disabled people.

D. The Normality of Disability
   1. H.G. Wells' 1927 short story, "The Country of the Blind."
   2. Wells' tale makes blindness seem normal, which comes close to the way disabled people today think of their disabilities – not as a form of deviance but as a different form of normality.
   3. The idea of normality of disability partly supplanted the rehabilitation ideal that originated in the 19th century. Disabled people participated little in efforts to improve the conditions of their existence, which began to change when they organized in the 1960s. Since the 1980s, disabled people have begun to assert their autonomy and the "dignity of difference." Rather than see disability as a personal tragedy, they see is as a social problem, and rather than see themselves as deviant, they think of themselves as inhabiting a different but normal world.
   4. The deaf community typifies the new challenge to ablism.

IV. AGING

A. Sociology of Aging
   1. Sociologists see aging in a complex light. Aging is a process of socialization, and its significance varies from one society to the next.
   2. Aging and the Life Course
      a. The **life course** refers to the distinct phases of life through which individuals pass.
      b. These stages tend to be marked by **rites of passages**, or rituals signifying the transition from one life stage to another, such as circumcision, baptism, confirmation, the bar and bat mitzvah, college convocation, the wedding ceremony, and the funeral.
      c. From one society and historical period to another, each stage of life varies in duration, number of life stages, and significance cultures attach to them. Increased **life expectancy** and the need for a highly educated labor force made childhood possible and necessary.
   3. Age Cohort
      a. An **age cohort** is a category of people born in the same range of years, and have learned common patterns of behavior.
      b. **Age roles** are patterns of behavior that are expected of people in different age cohorts, and from an important part of our sense of self and others.
      c. Sociologists consider youth culture as a distinct *subculture* because differences across age cohorts are large in the United States.
   4. Generation

    a. A **generation** is a special type of age group composed of members of an age cohort who have unique and formative experiences during their youth, such as "baby boomers," and "Generation X."

    b. Generationally defined moments sometimes crystallize the feeling of being a member of particular generations, such as the assassination of President Kennedy and September 11, 2001.

    c. The crystallization of a generational "we-feeling" among youth is a recent phenomenon.

    d. Generations sometimes play a major role in history, such as in revolutionary movements.

B. Aging and Inequality

  1. Age Stratification

    a. **Age stratification** refers to social inequality between age cohorts, which exists in all societies but varies across social contexts.

    b. The very young are often at the bottom of the stratification system.

    c. Even in rich countries, poverty is more widespread among children than adults. Childhood poverty exceeds poverty among adults by 71%. The United States has the highest child poverty rate among the world's two dozen richest countries.

  2. Gerontocracy

    a. **Gerontocracies** are societies in which elderly men ruled, earned the highest incomes, and enjoyed the most prestige, such as ancient China. People in some industrialized countries pay more attention to age than Americans do.

    b. Powerful, wealthy, and prestigious leaders are often mature, but not the oldest people in society. The United States, past and present, is typical in this regard.

    c. True gerontocracy in generally rare.

    d. The rule by youth is also uncommon.

  3. Theories of Age Stratification

    a. The Functionalist View

      i. *Functionalists* suggest that it came about when age cohorts were differentiated in the course of industrialization, and the cohort of retired elderly people grew.

      ii. Age stratification grew and developed because different age cohorts performed functions of differing value to society. In preindustrial societies, the elderly were important as a storehouse of knowledge and wisdom. With industrialization, their function became less important and their status declined.

      iii. Systems of age stratification "converge" under the force of industrialization.

    b. Conflict Theory

      i. *Conflict theorists* agree with functionalists regarding the needs of industrialization generating age categories. However, they disagree on two points:

      ii. They dispute that age stratification reflects the functional importance of different age cohorts. Instead, age stratification stems from competition and conflict. Power and wealth do not necessarily correlate with functional roles; competition and conflict may redistribute power and wealth between age cohorts.

      iii. Regarding convergence, conflict theorists suggest that political struggle can make a big difference in how much age stratification exists in a society.

    c. Symbolic Interactionist Theory

        i. *Symbolic interactionists* focus on the meanings people attach to age-based groups and age stratification. They stress the way people understand aging as a matter of interpretation. They have helped us understand the degree and nature of prejudice and discrimination against the elderly.

C. Social Problems and the Elderly

  1. The growth of the elderly population.

    a. In 1900, 4% of the U.S. population was 65 and over.

    b. Today, the figure is around 13%.

    c. By 2040, nearly 21% will be elderly.

    d. After 2040, the proportion of Americans 65 years and over will start to decline.

  2. Many sociologists of aging categorize elderly people as those who enjoy relatively good health, usually between 65 and 74 years, as the "young old"; 85 and older as the "old old."

  3. The reasons for the rising proportion of elderly people are clear. On the one hand, fertility rates have been declining in virtually all industrial societies. On the other hand, life expectancy has been increasing because of medical advances, better welfare provisions, and other factors.

  4. Aging and Poverty

    a. The proportion of young old is expected to decline, and the old old are expected to continue increasing, which raises many health concerns such as physiological decline, life-threatening diseases, social isolation, and poverty.

    b. Because the sex ratio imbalance (greater number of women than men) is most marked in the oldest age cohorts, poverty and related problems among the elderly is a gender issue.

    c. Economic inequality between elderly women and men is largely the result of women's lower earning power when they are younger. Women's employer pensions tend to be inferior to men's.

    d. In addition to the old old and women, the categories of elderly people most likely to be poor include African Americans, people living alone, and in rural areas. The elderly are sometimes socially segregated.

  5. A Shortage of Caregivers

    a. Among rich countries, the United States is a relatively young society. The proportion of the old to the total population is comparatively low.

  6. Ageism

    a. Especially in a society like the United States that highly values vitality and youth, being elderly is a *social stigma*. **Ageism** is prejudice about, and discrimination against, elderly people.

    b. Elderly people often do not conform to the negative stereotypes. Retirees are seldom a tangle of health problems and a burden on society.

    c. Their housing arrangements are not usually desolate and depressing.

  7. The Power and Wealth of the Elderly

    a. Many elderly people own assets.

    b. The elderly are well organized politically, which partly accounts for their relative economic security.

D. Death and Dying
1. The ultimate social problem the elderly face is their own demise. Why are death and dying *social* problems, rather than just religious, philosophical, and medical issues?
2. Attitude toward death and dying vary widely across time and place. In most traditional societies, most people accepted death, partly because dying was not isolated from other people, and was emotionally supported.
3. In contrast, in the United States today, dying and death are separated from everyday life. Our culture celebrates youth and denies death, which makes us less prepared for death than our ancestors were.
4. Our reluctance to accept death is evident in our many euphemisms.
5. Psychiatrist Elisabeth Kubler-Ross's analysis of the stages of dying suggests how reluctant we are to accept death.
    a. *Denial*
    b. *Anger*
    c. *Negotiation*
    d. *Acceptance*
6. Euthanasia
    a. The debate of euthanasia reveals the reluctance of many Americans to accept death.
    b. **Euthanasia**, also known as mercy killing or assisted suicide, involves a doctor prescribing or administering medication or treatment that is intended to end a terminally ill patient's life.
    c. The American Medical Association's Council on Ethical and Judicial Affairs (AMA-CEJA) says it is the duty of doctors to withhold life-sustaining treatment if that is the wish of a mentally competent patient.
    d. Public opinion polls show that about two thirds of Americans favor physician-assisted suicide euthanasia, and between 33% and 60% of American doctors say they would be willing to perform euthanasia if it were legal.
    e. The AMA-CEJA, the Catholic Church, some disabled people, and other groups oppose euthanasia.
    f. Euthanasia is legal in the Netherlands, is likely to become legal in some other countries in the next few years, and in Oregon, a physician-assisted suicide law took effect in October 1997. However, in a June 1997 ruling, the United States Supreme Court upheld constitutional state laws that bar assisted suicide.
    g. What is clear is that, "[t]hroughout the Nation, Americans are engaged in an earnest and profound debate about the morality, legality and practicality of physician assisted suicide."
7. The Business of Dying
    a. The way we die reflects the nature of our society and culture. In the United States, a capitalist, business-oriented society, funerals are big business – a $20 billion a year industry in 1999. The average undertaker's bill in the late 1990s was $4700, and adding other expenses the bill grew to $7800.
    b. Two main reasons why funerals are so expensive:
        i. Big corporations and supplanting small family operations in the funeral industry, in which the undisputed giant is Services Corp. International (SCI).

ii. The funeral industry takes full advantage of the vulnerability of people who lost their loved ones.

Study Activity: Applying The Sociological Compass

Review the concepts ablism and ageism in Chapter 12. Make a list of the similarities of these two forms of inequality. Explain each phenomenon from a functionalist, conflict, and symbolic interactionist perspective.

Similarities of Ablism and Ageism:

Explanations of Ablism:

Functionalist:

Conflict:

Symbolic Interactionist:

Explanations of Ageism:

Functionalist:

Conflict:

Symbolic Interactionist:

Infotrac College Edition Online Exercises

For the following exercises, log on to the online library of InfoTrac College Edition at http://www.infotrac-college.com/. Make note that InfoTrac has implemented a new registration system that will allow easier access to InfoTrac through the use of a personalized username and password. Once you've created your username and password you may proceed directly to the Log On page. To create an account, register your passcode packaged with your textbook, and create a username and password, by following the online prompt. After you are logged in, click on "Infotrac College Edition." You will arrive at a screen that enables you to search topics.

Keyword: **age appreciation**. Often the negative aspects of aging are stressed. Research the positive aspects of aging. What are the key factors that help one age in a positive way? What can our society do to show age appreciation?

Keyword: **euthanasia**. Arguments for and against euthanasia have become more vocal in our society. What are the policies in other countries about euthanasia and physician assisted suicide? How do their views compare with those in the United States?

Keyword: **health maintenance organizations (HMOs)**. HMOs focus on preventive health care for a fixed fee. Some HMOs are under the gun for not providing adequate care for their patients. What are some of the growing concerns of unhappy HMO subscribers? What solutions do they call for?

Keyword: **Medicare fraud**. Medicare pays some costs for people over age 65 in the United States. Sometimes the program is abused. Research the ways that Medicare is used in fraudulent ways. For example, is it abused more by the elderly or by health care providers?

Keyword: **Alzheimer's disease**. Look for articles that stress social support programs for Alzheimer's. What characteristics do they share in common? How do these programs relate to the sociological discipline?

Keyword: **Disability definition**. Search through these articles. How many definitions of disability can you find? What are the similarities and differences across definitions? Come up with your own definition.

Internet Exercises

Death is not commonly discussed in American Society. Go to: http://www.hospicenet.org/html/talking.html for some suggestions about talking to children about death. Read the short article. Make a list of their suggestions for discussing death with children. How can you implement their suggestions into your life?

You can find much information about the American with Disabilities Act (1990) at: http://www.usdoj.gov/crt/ada/adahom1.htm.

Identify the current key issues with the Act. Write a short paper stating what you have learned about the American with Disabilities Act from this short exercise.

Go to: http://www.msnbc.com/modules/quizzes/lifex.asp to find out your life expectancy. Answer the questions on your health risk. Write a short summary discussing your life expectancy, health risk, and if you are happy with your results.

The World Health Organization web site at: http://www.who.int/ is an excellent source of information on health issues. Browse the site. Research a particular health issue and then report back to the class.

Practice Tests

Fill-In-The-Blank

Fill in the blank with the appropriate term from the above list of "key terms."

1. A(An) _____ is an age group that has unique and formative historical experiences.

2. _____ is also known as mercy killing and assisted suicide.

3. _____ involves curing disabilities to the extent possible through medical and technological intervention.

4. The _____ refers to the distinct phases of life through which people pass.

5. _____ people are considered deficient in physical or mental capacity.

6. A(An) _____ is a society ruled by elderly people.

7. _____ is prejudice and discrimination against disabled people.

8. _____ is the average age at death of the members of a population.

Multiple-Choice Questions

Select the response that best answers the question or completes the statement:

1. People 30 years old or more tend to think of Bob Dole:
    a. as a political figure.
    b. as an advertising icon for Pepsi.
    c. as a poster boy for Viagra.
    d. All of these are correct.
    e. None of the above.

2. In the discipline of sociology, the human body is:
    a. a trivial topic, thus not highly interesting to most sociologists.
    b. a biological wonder, thus irrelevant to most sociological researchers.
    c. a subject of growing interest.
    d. All of these are correct.
    e. None of the above.

3. The relationship between the body and society is especially clear in the case of:
   a. sex.
   b. women.
   c. childhood.
   d. disability.
   e. gerontocracy.

4. Which statement is true of the height differences between the upper classes and lower classes?
   a. There are no average height differences between upper classes and lower classes within most countries.
   b. Class differences in height are greater than they were centuries ago.
   c. Middle-class members tend to be shorter than working-class members.
   d. All of these are correct.
   e. None of the above.

5. Short people experience _____ discrimination based on height.
   a. no
   b. blatant
   c. subtle
   d. reasonable
   e. ideological

6. Based on Figure 12.1, what is (are) the direct social cause(s) of height?
   a. Basic social causes such as income, inequality, and public health.
   b. Proximate social causes such as diet, disease, and work intensity.
   c. Social consequences such as life expectancy, health, and cognitive development.
   d. All of these are correct.
   e. None of the above.

7. The percentage of overweight people in the United States is:
   a. big and growing.
   b. small but growing.
   c. big but decreasing.
   d. small and decreasing.
   e. None of the above.

8. Which statement is true regarding the body and society?
   a. The practice of affecting body shape and appearance is unique to urban, industrial societies.
   b. Most social distinctions such as social class and race, have little to do with the ways in which people have shaped the appearance of their bodies.
   c. People have always attempted to affect their body shape and appearance according to principles laid out by society.
   d. All of these are correct.
   e. None of the above.

9. As a so-called signifier of wealth and prestige in preindustrial societies, people generally favored:
   a. muscular female physiques.
   b. thin physiques.
   c. well-rounded physiques.
   d. All of these are correct.
   e. None of the above.

10. Obese children have a body mass index (BMI) at or above what percentile of the sex-specific BMI growth charts?
    a. 95th
    b. 80th
    c. 75th
    d. 65th
    e. 50th

11. From the 1920s to the 1970s, native North American women were forcefully sterilized due to being perceived as "disabled." What was their alleged disability?
    a. mental retardation
    b. being native North American
    c. physical impairment
    d. autism
    e. ablism

12. When architecture and urban planning do not provide nonstandard modes of mobility for disabled people, this is an example of which form of ablism?
    a. neglect
    b. active prejudice
    c. individual discrimination
    d. ideological ablism
    e. stereotyping

13. Which sociological theory focuses on the meanings people attach to age-based groups and age stratification?
    a. functionalism
    b. conflict
    c. symbolic interactionism
    d. feminism
    e. None of the above.

14. What percentage of the U.S. population was elderly in 2000?
    a. 42.2%
    b. 36.7%
    c. 20.7%
    d. 17.9%
    e. 12.8%

15. What is life expectancy?
    a. the average age of death of the members of a population.
    b. the maximum age of death of eldest members of a population.
    c. the minimum age of death of among members of a population who are defined as elderly.
    d. All of these are correct.
    e. None of the above.

16. Which factor about aging is emphasized by sociologists?
    a. Aging is a complex social phenomenon.
    b. Aging is a process of socialization.
    c. Aging is deeply rooted in society and culture.
    d. Aging involves the meanings societies attach to the life span.
    e. All of these are correct.

17. Rites of passage refer to:
    a. distinct stages of life through which individuals pass.
    b. rituals signifying the transition from one life stage to another.
    c. social systems in which elderly men rule.
    d. All of these are correct.
    e. None of the above.

18. Which is a viable projection about the growth of the elderly population in the United States in the future?
    a. By 2040, nearly 21% of the U.S. population will be elderly.
    b. After 2040, the proportion of elderly in the United States will start to decline.
    c. The growth of the elderly population in the future will present new and growing social problems.
    d. All of these are correct.
    e. None of the above.

19. Which is a precise example of a "generation" based on the definition provided in the textbook?
    a. baby boomers
    b. youth culture
    c. children ages 4 to 6.
    d. All of these are correct.
    e. None of the above.

20. Which category of elderly people is most likely to be poor?
    a. the "old old"
    b. African Americans
    c. people living alone
    d. people living in rural areas
    e. All of these are correct.

21. Which is NOT an example of ageism?
    a. segregation of the elderly
    b. stigmatization of the elderly
    c. meager Medicaid care provided to elderly
    d. discrimination of the elderly
    e. All of the answers are examples of ageism.

22. Regarding euthanasia, public opinion polls show that:
    a. About two thirds of Americans favor physician-assisted suicide.
    b. Sixty percent of American doctors would not be willing to perform euthanasia even if it were legal.
    c. The Catholic Church supports euthanasia for extenuating medical circumstances.
    d. All of these are correct.
    e. None of the above.

23. What is the sociological point of the movie Shallow Hall, according to Box 12.2?
    a. People with disabilities are ordinary folks.
    b. People with disabilities are inherently deviant.
    c. Many people are odd for not being able to see their normality.
    d. All of these are correct.
    e. None of the above.

24. Sociologists see aging as:
    a. a complex social phenomenon.
    b. a process of socialization.
    c. the learning of new roles appropriate to the latter stage of life.
    d. All of these are correct.
    e. None of the above.

25. Which term represents a category of people born in the same range of years?
    a. peer group
    b. age cohort
    c. age-range community
    d. census generation
    e. birth date enclave

True False

1. The example of Bob Dole is sociologically amusing and trivial.

   TRUE or FALSE

2. The social consequences of body stature are sociologically important.

   TRUE or FALSE

3. Socially, what is true of height is also true of weight.

   TRUE or FALSE

4. The assertion in the United States, that people ought to have rights over their own bodies, has seldom been challenged throughout U.S. history.

   TRUE or FALSE

5. Ablism involves more than active prejudice and discrimination.

   TRUE or FALSE

6. Over time, life expectancy has declined because of the depletion of our environment.

   TRUE or FALSE

7. The United States is a typical example of gerontocracies.

   TRUE or FALSE

8. The elderly are sometimes socially segregated.

   TRUE or FALSE

Short Answer

1. Define age cohort and place yourself in an appropriate context. (p. 357)

   _____

   _____

2. Explain how the deaf community typifies the new challenge to ablism. (p. 356)

   _____

   _____

3. Provide two sociological reasons why funerals are so expensive in the United States. (p. 366-368)

   _____

   _____

4. Define and explain the connections between the terms "age cohort" and "generation." Provide an example. (p. 357-358)

   _____

   _____

5. Define and explain ageism, with examples. (p. 363)

   _____

   _____

Essay Questions

1. Explain age stratification from functionalist, conflict, and symbolic interactionist perspectives. (p. 358-364)

2. With the current and projected growth of the elderly population in the United States, sociologists anticipate a host of social problems of the elderly. Identify and explain three social problems. (p. 361-364)

3. Explain how social class, gender, and race are connected in the process of ablism. (p. 352-356)

4. Explain the idea of *the normality of disability*. How has this idea evolved over the last century, and how is it related to the idea of rehabilitation? Use concrete examples to illustrate your points. (p. 352-356)

5. Describe the ways that body characteristics (height, weight, etc.) and social status are related. Use clear examples. Have you seen evidence of these relationships in your own life? (p. 347-352)

Solutions

Practice Tests

Fill-In-The-Blank

1. generation
2. Euthanasia
3. Rehabilitation
4. life course
5. Impaired
6. gerontocracy
7. Ablism
8. Life expectancy

Multiple-Choice Questions

1. A, (p. 347)
2. C, (p. 348)
3. D, (p. 348)
4. E, (p. 348-349)
5. C, (. 348-349)
6. B, (p. 349)
7. A, (p. 350)
8. C, (p. 351)
9. C, (p. 350)
10. A, (p. 351)
11. B, (p. 353)
12. A, (p. 353-354)
13. C, (p. 361)
14. E, (p. 361)
15. A, (p. 357)
16. E, (p. 356-361)
17. B, (p. 356)
18. D, (p. 361-362)
19. A, (p. 357)
20. E, (p. 361-363)
21. C, (p. 363-364)
22. A, (p. 366)
23. C, (p. 355)
24. D, (p. 356)
25. B, (p. 357)

True False

1. F, (p. 347-348)
2. T, (p. 349)
3. T, (p. 349)
4. F, (p. 351)
5. T, (p. 353)
6. F, (p. 356-357)
7. F, (p. 359-360)
8. T, (p. 358-363)

WORK AND THE ECONOMY

Student Learning Objectives

After reading Chapter 13, you should be able to:

1. Describe the development of the three sectors of the economy.

2. Discuss the current and future prospects for "good" and "bad" jobs in terms of competing perspectives on work and labor.

3. Discuss the significance of worker resistance and labor unions to the organization and operation of work in a comparative context.

4. Explain the difficulty of moving from the primary to secondary labor market, and describe the demographic characteristics of the barriers in regards to gender, race, ethnicity, and union membership.

5. Describe current conditions of work in general in terms of the time-crunch, and explain the main reasons for the frantic pace of work and social life in the United States.

6. Describe and explain the problem of social inequality in the labor market in general, and in relation to corporate control and globalization, in particular.

7. Contrast the neoclassic school of economics and economic sociology.

8. Contrast capitalism and communism as ideal-types of economic systems in a comparative context.

9. Explain the impact of globalization on American workers and less developed countries.

10. Describe the working conditions of workers in multinational corporations in less developed countries.

11. Discuss future trends of work and the economy based on prevailing patterns involving the implementation of new technologies and the organization of work.

Key Terms

economy (373)

productivity (373)

markets (373)

division of labor (373)

deskilled (375)

scientific management (375)

labor market segmentation (379)

primary labor market (379)

secondary labor market (379)

human relations school of management (380)

quality of work life (381)

codetermination (382)

unions (382)

internal labor markets (382)

professionals (382)

free market (387)

regulated market (387)

capitalism (390)

corporations (390)

communism (391)

oligopolies (393)

democratic socialist (393)

conglomerates (394)

interlocking directorates (394)

Detailed Chapter Outline

I.   THE PROMISE AND HISTORY OF WORK

   A.   Salvation or Curse?

     1.   Two strikingly different images of work that form the questions of the sociology of work:

       a.   The information age, computerization, work automation, and standardization are degrading and inhumane processes.

       b.   Bill Gates argues that computers:

         i.   Reduce work hours.

         ii.   Make goods and services less expensive;

         iii.   Allow us to enjoy our leisure time more.

   B.   Three Revolutions

     1.   The **economy** is the institution that organizes the production, distribution, and exchange of goods and services. Three sectors of the economy:

       a.   *Primary* or "agricultural" sector, which includes farming, fishing, logging, and mining.

       b.   *Secondary* or "manufacturing" sector, which turns raw materials into finished goods, and manufacturing takes place.

       c.   *Tertiary* or "service" sector, where services are bought and sold.

     2.   Each sector rose to dominance respectively within three revolutionary events—the agricultural revolution, the industrial revolution, and the revolution in services.

     3.   The Agricultural Revolution

       a.   With the invention of the plow, cultivation of land and **productivity** (the amount produced for every hour worked) increased.

     4.   The Industrial Revolution

       a.   Since the 15th century, international exploration, trade, and commerce stimulated the growth of **markets**, social relations that regulate the exchange of goods and services.

       b.   About 225 years ago, the steam engine, railroads, and other technological innovations increased the ability of producers to supply markets, which marked the era of the Industrial

Revolution that began in Western Europe and spread to other parts of Europe, North America, Russia, and Japan within a century.

5. The Revolution in Services
   a. As productivity increased, service-sector jobs proliferated.
   b. The computer is responsible for the rapid change in the composition of the labor force during the final decades of the 20ᵗʰ century.
6. The Social Organization of Work
   a. These three revolutions increased the **division of labor**, or specialized work tasks, which consequently created new skills. Social relations among workers became more hierarchical.

II. THE QUALITY OF WORK: "GOOD" VERSUS "BAD" JOBS

A. John Lie's experiences with "good" and "bad" jobs.

B. What can we say about the overall mix of jobs in the United States? Are there more good jobs than bad jobs? What does the future hold? Are good or bad jobs likely to become more plentiful? What are your job prospects?

C. The Deskilling Thesis
1. Thirty years ago, Harry Braverman argued that because capitalists are eager to organize work to maximize profits, and break complex tasks into simple routines (replacing labor with machines and managerial control over workers), work tends to become **deskilled** overtime.
   a. In the early 1900s, Henry Ford introduced the assembly line.
   b. In the 1900s, Frederick W. Taylor developed the principles of **scientific management**, a system of improving productivity and efficiency by eliminating unnecessary actions.
2. Criticisms of Braverman's deskilling thesis:
   a. Although his characterization of factory work was in some respects accurate, factory workers represent only a small proportion of the labor force.
   b. The manufacturing sector is declining, and the service sector is expanding, which accounts for three quarters of U.S. jobs today.
3. In the 1980s and 1990s, some analysts feared that good jobs in manufacturing were being replaced by bad jobs in services.
4. Part-Time Work
   a. The proportion of part-time workers in the United States labor force increased 46% between 1957 and 1996.
   b. Although the growth of part-time jobs is not necessarily a problem (and can be advantageous) for voluntary part-time workers or people who have good part-time jobs, an increasingly number of people rely on part-time work to meet the needs of full-time living. The fastest growing category of part-time workers is *involuntary* part-timers, who often want and need to work more hours.
   c. The problem of part-time work also involves issues of self-respect in the face of low pay, benefits, security, status, and creativity.
      i. Katherine Newman's research on fast-food workers in Harlem, as a case of the indignity endured by part-time workers.
      ii. The likelihood of "temps" to be sexually harassed.

D. A Critique of the Deskilling Thesis
  1. Braverman's and Zuboff's analyses are too narrowly focused on jobs near the bottom of the occupational hierarchy, therefore contributes little explanation of what is happening to the occupational structure as a whole.
  2. The decline of the manufacturing sector, and the rise of the service sector do not necessarily imply a decline of the entire labor force, so one should avoid generalizing about the entire sector from case studies of a few job categories.
  3. Although the least skilled service jobs are dull and pay poorly, they are not as "dead-end" as they are made out to be.
      a. The least skilled workers tend to be under the age of 25 and hold their jobs only briefly.
      b. After working in least skilled entry-level jobs, most men move on to blue-collar jobs, and most women move on to clerical jobs.
  4. Braverman and Zuboff exaggerated the downward slide of the U.S. labor force because they underestimated the continuing importance of skilled labor in the economy. So rather than involving a downward shift in the entire labor force, a polarization between good jobs and bad jobs seems more accurate.
      a. Job polarization in the Silicon Valley.
      b. The top of the service sector is growing, the bottom is growing more rapidly, and the middle is growing more slowly.

E. Labor Market Segmentation: Three stages of labor market development in the United States according to David Gordon and his colleagues:
  1. Initial proletarianization:
      a. 1820–1890
      b. A large industrial working class replaced craft workers in small workshops.
  2. *Labor homogenization:* From the end of the 19th century to the start of World War II.
      a. Extensive mechanization and deskilling took place.
  3. **Labor market segmentation:** After World War II to the present.
      a. Job polarization is taking place. Work is found and experienced in different ways, in two labor markets. Social barriers make it difficult to move from one setting to another.
      b. The labor market has been divided into two distinct parts:
          i. (a) The primary labor market is composed mainly of highly skilled or well-educated white males, employed in large corporations that enjoy high levels of capital investment. Employment is secure, earnings are high, and fringe benefits are generous.
          ii. (b) The secondary labor market contains a disproportionately large number of women and members of racial minorities, particularly African and Hispanic Americans. Employees tend to be unskilled and lack higher education. They work in small firms with low levels of capital investment. Employment is insecure, earnings are low, and fringe benefits are meager.

F. Worker Resistance and Management Response

1. An additional criticism of Braverman's analysis is that he inaccurately portrays workers as passive victims of management control. Workers often resist the imposition of task specialization and mechanization by managers.

2. Worker resistance has often forced management to modify its organizational plans (i.e. Henry Ford). In the 1920s, some employers started to treat their employees more humanely in hopes of improving the work environment and make employees more loyal and productive.

3. The **human relations school of management**, which advocated less authoritarian leadership, careful selection and training of personnel, and greater attention to human needs and job satisfaction, emerged in the 1930s as a challenge to Frederick W. Taylor's scientific management approach.

4. Over the next 70 years, owners and managers in all rich industrialized countries realized they had to make concessions to labor, such as higher wages, more decision-making authority, promotion policies, job design, product innovation, and company investments.

5. The biggest concessions to labor were made in countries with the most powerful trade union movements such as Sweden. A large proportion of workers are members of unions (over 80% in Sweden).

6. In contrast, in the United States less than 14% of eligible workers are union members. In comparison to other rich industrialized countries, Americans work more hours per week, have fewer paid vacation days per year, and lag behind Western Europe and Japan in industry-level decision making among workers.

7. Two main types of decision-making innovations since the early 1970s.

    a. *Reforms that give workers more authority on the shop floor* including those advanced by the **quality of work life** movement, which originated in Sweden and Japan, and involves small groups of workers and managers collaborating to improve the quality of goods and communication between workers and managers. Quality circles have been introduced in some American industries, but are less widespread in the United States than in Western Europe and Japan.

    b. *Reforms that allow workers to help formulate overall business strategy* give workers more authority than quality circles, which occurs in Western Europe and Germany. In Germany, the system is known as **codetermination**. There are a few American examples of this mostly in the auto industry, which were widely credited for the American auto sector's quality and productivity in the 1980s and 1990s. However, "worker participation programs have had a difficult birth in North America. No broad policy agenda guides their development, no systematic social theory light their way."

G. Unions and Professional Organizations

    1. **Unions** are organizations of workers that seek to defend and promote their member's interests,—and have succeeded in improvement of working conditions, higher wages, and more worker participation in decision making.

    2. Unions have helped develop systems of labor recruitment, training, and promotion, referred to as **internal labor markets** because they control pay rates, hiring, and promotions.

3. **Professionals** (people with specialized knowledge acquired through extensive higher education) have also created labor market shelters. They usually regulate themselves and enforce standards through professional associations.

H. Barriers Between the Primary and Secondary Labor Markets
   1. Three social barriers that make the primary labor market difficult to penetrate:
      a. Often, there are *few entry-level positions* in the primary labor market.
      b. A second barrier preventing people in the secondary labor market from penetrating the primary labor market is their *lack of informal networks* linking them to good job openings.
      c. Finally, mobility out of the secondary labor market is difficult because workers usually *lack the required training and certification* for jobs in the primary labor market.
   2. Men, non-Hispanic whites, and unionized workers tend to be concentrated in the primary labor market, while women, African and Hispanic Americans, and nonunion members tend to be concentrated in the secondary labor market.

I. The Time Crunch and Its Effects
   1. Overwork, lack of leisure have become features of both labor markets and our culture as a whole.
   2. All adults in most American households work full-time, many adolescents work part-time, and some people work two jobs.
   3. Many office workers, managers and professionals work 10, 12 or more hours a day due to tight deadlines, demands for high productivity, and a trimmed down work force.
   4. Compounded with family life, it is evident that stress, depression, aggression and substance abuse are on the rise in both labor markets. Work is the leading source of stress in the world.
   5. Three main reasons why leisure is on the decline and the pace of work is more frantic in the United States.
      a. Big corporations are in a position to invest enormous and increasing resources in advertising, which pushes Americans to consume goods and services at higher levels.
      b. More corporate executives apparently think it is more profitable to have employees work more hours rather than hire more workers and pay expensive benefits for new employees.
      c. American workers are not in a position to demand reduced working hours and more vacation time because few of them are unionized.

III. THE PROBLEM OF MARKETS

A. The secondary labor market is a relatively **free market**, which means that the supply of, and demand for, labor regulates wage levels and other benefits. The primary labor market is a more **regulated market**, whereby wage levels and other benefits are not just by supply and demand, but also by the power of workers and professionals, who exercise their influence to their advantage.

B. The freer the labor market, the higher the level of social inequality, which is why the secondary labor market cannot be entirely free.
   1. The federal government had to establish a legal minimum wage to prevent the price of unskilled labor from dropping below the point at which people are able to make a living.
   2. In late 18th—century England, a historical period that was most closely a free market, starvation became so widespread and the social instability so great the government had to establish "poor houses."

C. Although the question of whether free or regulated markets are better lies at the center of economic and political debates, many economic sociologists find the question too abstract.
   1. The structure of markets varies widely across cultures and historical periods.
   2. The degree and type of regulation depend on how power, norms, and values are distributed among various social groups.

D. The "neoclassical" school of contemporary economics.
   1. Neoclassical economists' approach is different from the economic sociologist's approach, in which they argue that free markets maximize economic growth.
   2. Example—Neoclassical claim that elimination of the minimum wage would benefit everyone.
   3. However, in the real world, resistance to the operation of free markets increases as you move down the social hierarchy. As the poorest and least powerful fight against falling wages, social instability can disrupt production and investment.

E. Economic Systems
   1. Capitalism
      a. **Capitalism** is the world's dominant economic system today, and has two distinct features.
         i. *Private ownership of property.* Individuals and **corporations** (legal entities) own nearly all the means of producing goods and services, and are free to buy and sell. Corporate ownership has the advantages of lower tax rates, and liability in the event of consumer harm or bankruptcy.
         ii. *Competition is the pursuit of profit.* A purely capitalist economy is often called a *laissez-faire* system, whereby the government does not interfere with the operation of the economy.
      b. In reality, no economy is purely laissez-faire.
         i. The state had to intervene to create markets in the first place (European conquest and land acquisition from native peoples in what is now North America, to transform the land into a marketable commodity).
         ii. Today, governments must intervene in the economy to keep the market working effectively (economic infrastructure involving roads and ports, laws governing minimum wage, occupational health and safety, child labor, and industrial pollution).
         iii. Governments play an influential role in establishing and promoting many leading industries, especially those that require large outlays on research and development.
      c. Which capitalist economies are most free and which are least free?
         i. Based on an annual index of competitiveness of 47 capitalist countries published by the International Institute for Management Development in Switzerland, the United States is the most competitive economy in the world. Russia ranks last.
   2. Communism
      a. Like laissez-faire capitalism, **communism** is an ideal, and the term Karl Marx gave to the classless society that, according to him, is bound to develop out of capitalism. *Socialism* is the name Marx gave to the transitional phase between capitalism and communism.
      b. While no country is pure communist, about two dozen in Asia, South America, and Africa consider themselves socialist, including China, North Korea, Vietnam, and Cuba.

    c. As an ideal, communism is an economic system with two distinct features.

        i. *Public ownership of property*.

        ii. *Government planning*.

    d. In perhaps the most surprising and sudden change in modern history, the countries of Central and Eastern Europe (formerly single-party, socialist societies) namely the Soviet Union introduced capitalism and started holding multiparty elections in the late 1980s and 1990s. Factors for the collapse socialism in Central and Eastern Europe:

        i. The citizens of the region enjoyed few civil rights.

        ii. Their standard of living was only about half as high as that of people in the rich industrialized countries of the West.

        iii. Dissatisfaction was widespread and expressed in many ways including strikes and political demonstrations.

    e. In the 1990s, some Central and East European countries were more successful in introducing elements of capitalism and raising the standard of living. Example—the Czech Republic versus Russia.

        i. The most important factor for the different success rates lies in the way different countries introduced reforms.

        ii. The Czechs introduced private property and competition, while Russia introduced private property without competition.

        iii. Russia's level of socioeconomic inequality became among the highest in the world.

        iv. Russia is full of oligopolies, giant corporations that control a part of an economy, are few in number, compete against one another, and set prices at levels that are most profitable for them, which constrains innovation.

3. Democratic Socialism

    a. Several prosperous and highly industrialized countries are **democratic socialist societies**, including Sweden, Denmark, Norway, and to a lesser degree France and Germany.

    b. Such societies have two distinctive features:

        i. *Public ownership of certain basic industries*. These industries include telecommunications, electricity, railways, airlines, and steel. The level of public ownership is still not high. The great bulk of property is privately owned, and competition in pursuit of profit is the main motive of business.

        ii. *Government Intervention in the market*. These countries enjoy regular, free, multiparty elections, like the United States. Unlike the United States, political parties backed by a strong trade union movement have formed governments in democratic socialist countries for much of the post – World War II period. The governments that these unions back intervene strongly in the operation of markets for the benefit of ordinary workers. Taxes are considerably higher than in capitalist countries. Social services are much more generous. The gap between the rich and the poor is much narrower than in capitalist countries. On average, workers earn more, work fewer hours, and enjoy more paid vacation days.

    c. Since the 1980s, the democratic socialist countries have moved in a somewhat more capitalist direction, but still retain their distinctive approach to governments and markets, which is why democratic socialism is sometimes called a "Third Way" between capitalism and socialism.

F. The Corporation
1. Giant Corporations
    a. In the United States and other Western countries, "antitrust" laws limit the growth of oligopolies (i.e. 1890 Sherman Antitrust Act and the 1914 Clayton Act in the United States).
    b. The law has been only partly effective in stabilizing its growth, especially when the four largest corporations in an industry make up 4 out of every 10 dollars in sales—and when the top 500 corporations control over two thirds of business resources and profit.
2. The Growth of Conglomerates
    a. United States antitrust law encourages companies to diversify; that is move into new industries. Big companies that operate in a number of industries simultaneously are referred to as **conglomerates**. Conglomerates are growing rapidly in the United States.
3. Interlocking Directorates
    a. **Interlocking directorates** is a way corporations may be linked, and are formed when an individual sits on the board of directors of two or more noncompeting companies (1997 Board of Directors of IBM).
4. Small Firms and Big Corporations
    a. Smaller businesses continue to exist, and are especially important in the service sector.
    b. Most of the United States labor force works in large corporations.

G. Globalization
1. In the 1980s and early 1990s, the United States was hit by a wave of corporate "downsizing," that consequently resulted in hundreds of thousands of blue-collar workers and middle managers being unemployed. Some blamed the government, others blamed the unemployed, and others blamed the corporations.
2. In the 1980s, workers, governments, and corporations got involved as unequal players in the globalization of the world economy. On the largest scale ever, American and Japan-based multinationals, along with other highly industrialized countries, built plants in many countries to take advantage of inexpensive labor and low taxes.
3. Because workers were rooted in their communities, and governments in the nation-states, multinationals had a big advantage in the globalization game, with the threat of relocation and by playing governments off one another.
4. Today, it is clear that the winners are the stockholders of multinational corporations, and the losers at least initially were blue-collar workers (i.e. General Motors' labor force cuts since 1980). In 2000, unemployment in the United States hit a 38-year low.
5. By 2003, new worries surfaced because of the rise of China as an economic powerhouse.
    a. Since 1979, Chinese economic growth has averaged 10% a year, transforming China into the world's 7th largest economy.
    b. Chinese wages are low, so manufactured goods are inexpensive.
    c. Manufacturing employment fell to 13% of total employment in the United States, half of what it was in 1970.

d. Free trade supporters argue that decline in manufacturing jobs is due not to Chinese competition, but increased productivity of the American worker.

e. Free trade defenders also note that most displaced American workers eventually find other jobs.

  i. Free trade supporters fail to note that increased productivity is a response to competition from abroad, and that displaced workers often find inferior new jobs.

f. Many mainstream economists prefer to ignore the negative effects of free trade on average income and class inequality in the United States.

6. Globalization in the Less Developed Countries

a. On the one hand, many workers in the regions of the world where branch plants of multinationals have sprung up rush to fill factory jobs in hopes of better work for future generations (i.e. rural Indonesian women workers, Mexico's northern border with the United States). On the other hand, there are negative consequences.

b. Governments in developing countries impose few in any pollution controls, which have dangerous effects on the environment.

c. Fewer jobs are available than the number of workers drawn from the countryside, resulting in growth of urban slums.

d. Branch plants create few jobs involving design and technical expertise, because of the importation of high-value components.

e. Some branch plants, especially clothing and shoe factories in Asia, exploit children and women, by requiring them to work long workdays for meager wages in unsafe conditions.

  i. Nike has been at the forefront of moving production overseas to places like Vietnam and Indonesia. Workers are paid 10 cents an hour in Indonesia, which accounts for only about 4% of the price for a pair of Nike shoes. Workdays are up to 16 hours. Substandard air quality and excessive exposure to toxic chemicals is normal.

  ii. The Gap invested heavily in the Northern Mariana Islands near Guam. Although the Marianas are not a poor foreign country due to their commonwealth territorial status with the United States similar to that of Puerto Rico, garment manufacturing is the biggest source of income on the islands and the Gap is the largest employer. Below U.S. minimum-wage rates, duty-free access to the United States markets, and the right to sew "Made in U.S.A" labels manufactured there, attracts The Gap. (which also owns Banana Republic and Old Navy). Working conditions are horrific.

H. The Future of Work and the Economy

1. In the future, it is likely that businesses will continue to introduce new technologies and organize the workplace in more efficient ways, to cut costs and maximize profits.

2. It is less predictable how these practices will be implemented.

a. Possibilities

  i. Complement the abilities of skilled workers, or "competing with the high end of the wage scale".

  ii. Replacement of workers, deskilling of jobs, and employment of low-cost labor (mainly women and minority group members) on a large scale, or "competing with the low end of the wage scale".

3. The future of work in the United States will be closely tied to which end of the wage scale we decide to compete with.
   a. The future of work and the economy is up for grabs.

Study Activity: Applying The Sociological Compass

Consider your role in society as a consumer. In other words, consider all the clothes, shoes, and other things you purchase at the malls and shopping centers. First, make a list of the garments and shoes you purchase, and identify the name brand and estimated cost of each item. Second, consider the likely cost of labor to produce the product. For the purpose of this activity, calculate the labor cost based on Nike's 4% rate that labor is accounted for in the retail price for a pair of shoes. For instance, it costs Nike $4 in labor for a pair of Air Jordan's that sells for over $100 retail. Third, explain why and how such a huge disparity exists between labor cost and retail price. Finally, consider how globalization trends will affect the job prospects and conditions in your occupational field of aspiration in terms of the "deskilling" and "labor market segmentation" perspectives, respectively.

Garments and Shoes

Items:

1.

2.

3.

4.

Brands:

1.

2.

3.

4.

Estimated Cost:

1.

2.

3.

4.

Estimated Labor Cost by Item

1.

2.

3.

4.

Explanation of the Disparity between Labor Cost and Retail Price

Your Job Prospects and Foreseeable Conditions in the Future

Deskilling thesis:

Labor segmentation perspective:

Infotrac College Edition Online Exercises

For the following exercises, log on to the online library of InfoTrac College Edition at http://www.infotrac-college.com/. Make note that InfoTrac has implemented a new registration system that will allow easier access to InfoTrac through the use of a personalized username and password. Once you've created your username and password you may proceed directly to the Log On page. To create an account, register your passcode packaged with your textbook, and create a username and password, by following the online prompt. After you are logged in, click on "Infotrac College Edition." You will arrive at a screen that enables you to search topics.

Keyword: **self-employment**. Self-employment was once very common in the United States. Self-employment decreased significantly with the industrial revolution. The number of people who are self-employed continues to increase. Search for articles that discuss the increase in self-employment and the special problems the self-employed face. Write a short paper stating your findings.

Keyword: **Age discrimination**. Search for articles that deal with age discrimination in the labor force. How common is this problem. Is there any evidence that employers are willing to hire older persons?

Keyword: **sweatshops**. Look for articles dealing with the sweatshop industry. How widespread is the sweatshop industry?

Keyword: **absenteeism**. Search for articles that discuss the causes of absenteeism. How much of absenteeism can be explained by lack of day care or worker alienation?

Keyword: **disabled workers**. Search for articles that discuss the issues facing disabled workers. Make a list of all the issues you find. Come prepared to class to discuss your list in small groups.

Internet Exercises

Go to: http://www.leftbusinessobserver.com/Work.html and read about work in the future. Write a short about your feelings concerning this provocative analysis of the future of work. Compare this analysis of future work with some key concepts in Chapter 13.

Go to: http://www.bls.gov/oco/home.htm to access the *Occupations Outlook Handbook*. You will arrive at a screen that lists a few ways to explore job outlooks. Continue to navigate through the Web site until you arrive at your occupation of choice. The *Occupations Outlook Handbook* summarizes the nature of work, working conditions, employment, training, qualification and advancement, job outlook, earnings, and related occupations. Read the outlook description of the occupation of your choice, paying particular attention to "job outlook." Summarize the main implications, suggestions, and projections for your occupational aspiration. Are the projections consistent with your perceptions based on what your learned in Chapter 13?

A report put out by the Educational and Employment Statistics Division of American Physics (AIP) discusses the underemployment of physics post-doctorates. You will find this report at: http://www.aip.org/statistics/trends/reports/underemp/under.htm.

How can you explain the fact that people with such high academic degrees are underemployed? Is it because of individual factors or social structure factors? Write a short paper with your response.

To find information on employment for the disabled, look at: http://dis-rights.info/dr-follow.html. It discusses all aspects of employment for the disabled. Investigate some area of employment and report back to the class. Be sure to tie it in with some information from Chapter 12 that discusses disabilities.

For information about the stock market, go to: http://library.thinkquest.org/3088/

Research stocks and learn about stocks. Complete the online tutorial. Write a short paper discussing your experience.

The United for a Fair Economy website deals with different aspects of our economy. Visit the site at: http://www.ufenet.org/

Write a short comparison paper comparing what was presented in class lecture with the information found at this Website.

Practice Tests

Fill-In-The-Blank

Fill in the blank with the appropriate term from the above list of "key terms."

1. The _____ is the institution that organizes the production, distribution, and exchange of goods and services.

2. The _____ refers to specialization of works tasks.

3. _____ is a system of improving productivity developed in the 1910s by Frederick W. Taylor.

4. The _____ contains a disproportionately large number of women and members of racial minorities, who tend to be unskilled and lack higher education.

5. The _____ movement involves small groups of a dozen or so workers and managers collaborating to improve both the quality of goods produced and communication between workers and managers.

6. _____ are organizations that seek to defend and promote their members' interests.

7. In a(an) _____ , prices are only by supply and demand.

8. _____ are legal entities that can enter into contracts and own property.

9. In _____ countries, democratically elected governments own certain basic industries entirely or in part and intervene vigorously in the market to redistribute income.

10. _____ are large corporations that operate in several industries at the same time.

Multiple-Choice Questions

Select the response that best answers the question or completes the statement:

1. Which sector of the economy involves the purchase and sell of services?
   a. primary sector
   b. secondary sector
   c. tertiary sector
   d. manufacturing sector
   e. None of the above.

2. With the development of the three sectors of the economy, and as social relations among workers became more hierarchical, specialized work tasks increased and became more complex. The increase in specialized work tasks and its complexity is referred to as:
   a. the division of labor
   b. labor market segmentation
   c. internal labor markets
   d. the deskilling of the workforce
   e. scientific management

3. Which is NOT a criticism of the deskilling thesis?
   a. It is too narrowly focused on jobs toward the top of the occupational hierarchy.
   b. Lacks explanation of the processes of the occupational structure as a whole.
   c. Exaggerates the downward slide of the U.S. labor force.
   d. It overlooks the polarization between "good" and "bad" jobs.
   e. All of the answers are critiques.

4. Which part of the labor market contains disproportionately large numbers of women and members of racial minorities?
   a. the primary labor market
   b. the secondary labor market
   c. the tertiary labor market
   d. All of these are correct.
   e. None of the above.

5. Which is a decision-making innovation that has improved communication between workers and managers, and the quality of goods?
   a. The emergence of interlocking directorates.
   b. The growth of socialist revolutions.
   c. Reforms that give workers more authority on the shop floor.
   d. All of these are correct.
   e. None of the above.

6. _____ are social mechanisms for controlling pay rates, hiring, and promotions within corporations.
   a. Decision-making innovations
   b. Quality circles
   c. Oligopolies
   d. Internal labor markets
   e. None of the above.

7. Which is NOT a barrier between primary and secondary labor markets?
   a. There are few entry-level positions in the primary labor market.
   b. There is a lack of informal networks in the secondary labor market to good job openings.
   c. Workers usually lack the required training and certification for jobs in the primary labor market.
   d. There are too many union members in the primary labor market.
   e. All of the answers are barriers.

8. Why is leisure on the decline and work more fast-paced in the United States?
   a. Big corporations push Americans to consume more goods and services, through enormous investment in advertising.
   b. More corporate executives seem to think it is more profitable to have employees work more hours, rather than to hire more workers.
   c. American workers are not in a position to demand reduced working hours and more vacation because few are unionized.
   d. All of these are correct.
   e. None of the above.

9. The fact that the federal government had to establish a legal minimum wage in the United States to prevent the price of unskilled labor from dropping below the point at which people could not make a living illustrates:
   a. why the primary labor market is so tightly controlled.
   b. why the secondary labor market cannot be entirely free.
   c. the problem of capitalism for workers.
   d. the prevalence of labor union membership among workers.
   e. All of these are correct.

10. Which statement is INCONSISTENT with neoclassic economists?
    a. Resistance to the operation of free markets increases the likelihood that you will move down the social hierarchy.
    b. Free markets maximize economic growth.
    c. The elimination of the minimum wage would benefit everyone.
    d. All of the answers are inconsistent statements.
    e. None of the above are inconsistent statements.

11. Which economic system is dominant in the world?
    a. communism
    b. socialism
    c. communalism
    d. oligopolies
    e. capitalism

12. Marx gave the name _____ to the transitional phase between capitalism and communism.
    a. communalism
    b. socialism
    c. bourgeoisie
    d. proletariat
    e. petite bourgeoisie

13. Which is a characteristic of communism in its ideal form?
    a. private ownership of property
    b. public ownership of property
    c. laissez-faire
    d. competition in the pursuit of profit
    e. free market

14. Which is NOT a negative consequence of multinationals to less developed countries?
    a. Fewer jobs are available than the number of workers.
    b. Governments in developing countries lack pollution controls.
    c. Branch plants create few jobs involving design and technical expertise.
    d. Exploitation of children and women.
    e. Workers of multinationals rarely resist exploitation.

15. Although the future of work and the economy is difficult to predict, which is a likely pattern in the future?
    a. Businesses will likely export capital to take advantage of inexpensive labor abroad.
    b. Businesses will likely continue to introduce new technologies.
    c. Businesses will likely continue to attempt to cut costs and maximize profits.
    d. Businesses will likely attempt to organize the workplace in more efficient ways.
    e. All of these are correct.

16. Interlocking directorates are one way _____ may be linked.
    a. workers
    b. workers and managers
    c. unions and workers
    d. corporations
    e. multinationals and workers in developing countries

17. Which country is the best example of an oligopoly?
    a. Czechoslovakia
    b. Russia
    c. The United States
    d. Sweden
    e. Japan

18. The fact that the state had to intervene to create markets to begin with shows that in reality, no economy is purely:
    a. communist
    b. socialist
    c. democratically socialist
    d. laissez-faire
    e. None of the above.

19. Katherine Newman's research on fast-food workers in Harlem illustrates:
    a. the importance of indignity endured by part-time workers.
    b. that labor exploitation in an inner-city setting is like worker conditions in less developed countries.
    c. that issues of self-respect are trivial in comparison to more compelling issues of absolute poverty.
    d. All of these are correct.
    e. None of the above.

20. The World Trade Organization (WTO) was set up by the governments of 134 countries in 1994 to:
    a. encourage and referee global commerce.
    b. discourage and referee global commerce.
    c. monitor and referee protests globally.
    d. All of these are correct.
    e. None of the above.

21. What is the percentage of workers in the United States earning poverty-level wages?
    a. 5%
    b. 12%
    c. 29%
    d. 21%
    e. 48%

22. Which occupation is expected to grow the most in the United States, in terms of estimated number of new jobs, between 1998 and 2008?
    a. systems analyst
    b. registered nurses
    c. secondary school teachers
    d. office and administrative support supervisors and managers
    e. security guards

23. What country has the greatest hours worked per week on average, and many fewer paid vacation days per year, than workers in other rich industrialized countries?
    a. Norway
    b. Germany
    c. The United States
    d. Canada
    e. Japan

24. The movie *Roger and Me* was released when many scholars and politicians were expressing fears that the U.S. labor force was on a downward slide due to:
    a. deindustrialization.
    b. scientific management.
    c. the shrinking division of labor.
    d. labor union organization.
    e. global protests against globalization.

25. Which group of workers was furthest from the economic core and displayed the lowest median weekly earnings of full-time wage and salary workers in the United States in 1995?
    a. White male unionized
    b. White female unionized
    c. Latino (male) unionized
    d. Latina (female) unionized
    e. Latina (female) non-unionized

True False

1. New opportunities make it relatively easy to move from the secondary labor market to the primary labor market.

   TRUE or FALSE

2. The human relations school of management advocated more efficient authoritarian leadership, and greater attention to strategies to maximize profit.

   TRUE or FALSE

3. The secondary labor market cannot be entirely free.

   TRUE or FALSE

4. Professionals are people with specialized knowledge acquired through extensive higher education.

   TRUE or FALSE

5. The central assumption of most economic sociologists is that free markets maximize economic growth, which is beneficial to society at large.

   TRUE or FALSE

6. The number of conglomerates is declining in the United States.

   TRUE or FALSE

7. Some workers in the least developed countries, where multinationals have sprung up, rush to fill low-wage factory jobs.

   TRUE or FALSE

8. Since the Northern Mariana Islands are a commonwealth territory of the United States, The Gap factories in the Marianas are characterized by exceptionally healthy and humane working conditions, unlike most transnationals.

   TRUE or FALSE

9. The primary labor market is more regulated than the secondary labor market.

   TRUE or FALSE

Short Answer

1. List and define the three sectors of the economy. (p. 373)

   _____

   _____

2. What is an oligopoly and conglomerate? Provide an example for each term. (p. 393-394)

   _____

   _____

3. List and describe three negative consequences of multinationals on less developed countries. (p. 397)

   _____

   _____

4. Define the term "laissez-faire." Are there any purely laissez-faire economies? Why or why not? (p. 390)

   _____

   _____

5. Identify the three stages of labor market development in the United States according to David Gordon and his colleagues. (p. 379-380)

   _____

   _____

Essay Questions

1. Describe and explain the historical development of the agricultural, manufacturing and service sectors. (p. 371-374)

2. Contrast the deskilling thesis and the labor segmentation perspective. (p. 375-380)

3. Define the terms capitalism, communism, and democratic socialism. Explain the differences between the three economic systems in their ideal and real forms. (p. 387-394)

4. What is the difference between the neoclassical school of economics and the economic sociological view of markets? Make sure you explain the difference between the primary and secondary labor markets, and a free or regulated market. (p. 387-390)

5. Explain what it means to compete on the low or high end of the wage scale. How do you see the United States responding to new and powerful global economic challenges in the future? (p. 394-399)

Solutions

Practice Tests

Fill-In-The-Blank

1. economy
2. division of labor
3. Scientific management
4. secondary labor market
5. quality of work life
6. Unions
7. free market
8. Corporations
9. democratic socialist
10. Conglomerates

Multiple-Choice Questions

1. C, (p. 373)
2. A, (p. 373)
3. A, (p. 377-379)
4. B, (p. 379)
5. C, (p. 380-382)
6. D, (p. 382)
7. D, (p. 383-385)
8. D, (p. 386-387)
9. B, (p. 387-390)
10. A, (p. 389-390)
11. E, (p. 390)
12. B, (p. 391-392)
13. B, (p. 392)
14. E, (p. 395-396)
15. E, (p. 398)
16. D, (p. 394)
17. B, (p. 393)
18. D, (p. 390)
19. A, (p. 376-377)
20. A, (p. 396)
21. C, (p. 388)
22. A, (p. 379)
23. C, (p. 380-381)
24. A, (p. 384)
25. E, (p. 385)

True False

1. F, (p. 383)
2. F, (p. 380)
3. T, (p. 387)
4. T, (p. 382)
5. F, (p. 387-390)
6. F, (p. 394)
7. T, (p. 397)
8. F, (p. 397-398)
9. T, (p. 387)

POLITICS

Student Learning Objectives

After reading Chapter 14, you should be able to:

1. Discuss democracy as an ideal versus a reality.

2. Define the core concepts of politics.

3. Identify and define the three bases of authority according to Max Weber.

4. Compare and contrast competing theories of democracy.

5. Understand and apply theories of democracy to discuss the future of democracy.

6. Explain the role of elites in the decision-making processes that govern society according to C. Wright Mills' elite theory.

7. Describe the philosophies of liberal and conservative policy groups.

8. Discuss the patterns of party support in terms of social class, religion, race, and gender.

9. Explain the role of organization as a source of the distribution of political power.

10. Describe how the state can structure political life, regardless of how power is distributed.

11. Explain why the U.S. political system is ironically less responsive to the needs of the disadvantaged, in spite of its promotion of democracy.

12. Describe historical examples of periods of political shock in the United States.

13. Explain the contradictions of democracy in terms of its formal and liberal forms.

14. Describe the three waves of democratization that swept the world since 1828.

15. Describe the social preconditions for democracy to be fully recognized.

16. Discuss the future prospects for democracy in light of electronic and Internet technology.

17. Explain and critique postmaterialist views of current and future political life.

18. Discuss the concepts of war and terrorism, and explain what is meant by "politics by other means".

Key Terms

power (405)

authority (405)

legitimate (405)

traditional authority (406)

legal-rational authority (406)

charismatic authority (406)

political revolution (406)

state (406)

civil society (406)

autocracy (406)

authoritarian state (406)

democracy (406)

political parties (408)

lobbies (408)

mass media (408)

public opinion (408)

social movements (408)

pluralists (408)

elite theorists (409)

elites (409)

ruling class (409)

political action committees-PACs (411)

power-resource theory (413)

state-centered theorists (416)

formal democracies (423)

liberal democracies (423)

postmaterialists (426)

war (428)

terrorism (432)

Detailed Chapter Outline

I. INTRODUCTION

   A. The Tobacco War

     1. In the spring of 1998, Congress was ready to pass an antitobacco bill that would be very costly for tobacco companies.

     2. The public seemed eager to support the legislation, due to compelling health concerns, and the targeting of teenagers by the tobacco industry.

     3. Representatives of the tobacco industry mobilized enormous resources against the bill (i.e. lobbyists and ad campaigns), and wined, dined, and cajoled members of Congress to vote against the bill.

     4. Ads swayed public opinion, and influenced the belief that the bill's real aim was to raise revenue.

     5. The bill was defeated in June 1998.

     6. The defeat of the federal bill raises important political questions.

       a. Does the outcome of the tobacco war illustrate the operation of "government of the people, by the people, for the people" as Abraham Lincoln defined democracy?

       b. Does big business's access to economic and political resources lead one to doubt Lincoln's definition of democracy?

     7. The tobacco war raises the core question of political sociology.

       a. What accounts for the degree to which a political system responds to the demands of all its citizens?

    b. Political sociologists examine the effects of social structures, especially class structures, on politics.

    c. An adequate theory of democracy requires that we also examine how state institutions and laws affect political processes.

II. POWER AND AUTHORITY

A. Politics is a machine that determines "who gets what, when, and how." **Power**, the ability to control others against their will, fuels the machine.

    1. Power:

        a. May involve force.

        b. Becomes **authority** when it is legitimate and institutionalized to the point that people basically agree or accept how the political machine is run, even if grudgingly.

        c. Is **legitimate** when regarded as valid or justified.

        d. Is institutionalized when the norms and statuses of social organizations govern its use.

B. Types of Authority

    1. **Traditional authority** is the norm in tribal and feudal societies, involves the inheritance of authority by rulers through family or clan ties. The right of a family or clan to monopolize leadership is deemed justified by religion.

    2. **Legal-rational authority** is typical of modern societies, and is derived from respect for law. Laws are generally believed to be rational. And if someone achieves office by following laws, their authority is respected.

    3. **Charismatic authority** is based on belief in the claims of extraordinary individuals to be inspired by God of some higher principle. Charismatic figures sometimes emerge during a **political revolution**, an attempt by many people to overthrow existing political institutions and establish new ones.

C. Types of Political System

    1. Political sociology is concerned with institutions that *specialize* in the exercise of power and authority, which form the **state**, institutions that formulate and carry out a country's laws and public policies.

    2. **Civil society** is the private sphere of social life, where the state regulates citizens.

    3. Autocracies and Authoritarian States

        a. In an **autocracy**, absolute power rests in the hands of a single person or party.

        b. **Authoritarian state** is characterized by sharp restrictions on citizen control.

        c. In a totalitarian state, citizens lack almost any control of the state, but is virtually nonexistent.

    4. Democracies

        a. **Democracy** involves citizens exerting a relatively high degree of control over the state, partly by choosing representatives in regular, competitive elections.

        b. Citizens control the state indirectly through several organizations in modern democracies.

        c. **Political parties** compete for control of government in regular elections. They give voice to policy alternatives and rally adult citizens to vote.

        d. **Lobbies** are formed by special interest groups to advise and influence politicians.

e. The **mass media** help to keep the public informed about the quality of government by keeping a critical eye on the state.

f. Public opinion 316 refers to values and attitudes of the adult population as a whole, and is expressed in polls and letters to lawmakers.

g. **Social movements** are collective attempts to change all or part of the political and social order.

III. THEORIES OF DEMOCRACY

A. Pluralist Theory

1. Social scientists who studied New Haven politics in the 1950s, during a time when New Haven was in decline (white middle-class settled in suburbs, erosion of city's tax base, slums), were known as **pluralists**, who concluded that no single group exercised disproportionate power in New Haven.

2. Pluralists believe:

   a. Politics operate democratically in the United States as a whole.

   b. America is a heterogeneous society with many competing interests and centers of power, none of which can consistently dominate.

   c. Politics involves negotiation and compromise between competing groups.

   d. Democracy is guaranteed.

B. Elite Theory

1. **Elite theorists** sharply disagreed with pluralists.

2. C. Wright Mills is a chief elite theorist.

   a. **Elites** are small groups that occupy the command posts of America's most influential institutions, including the two or three biggest corporations, the executive branch of government, and the military.

   b. Elites make decisions that profoundly affect all members of society, yet without regard for elections or public opinion.

   c. However, connections between the corporate, state, and military elites, do not turn the three elites into a **ruling class**, a self-conscious group of cohesive people.

C. A Critique of Pluralism

1. In light of the well-established existence of wealth-based inequalities in political influence and participation, most political sociologists today question the pluralist perspective, and are more sympathetic to the elitist view.

2. Political Participation

   a. "Citizen Participation Study" found that people with higher incomes are more politically active.

   b. The rich contribute 17.5 times more money to election campaigns than the poor, although the poor are six times more numerous.

3. Political Influence and PACs

   a. **Political action committees** (PACs) are organizations that raise funds for politicians that support particular issues.

  b. Dawn Clawson's research on PACs found a split between a unified business-Republican group and a labor-women-environmentalist-Democratic group.
  c. PACs tend to favor Republican candidates.

D. Power Resource Theory
  1. While elite theorists believe it makes little difference whether Republicans or Democrats are in power, and the victory of one party doesn't deserve much sociological attention, **power resource theory** focuses on how *variations* in the distribution of power affect the fortunes of parties and policies.
  2. American voters cluster in two policy groups.
      a. *Liberal* or left-wing voters promote extensive government involvement in the economy, a strong "social safety net" of health and welfare benefits to help less fortunate members of society, and equal rights for women, racial and sexual minorities.
      b. *Conservative* or right wing voters favor a reduced role for government in the economy, a smaller welfare state, individual initiative in promoting economic growth, and traditional social and moral values.
  3. There are different patterns of support for one party over another by social class, religious groups, races, and other groups.
  4. Political Parties and Class Support
      a. In most Western societies, *class* support is a main factor, but varies by country. The strength of this class tendency depends on how socially organized or cohesive classes are.
  5. Organization and Power
      a. The main insight of power resource theory—Organization is a source of power. Change in the distribution of power between major classes partly accounts for the fortunes of different political parties and different law and policies.
  6. Other Party Differences: Religion, Race, and Gender
      a. *Religion*, *race*, and *gender* are also important factors that distinguish parties.
      b. The 2002 GSS indicates that 60% of male voters but only 48% of female voters chose Bush in the 2000 presidential race. 40% of male voters and 52% of female voters preferred Gore.

E. State-Centered Theory
  1. Theda Skocpol and other **state-centered theorists** show how the state itself can structure political life, regardless of how power is distributed.
  2. Example—American voter registration law and nonvoting in the United States.
      a. The United States has a proportionately smaller pool of eligible voters than other democracies.
      b. Only in the United States do individual citizens have to take the initiative to register themselves in voter registration centers, however many Americans are unable or unwilling to register.
      c. The poor, less educated, and members of racial minority groups are less likely to vote.
  3. Voter Registration Laws
      a. American voter registration law is a pathway to democracy for some, and a barrier for others.

4. Two reasons why the American political system is less responsive than other rich democracies to the needs of the disadvantaged.
    a. The working class is nonunionized and weak.
    b. The law requires citizen-initiated voter registration.
5. Changing State Structures
    a. State structures resist change.
    b. *Constitutions* anchor their foundations.
    c. *Laws* surround the upper stories of state structures.
    d. *Ideology* reinforces the entire structure.
6. Shocks periodically reorient American public policy and cause major shifts in voting patterns.
    a. 1890s industrial unrest
    b. 1930s Great Depression
    c. 1970s another economic crisis
7. Summing Up
    a. Political sociology has made good progress since the 1950s.

IV. THE FUTURE OF DEMOCRACY

A. Two Cheers for Russian Democracy
1. In 1989 in the midst of the decline of totalitarianism, the prospects for democracy, capitalism, and multiparty elections seemed in place.
2. The great questions of Russian politics remain unanswered, and democracy enabled a few people to enrich themselves at the expense of most Russians. Many citizens equated democracy with distress, not freedom.
    a. Russia held multiparty elections in 1991.
    b. Although most Russians favored democracy, support soon fell because the economy collapsed.
    c. Formerly fixed prices were allowed to rise by the market, which led to consumer goods costing 10 or 12 times more than just a year earlier.
    d. Unemployment grew.
    e. Workers were paid irregularly.
    f. 39% of the population lived below the poverty line in 1999.
    g. Profitable businesses and valuable real estate formerly owned by the government were sold to private individuals and companies, and disproportionately went to members of the Communist and organized crime syndicates.
    h. By 1994, the richest 10% of Russians earned 15 times more than the poorest 10%.
3. What are the necessary social conditions for a country to become fully democratic?
B. The Three Waves of Democracy
1. The first wave
    a. Began in the 1828 presidential election, when over half the white adult male population in the United States became eligible to vote.
    b. By 1926, 33 countries enjoyed minimally democratic institutions.

    c. A democratic reversal occurred between 1922 and 1942, when fascist, communist, and militaristic movements caused two-thirds of the world's democracies to fall under authoritarian and totalitarian rule.

2. The second wave
   a. 1943 to 1962
   b. Allied victory in World War II returned democracy to many of the defeated powers (i.e. West Germany and Japan).
   c. The beginning of the end of colonial rule in some state in Africa and elsewhere.
   d. In the 1950s the world was in the midst of a second democratic reversal. A third of the democracies in 1958 were authoritarian by the mid-1970s.

3. The third wave
   a. The biggest wave that began in 1974 with the overthrow of dictatorships in Portugal and Greece.
   b. In the early 1990s, a series of authoritarian regimes fell in Southern and Eastern Europe, Latin America, Asia, and Africa.
   c. In 1991, Soviet communism collapsed.
   d. By 1995, 117 of the world's 191 countries (61% of the world's countries and 55% of the world's population) were democratic in terms of representative elections.

4. While these countries are **formal democracies** (hold regular, competitive elections), many are not **liberal democracies** (lack the freedoms and constitutional protections that make political participation and competition meaningful), such as Russia.

C. The Social Preconditions of Democracy
   1. Liberal democracies emerge and endure when countries enjoy considerable economic growth, industrialization, urbanization, the spread of literacy, and decrease in economic inequality.
   2. When a country's military is as powerful as its middle and working classes, democracy is precarious and often merely informal.
   3. Favorable external political and military circumstances help liberal democracy endure.
   4. Liberal democracy will spread in the less economically developed countries only if they prosper and enjoy support from a confident United States and European Union, the world centers of liberal democracy.
   5. However, the United States is not always a friend to democracy,—and just because the United States promotes democracy in many parts of the world, we should not assume that liberal democracy has been fully actualized in this country.

D. Electronic Democracy
   1. Like the vision of public opinion polls by George Gallup in 1935, some people view new technology with enthusiasm today.
   2. Computers connected to the Internet could:
      a. allow citizens to debate and vote on issues.
      b. give politicians a clear signal of how public policy should be conducted.
      c. revive American democracy.
   3. Flaws of this grand vision

    a. If electronic public meetings were accessible to everyone, disinterest would limit participation.

    b. If electronic democracy becomes widespread, it will probably reinforce the same inequalities in political participation, in the form of a digital divide.

E. Postmaterialism

  1. **Postmaterialists** believe that economic or material issues are becoming less important in American politics, and claim that growing inequality and prosperity in rich industrialized countries have resulted in a shift from class-based to value-based politics.

  2. Postmaterialists:

    a. Argue that in liberal democracies, the gap between the rich and the poor is less extreme and society as a whole is more prosperous.

    b. Are more concerned with women's rights, civil rights, and the environment.

    c. Conclude that the old left-right political division, based on class differences and material issues, is being replaced by a new left-right division based on age and postmaterialist value-based issues.

  3. Criticisms of postmaterialism:

    a. Affluence is not universal in America.

    b. Inequality is not decreasing.

    c. Poverty is especially widespread among youth, the very people postmaterialists view as the most affluent and least concerned with material issues.

    d. Research shows that social class issues are just as important in voting as it was 40 years ago.

  4. The big dilemma of American politics

    a. Problems of economic inequality are still prevalent, but will likely not be seriously addressed unless disadvantaged Americans become more politically involved. Yet unequal political participation is a likely future pattern.

    b. A solution may have to emerge outside normal politics, such as another historical shock, for democracy to be fully actualized.

V. POLITICS BY OTHER MEANS

A. War

  1. War Deaths

    a. A **war** is a violent armed conflict between politically distinct groups who fight to protect or increase their control of territory.

    b. Humanity has spent much of its history preparing for, fighting, and recovering from war.

    c. Wars have become more destructive over time with "improvements" in the technology of human destruction. The 20th century represents 2.6% of the time since the beginning of recorded war history but accounts for roughly 3.3% of the world's war deaths, military and civilian.

  2. The Business of War

    a. War is an expensive business, and the United States spends far more than any other country financing it.

    b. With 4.5% of the world's population, the United States accounts for about a third of total military expenditures in the world. The United States is by far the largest exporter of arms, accounting for nearly 60% of world arms exports. Nearly all the big arms importers are U.S. allies, such as our best customers Saudi Arabia and Taiwan.

3. Types of War
    a. Wars may take place between countries (interstate wars) and within countries (civil or societal wars).
    b. A special type of interstate war is the colonial war, which involves a colony engaging in armed conflict with an imperial power to gain independence.
    c. Since the mid-1950s, most armed conflict in the world has been societal rather than interstate.
    d. Wars like the 2003 U.S.-Iraq war account for little of the total magnitude of armed conflict although loom large in the mass media. Wars like the ongoing conflict in the Democratic Republic of the Congo account for most of the total magnitude of armed conflict yet are rarely mentioned in the mass media.

4. The Risk of War
    a. War risk varies from one country to the next, but what factors determine the risk of war on the territory of a given country?
    b. Type of government affects risk of war. Government types include democracy, autocracy, and "intermediate" forms. Intermediate-type countries are at highest risk of war, especially civil war, because these types of countries have neither high legitimacy as in democratic countries, nor iron rule as in autocratic countries.
    c. Level of prosperity affects risk of war. A country's gross domestic product per capita (GDPpc) indicates level of prosperity. Democracy is more common and autocracy is less common in prosperous countries.

B. Terrorism and Related Forms of Political Violence
1. Why has societal war largely replaced interstate warfare since World War II? A historical perspective is useful.
2. Historical Change
    a. From the rise of the modern state in the 17th century until World War II, states increasingly monopolized the means of coercion in society. This had three important consequences.
    b. As various regional, ethnic, and religious groups came under the control of powerful central states, regional, ethnic, and religious wars declined and interstate warfare became the norm.
    c. Because states were powerful and monopolized the means of coercion, conflict became more deadly.
    d. Civilian life was pacified because the job of killing for political reasons was largely restricted to state-controlled armed forces.
    e. All this changed after World War II (1939-1945). Since then, there have been fewer interstate wars and more civil wars, guerrilla wars, massacres, terrorist attacks, and instances of attempted ethnic cleansing and genocide perpetrated by militias, mercenaries, paramilitaries, suicide bombers, and the like. Large-scale violence has increasingly been visited on civilian rather than military populations.

3. Reasons for Change
   a. This change in the form of collective action came about for three main reasons.
      i. Decolonization and separatist movements roughly doubled the number of independent states in the world, and many of these new states, especially in Africa and Asia, were too weak to control their territories effectively.
      ii. Especially during the Cold War (1946-1991), the United States, the Soviet Union, Cuba, and China often subsidized and sent arms to domestic opponents of regimes that were aligned against them.
      iii. The expansion of international trade contraband provided rebels with new means of support.
4. Al Qaeda and Contemporary Warfare
   a. From this perspective, al Qaeda is a typical creature of contemporary warfare.
   b. **Terrorism** is defined by American law as premeditated, politically motivated violence against noncombatant targets.
   c. Al Qaeda originated in Afghanistan, a weak and dependent state.
   d. The United States supported al Qaeda's founders militarily in their struggle against the Soviet occupation of Afghanistan in the 1980s.
   e. Al Qaeda organized international heroin, diamond, and money laundering operations.
   f. It established a network of operatives around the world.
   g. International terrorists of the type the U.S. Department of State collects data on, typically demand autonomy or independence from some country, population, or region.
   h. For example, among al Qaeda's chief demands are Palestinian statehood and the end of U.S. support for the wealthy regimes in Saudi Arabia, Kuwait, and the Gulf States. Al Qaeda has turned to terror as a means of achieving these goals because other ways are largely closed off.
   i. The United States considers support for the oil-rich Arab countries to be of national interest. It has so far done little to further the cause of Palestinian statehood, although this may change with President Bush's Roadmap to Peace, which envisages Palestinian statehood by 2005.
   j. Staunch opponents of American policy cannot engage in interstate warfare with the United States because they lack states of their own. Because the existing structure of world power closes off other possibilities for achieving political goals, terror emerges as a viable alternative for some desperate people.

Study Activity: Applying The Sociological Compass

Interview a handful of family members, friends, and associates. Ask them to identify their political party preferences in general, and the reasons for their preferences. Make a list of responses by party preference and reasons. Analyze the list by applying power resource theory. In other words, explain your findings according to power resource theory.

Party Preferences and Reasons

Power Resource Theory

Infotrac College Edition Online Exercises

For the following exercises, log on to the online library of InfoTrac College Edition at http://www.infotrac-college.com/. Make note that InfoTrac has implemented a new registration system that will allow easier access to InfoTrac through the use of a personalized username and password. Once you've created your username and password you may proceed directly to the Log On page. To create an account, register your passcode packaged with your textbook, and create a username and password, by following the online prompt. After you are logged in, click on "Infotrac College Edition." You will arrive at a screen that enables you to search topics.

Keyword: **Voting**. Identify and read an article that discusses patterns of voting. Read the article. Are voting patterns discussed in the article consistent with what you learned in Chapter 14? Why or why not? Are there any implications for electronic democracy that you can draw from the article? If so, what are they?

Keyword: **Osama bin Laden**. Search for articles that discuss the political philosophy of bin Laden's involvement with politics. What key political ideas do they support and reject?

Keyword: **whistle blowers**. Search for articles on the political involvement of whistle blowers. How common is this? Does it seem that there is some type of political involvement or not? Write a short paper expressing your findings.

Keyword: **Political Action Committees (PACs)**. Political Action Committees work with special interest groups to support specific political ideas. Search for articles that discuss what type of political ideas PACs support. Make a list of the ideas.

Keyword: **National Rifle Association (NRA)**. The National Rifle Association has a lot of political influence. Search for articles that discuss how the NRA is involved with politics. Do you think that the NRA should be involved with politics? Write a short paper stating your opinion.

Keyword: **voter apathy**. Not very many eligible voters vote in American elections. Search for articles that discuss why people are indifferent about voting? What reasons are there for voter apathy?

Internet Exercises

Go to: http://www.theatlantic.com/issues/96jul/gender/gender.htm

for a discussion on the growing gap in politics in American society. There is a growing trend for women to move to the Democratic party and men to move to the Republican party. Read this short article. Make a list of the reasons why there appears to be this gender polarization in politics. Do you agree or disagree with this phenomenon?

A good site for research about politics can be found at The Center for Responsible Politics homepage: http://www.opensecrets.org/

You can access political profiles, political participation, voter perceptions and subjects along these lines.

With the 2000 presidential election so close in history, students may want to look at The Federal Election Commission website at: http://www.fec.gov/ and look over the process of voting and the elections in general.

For general information about the United States political system go to: http://www.trinity.edu/~mkearl/polisci.html

Explore the different links. What conclusions can you make about the U.S. political system from this short exercise?

Compare and contrast one major issue of concern (for example, tax reform) on all of the following four political party websites:

http://www.democrats.org/index.html

http://www.greenparty.org/

http://rnc.org/

http://www.reformparty.org/

Practice Tests

Fill-In-The-Blank

Fill in the blank with the appropriate term from the above list of "key terms."

1. A(An) _____ is the overthrow of political institutions by an opposition movement and replacement by new institutions.

2. _____ is derived from respect for the law, and is typical of modern societies.

3. _____ is composed of the values and attitudes of the adult population as a whole.

4. In a(an) _____ , citizens exercise high degree of control over the state, mainly by choosing representatives in regular, competitive elections.

5. Civil society _____

6. _____ are organizations formed by special interest groups to advise and influence politicians.

7. A(An) _____ is a group that controls the command posts of an institution.

8. A(An) _____ is a country whose citizens enjoy regular competitive elections and the freedoms and constitutional protections that make political participation and competition meaningful.

9. A(An) _____ is a violent conflict between politically distinct groups who fight to protect or increase their control of a territory.

10. _____ is defined by American law as premeditated, politically motivated violence against noncombatant targets including unarmed or off-duty military personnel by subnational groups.

Multiple-Choice Questions

Select the response that best answers the question or completes the statement:

1. Which is a question of political sociology?
   a. What accounts for the degree to which a political system responds to the demands of all its citizens?
   b. What are the effects of social structures, especially class structures, on politics?
   c. How do state institutions and laws affect political processes?
   d. All of these are correct.
   e. None of the above.

2. When power is legitimized and institutionalized to the point that people generally accept political operations, power becomes:
    a. totalitarian
    b. monopolized
    c. authority
    d. ideology
    e. normalized

3. According to Max Weber, which type of authority is derived from respect for law, and is typical of modern societies?
    a. traditional authority
    b. legal-rational authority
    c. charismatic authority
    d. totalitarian authority
    e. formal democracy

4. In _____ states, citizen control is sharply restricted.
    a. authoritarian
    b. totalitarian
    c. democratic
    d. egalitarian
    e. traditional

5. Political organizations are utilized by citizens to control the state in modern democracies. Which organization is formed by special interest groups to advise and influence politicians?
    a. political parties
    b. lobbies
    c. political action committees
    d. advisory boards
    e. Washington think tanks

6. Sometimes _____ leaders emerge when the historical conditions are ripe for extraordinary leadership such as a political revolution.
    a. traditional
    b. totalitarian
    c. charismatic
    d. All of these are correct.
    e. None of the above.

7. Which is NOT considered part of the elite according to elite theory?
    a. corporations
    b. the executive branch of government
    c. the military
    d. the police
    e. All of the above are considered elites.

8. Which policy group tends to support a smaller welfare state?
   a. liberals
   b. radicals
   c. conservatives
   d. revolutionaries
   e. None of the above.

9. Which is NOT a source of resistance to change among state structures?
   a. constitutions
   b. laws
   c. ideology
   d. civil unrest
   e. All of the answers are sources of resistance among state structures.

10. Which historical event shocked American public policy and caused major shifts in voting patterns?
    a. Industrial unrest in the 1890s
    b. The Great Depression
    c. The economic crisis of the 1970s
    d. All of these are correct.
    e. None of the above.

11. From the late 1980s to the early 1990s, many Russian citizens equated democracy with:
    a. freedom
    b. distress
    c. revitalization
    d. rebirth
    e. economic well-being

12. Which wave of democracy is considered the biggest wave based on the proportion of countries and the world's population that were considered democratic?
    a. first wave
    b. second wave
    c. third wave
    d. fourth wave
    e. fifth wave

13. Which is a social precondition for liberal democracy to be actualized in a country?
    a. The enjoyment of considerable economic growth.
    b. Favorable external political and military connections.
    c. Support from the world centers of democracy.
    d. Decrease in economic inequality and the spread of literacy.
    e. All of these are correct.

14. Which is NOT a postmaterialist argument?
    a. The gap between the rich and the poor is less extreme and society as a whole is more prosperous in liberal democracies.
    b. Women's rights, civil rights, and the environment are primary concerns.
    c. The old left-right political division, based on class differences and material issues, is being replaced by a new left-right division based on age and postmaterialist value-based issues.
    d. Poverty is widespread among youth.
    e. All of the answers are postmaterialist arguments.

15. Staunch opponents of American policy who engage in terrorism:
    a. can engage in interstate warfare with the United States.
    b. use terror because the existing structure of world power closes off other possibilities for achieving political goals.
    c. have states of their own, but suffer from insufficient infrastructure.
    d. All of these are correct.
    e. None of the above.

16. A special type of interstate war, which involves a colony engaging in armed conflict with an imperial power to gain independence, is referred to as:
    a. colonial war.
    b. civil war.
    c. societal war.
    d. All of these are correct.
    e. None of the above.

17. With 4.5% of the world's population, the United States accounts for about how much total military expenditure in the world?
    a. one-eighth
    b. one-fourth
    c. one-third
    d. three-fourths
    e. nearly all of military expenditures in the world.

18. Which factor(s) determine the risk of war on the territory of a given country?
    a. type of government.
    b. level of national prosperity.
    c. level of higher educational access.
    d. both 'a' and 'b'.
    e. both 'a' and 'c'.

19. Which type of government places a country at the greatest risk for war?
    a. democratic governments
    b. autocratic governments
    c. intermediary governments
    d. All of these are correct.
    e. None of the above.

20. According to Figure 14.1 "The Institutions of State and Civil Society" in Chapter 14, which is part of civil society?
    a. political parties
    b. lobbies
    c. mass media
    d. social movements
    e. All of these are correct

21. Vladimir Lenin, the Bolshevik leader of the Russian Revolution of 1917, represents an example of which type of authority according to Max Weber?
    a. traditional authority
    b. charismatic authority
    c. legal-rational authority
    d. All of these are correct.
    e. None of the above.

22. In a nationally televised address on January 17, 1961, President Eisenhower sounded much like _____ when he warned of the "undue influence" of the "military-industrial complex" in American society.
    a. Theda Skocpol
    b. Richard Lee
    c. Max Weber
    d. Emile Durkheim
    e. C. Wright Mills

23. What is the primary issue discussed in Box 14.1, Social Policy: What Do You Think, "Financing Political Campaigns", in Chapter 14 of the textbook?
    a. the strength of lobbyists.
    b. political campaign financing.
    c. bipartisan politics.
    d. All of these are correct.
    e. None of the above.

24. According to Fig. 14.3 in the textbook, the degree of business unity or lack thereof in terms percent of electoral races, indicates that big business was:
    a. divided.
    b. predominant.
    c. unified.
    d. All of these are correct.
    e. None of the above.

25. In the 2004, presidential election, which social group tended to support George W. Bush?
    a. low-income earners
    b. African Americans
    c. Hispanic Americans
    d. male voters
    e. supporters of reproductive choice

26. Today's unemployment insurance, old-age pension, and public assistance programs all originated in:
    a. the Roosevelt administration.
    b. the Eisenhower administration.
    c. the Kennedy administration.
    d. the Reagan administration.
    e. the Clinton administration.

27. Which sociological theory of democracy views power as being dispersed as opposed to concentrated?
    a. pluralist theory
    b. elitist theory
    c. power resource theory
    d. state-centered theory
    e. All of these are correct.

True False

1. In June 1998, an antitobacco bill was passed in response to public health concerns.

    TRUE or FALSE

2. Civil society is the private sphere of social life, where the state regulates citizens.

    TRUE or FALSE

3. The mass media help to keep the public informed by keeping a critical eye on the state.

    TRUE or FALSE

4. Elite theorists tend to agree with pluralists on key political issues.

    TRUE or FALSE

5. Corporations, the State, and the military are not a self-conscious cohesive ruling class.

    TRUE or FALSE

6. The strength of social classes to influence political processes depends on how socially organized and cohesive they are.

   TRUE or FALSE

7. Despite the social ills of the United States, in comparison to most other rich industrialized countries, it is reasonable to assume that liberal democracy has been fully actualized in the United States.

   TRUE or FALSE

8. Humanity has spent much of its history preparing for, fighting, and recovering from war.

   TRUE or FALSE

9. Wars have become less destructive with the improvements in the technology to aid humanity.

   TRUE or FALSE

10. Pluralist theory portrays politics as a neatly ordered game of negotiation and compromise in which all players are equal.

    TRUE or FALSE

Short Answer

1. List and define three bases of authority according to Max Weber. (p. 406)

   _____

   _____

2. List and describe the three historical waves of democracy. (p. 423-424)

   _____

   _____

3. What are three criticisms of postmaterialism? (p. 426-427)

   _____

   _____

4. Explain the big dilemma of American politics according to Robert Brym and John Lie. (p. 427)

   _____

   _____

5. Describe the future implications of electronic democracy based on the issues raised in the textbook. (p. 425-426)

   _____

   _____

Essay Questions

1. Compare and contrast pluralist and elite theories of democracy. (p. 407-411)

2. Consider the 2000 and 2004 presidential elections. Compare and contrast power resource theory and state-centered theory in relation to these events. (p. 413-420)

3. Explain the necessary conditions for a country to become fully democratic based on what you learned in Chapter 14. (p. 421-427)

4. Explain power resource theory and describe the differences between *liberal* and *conservative* labels in American politics. How does power resource theory aid our understanding of the success or failure of different parties and policies in different times and places? (p. 413-416)

5. From the rise of the modern state in the 17th century until World War II, states increasingly monopolized the means of coercion in society. Explain three important consequences that this had in terms of the replacement of interstate war with societal war since World War II. (p. 431-432)

Solutions

Practice Tests

Fill-In-The-Blank

1. political revolution
2. Legal-rational authority
3. Public opinion
4. democracy
5. is the private sphere of social life.
6. Lobbies
7. elite
8. liberal democracy
9. war
10. Terrorism

Multiple-Choice Questions

1. D, (p. 405)
2. C, (p. 405)
3. B, (p. 406)
4. A, (p. 406)
5. B, (408)
6. C, (p. 406)
7. D, (p. 409)
8. C, (p. 413-414)
9. D, (p. 418)
10. D, (p. 418-420)
11. B, (p. 421-422)
12. C, (p. 423)
13. E, (p. 424-425)
14. D, (p. 426-427)
15. B, (p. 431-432)
16. A, (p. 428)
17. C, (p. 428)
18. D, (p. 429-431)
19. C, (p. 429)
20. E, (p. 406)
21. B, (p. 407)
22. E, (p. 410)
23. B, (p. 412)
24. C, (p. 412)
25. D, (p. 414)
26. A, (p. 418)
27. A, (p. 408-409)

True False

1. F, (p. 403-405)
2. T, (p. 406)
3. T, (p. 408)
4. F, (p. 407-411)
5. T, (p. 409)
6. T, (p. 413-416)
7. F, (p. 423-424)
8. T, (p. 428)
9. F, (p. 428)
10. T, (p. 407-411)

CHAPTER **15**

FAMILIES

Student Learning Objectives

After reading Chapter 15, you should be able to:

1. Identify and describe prevailing forms of families in the United States.

2. Discuss the historical trends of the traditional nuclear family in the United States since the 1940s.

3. Compare and contrast functionalist, conflict and feminist perspectives on families.

4. Identify the main functions of marriage and the nuclear family.

5. Describe the transformation of the family from foraging societies to the post World War II era in the United States.

6. Describe the trends of the nuclear family since the second half of the 19th century in terms of divorce, marriage, and fertility rates.

7. Discuss the relationship between love and marriage in a historical perspective.

8. Identify and explain the forces underlying marital satisfaction.

9. Explain the economic and emotional effects of divorce on women and children.

10. Discuss the gender, political and ethical issues surrounding reproductive choice and reproductive technologies.

11. Describe the prevailing patterns of housework, childcare, and wife abuse in the context of gender inequality.

12. Discuss the prevailing patterns and controversies of family diversity in terms of cohabitation, same-sex marriage, single-motherhood and race.

13. Describe the patterns of alternative family forms in terms of adaptations to poverty by racial minority families.

14. Compare and contrast family policy in the United States and Sweden.

Key Terms

nuclear family (436)

traditional nuclear family (436)

polygamy (438)

extended family (438)

marriage (438)

divorce rate (441)

marriage rate (442)

total fertility rate (442)

Detailed Chapter Outline

I. INTRODUCTION

   A. Our most intense emotional experiences are bound up with our families.
      1. Seemingly trivial issues hold deep meaning and significance for family members.
      2. Family issues are central to political debate in the United States.

II. IS THE FAMILY IN DECLINE?

   A. As a subject of sociological inquiry, family issues generate much debate, which centers on the question, "Is the family in decline, and if so, what should be done about it?"
      1. Whenever the family changes rapidly, and when the divorce rate increases, this crisis alarm is sounded.

   B. Two forms of families when people speak of "the decline of the family."
      1. The **nuclear family** is composed of a cohabiting man and woman who maintain a socially approved sexual relationship and have at least one child.
      2. The **traditional nuclear family** is a nuclear family in which the wife works in the home without pay while the husband works outside the home for money, making him the "primary provider and ultimate authority."

   C. Trends of the traditional nuclear family.
      1. In the 1940s and 1950s, much of the American public and many sociologists considered the traditional nuclear family the most widespread and ideal family form.
      2. Between 1960 and 2000, the percentage of married-couple families with children living at home fell from about 44% to 24%.
      3. Between 1960 and 2000, the percentage of women over the age of 16 in the paid labor force increased from approximately 38% to 60%.

   D. Functionalist sociologists view the decreasing prevalence of the married-couple family and the rise of the "working mother" as sources of various social problems (i.e. crime rates, illegal drug use, poverty, and welfare dependency)—and call for legal and cultural reforms to shore up the traditional nuclear family.

   E. Conflict and feminist sociologists disagree with functionalists. They argue that:
      1. It is inaccurate to talk about *the* family as this assumes one single form.
      2. Changing family forms do not necessarily represent deterioration, but often represent improvement in the way people live.

3. Various economic and political reforms, such as an affordable nationwide day-care system, would eliminate most of the negative effects of single-parent households.

III. FUNCTIONALISM AND THE NUCLEAR IDEAL

A. Functional Theory
1. Since the 1940s, functionalists have argued that the nuclear family is ideally suited to meet the challenges of reproducing and raising offspring in an emotionally supportive environment so that they may operate as productive adults.
2. Other family forms.
   a. **Polygamy** expands the nuclear unit "horizontally" by adding one or more spouses (almost always wives) to the household. The overwhelming majority of families are *monogamous*.
   b. The **extended family** expands the nuclear family "vertically" by adding another generation (one or more of the spouse's parents) to the household. This family form was common throughout the world, and still is in some places.
3. **Marriage** is a socially approved, presumably long-term, sexual and economic union between a man and a woman, and involves rights and obligations.

B. Functions of the Nuclear Family
1. Five main functions of marriage and the nuclear family:
   a. *Sexual regulation*;
   b. *Economic cooperation*;
   c. *Reproduction*;
   d. *Socialization*; and
   e. *Emotional support*.

C. Foraging Societies
1. Although a gender division of labor exists among foragers (most men hunt and most women gather), men have few if any privileges over women, and relative gender equality is based on women's contribution of up to 80% of food. Also men tend to babies and children.
2. It is the mobile band, as opposed to the nuclear family, that enables social organization and provides subsistence.
3. Children are considered an investment in the future, however too many children are a liability.
4. Life is highly cooperative.
   a. Men and women care for each other's children.
   b. Socialization is a public matter.
5. Research on foraging societies calls into question many of the functionalist generalizations about the family.

D. The American Middle Class in the 1950s
1. After World War II, functionalists recognized the decline in importance of the productive functions and greater importance of socialization and emotional functions of the family.
2. In the typical urban or suburban nuclear family, for the most part, the husband was the breadwinner, and strong normative pressures reinforced women to marry, have babies, and stay home to raise them.

3. Functionalism had its merits in the 15 years after World War II, a time when Americans wanted to enjoy the security of family life in the wake of the Great Depression.
   a. Conditions were ideal for these family goals.
      i. Unparalleled optimism and prosperity.
      ii. Real per capita income rose.
      iii. Home-ownership jumped.
      iv. Government assistance supported and promoted the nuclear family.
   b. Thus people married younger, had more children, got divorced less, —and middle-class women engaged in an "orgy of domesticity."
4. However, the immediate postwar period was in many respects atypical as Andrew J. Cherlin shows.
5. Divorce, marriage, and childbearing trends show a gradual weakening of the nuclear family from the second half of the 19th century until the mid-1940s, and after the 1950s. Throughout the 19th century:
   a. The divorce rate rose.
      i. The **divorce rate** is the number of divorces that occur in a year for every 1000 people in the population.
   b. The marriage rate fell.
      i. The **marriage rate** is the number of marriages that occur in a year for every 1000 people in the population.
   c. The total fertility rate fell.
      i. The **total fertility rate** is the average number of children that would be born to a woman over her lifetime if she had the same number of children as women in each age cohort in a given year.

IV. CONFLICT AND FEMINIST THEORIES

A. Postwar families did not always operate like the smooth functioning, happy, white, middle-class, mother-householder, and father-breadwinner household portrayed in 1950s TV classics.

B. The nuclear family was often a site of frustration and conflict.
   1. Only about one-third of working-class, and two thirds of middle-class couples were happily married.
   2. Wives were less satisfied than husbands, reporting high rates of depression, distress, and feelings of inadequacy.
   3. Dissatisfaction among women workers during World War II, who were fired and downgraded after the war.

C. Poor families could not celebrate the traditional family unit.
   1. Some 40% of African-American women with small children had to work outside of the home in the 1950s, usually as domestics in upper-middle class and upper class white households.

D. Karl Marx and Friedrick Engels argued that the control of wives sexually and economically exists in the context of the emergence of the traditional nuclear family along side inequalities of wealth.
   1. Sexual control ensured that a man's property would be transmitted only to his offspring, particularly his sons.

E. Although Engels was correct to highlight the long history of male economic and sexual domination in the traditional nuclear family, he was incorrect in terms of other assumptions.
    1. He was wrong to think communism would eliminate gender inequality in the family.
    2. Gender inequality exists in noncapitalist and precapitalist societies.

F. Feminists
    1. Feminists view *patriarchy* (male dominance and norms justifying that dominance) as more deeply rooted in the economic, military, and cultural history of humankind than classical Marxists.
    2. Only a "genuine gender revolution" can alter gender inequality.

V. POWER AND FAMILIES

    A. Love and Mate Selection
        1. Although most Americans take for granted that marriage ought to be based on love, love has little to do with marriage in most societies throughout human history.
            a. Marriage:
                i.   Was typically arranged by a third party.
                ii. Was based mainly on maximizing prestige, economic benefits, and political advantages for families from which the bride and groom came from.
        2. The origins of love and marriage:
            a. The idea of love in the choice of marriage first gained currency in 18th-century England with the rise of liberalism and individualism.
            b. The intimate linkage between love and marriage that we are familiar with emerged in the early 20th century, when Hollywood and the advertising industry began to promote self-gratification and heterosexual romance on a grand scale.
        3. Social Influences on Mate Selection
            a. Love alone still does not determine mate selection. Three sets of social forces that influence falling in love and mate selection:
                i.   Marriage resources
                ii. Third parties
                iii. Demographic and compositional factors

    B. Marital Satisfaction
        1. As women became autonomous and free, marital stability came to depend more on marital satisfaction, than its mere usefulness.
        2. Factors that contributed to women's autonomy.
            a. The birth control pill allowed women to delay childbirth and to have fewer children.
            b. The entry of millions of women into higher education and the paid labor force made it easier for women to leave unhappy marriages. In the 1960s divorce laws were changed to make divorce easier and divide property more equitably.
        3. The Social Roots of Marital Satisfaction
            a. *Economic forces*;
            b. *Divorce laws*;
            c. *The family cycle*

      d. *Housework and child care*; *and*

      e. *Sex*

  4. Religion has little effect on the level of marital satisfaction, but influences the divorce rate.

C. Divorce

  1. Economic Effects

      a. After divorce, the most common pattern is a rise in the husband's income and a decline in the wife's, because husbands usually earn more, and children typically live with their mother, and support payments are often inadequate if paid at all. Consequently, while child support payments declined in the 1970s and 1980s, they rose in the 1990s.

  2. Emotional Effects

      a. Children of divorced parents:

          i. Tend to develop behavioral problems and do less well in school.

          ii. Are more likely to engage in delinquent acts and abuse drugs and alcohol.

          iii. Often experience an emotional crisis, particularly in the first 2 years after divorce.

      b. Adults of divorced parents are:

          i. Less likely to be happy than people of nondivorced parents.

          ii. More likely to suffer health problems.

          iii. More likely to depend on welfare.

          iv. More likely to earn low incomes.

          v. More likely to experience divorce themselves.

      c. Much of this research is based on families who seek psychological counseling.

  3. Factors Affecting the Well-Being of Children

      a. Researchers who utilize representative samples and control for the separate effects of many factors on children's well being, show otherwise. Amato and Keith showed that the overall effect of divorce on children's well being is not strong and declining over time. Three factors account for much of the distress among children of divorce.

          i. *A high level of parental conflict.*

          ii. *A decline in living standards.*

          iii. *The absence of a parent.*

D. Reproductive Choice

  1. As a result of the gender revolution, women now have more voice over having children, when they will have them, and how many they will have.

  2. Most women want to work in the paid labor force, many to pursue a career.

  3. Women's reproductive decisions are carried out by means of contraception and abortion.

  4. Americans have been sharply divided on the abortion issue since the 1970s.

      a. "Right-to-life" activists:

          i. Want to repeal laws legalizing abortion.

          ii. Tend to be homemakers in religious, middle-income families.

          iii. Argue that life begins at conception, thus abortion destroys human life.

          iv. Advocate adoption over abortion.

      b. "Pro-choice" activists:

        i.   Want laws legalizing abortion preserved.

       ii. Tend be women in pursuit of careers.

     iii. Are more highly educated, less religious, and better off financially.

     iv. Argue that every woman has a right to make choices about her body.

  c. Randall Collins and Scott Coltrane note, it seems likely that a repeal of abortion laws would return us to conditions of the 1960s. Abortions would be expensive, hard to obtain, pose more dangers to women's health, and poor women and their unwanted children would suffer most.

E. Reproductive Technologies

  1. Reproductive technologies facilitate pregnancy, as well as prevent pregnancy.

  2. Demand is strong for techniques to help infertile couples, some lesbian couples, and some single women have babies.

  3. Four main reproductive technologies:

    a. *Artificial insemination*;

    b. *Surrogate motherhood*;

    c. *In vitro fertilization*; and

    d. Various *screening techniques*

  4. Social, ethical, and legal issues:

    a. Discrimination

    b. Social and legal problems of rendering the terms "mother" and "father" obsolete.

F. Housework and Child Care

  1. Despite the gender revolution, the domain of housework, childcare, and senior care remains resistant to change.

  2. Arlie Hochschild showed that even women who work full-time in the paid labor force begin a "second shift" when they return home.

  3. Although some change has occurred as men have taken a more active role in household work, the change is modest.

  4. Men tend to do low-stress chores that have less pressing deadlines (i.e. mowing the lawn, repairing the car, and preparing income tax forms). Women tend to do higher stressed tasks that cannot wait (i.e. dressing children, preparing dinner, and washing clothes).

  5. Two main factors that shrink the gender gap in housework:

    a. The smaller the difference in income between husband and wife, the more equal the division of household labor.

    b. The more husband and wife agree there should be equality in the household , the more equality there is (attitude), which is linked to both spouses' having a college education.

G. Domestic Violence

  1. 1997 Gallup poll found that 22% of women, compared with 8% of men, reported physical abuse by a spouse or companion at least once in the past.

  2. There are three main types of domestic violence.

    a. *Common couple violence* occurs when partners have a specific argument and one partner lashes out physically at the other.

    b. *Intimate terrorism* is part of a general desire of one partner to control the other.

    c. *Violent resistance* is the defending of oneself against an individual who has engaged in intimate terrorism.

3. Severe forms of wife assault occur in all categories of the population, but is most common in lower class, less highly educated families, where men are more likely to believe that male domination is justified, and among couples who witnessed the abuse of their mothers as children.

4. Straus showed that high levels of wife assault are associated with gender inequality in the larger society. As gender equality increases, wife abuse declines.

5. For heterosexual couples, the incidence of domestic violence is highest:

    a. where a big power imbalance between men and women exist

    b. where norms justify the male domination of women

    c. where early socialization experiences predispose men to behave aggressively toward women.

VI. FAMILY DIVERSITY

A. Cohabitation

1. In the last three decades, the number of American couples who are unmarried and cohabiting increased more than 500%. Once considered a disgrace, cohabitation has gone mainstream.

2. Disapproval of cohabitation has decreased due to cultural and economic factors. In recent decades, the force of religious sanction has weakened, hence affecting opposition to premarital sex. The sexual revolution and growing individualism have allowed people to pursue intimate relationships outside of marriage. Because women have pursued higher education and entered the paid labor force in increasing numbers, their gender roles are not tied so closely to marriage as they once were.

3. Cohabitation is a relatively unstable relationship. Marriages that begin with cohabitation are associated with a higher divorce rate than marriages that begin without cohabitation.

4. Cohabitation and Marital Stability

    a. The most often cited and best supported explanation for the association between cohabitation and marital instability is that people who cohabit before marriage differ from those who do not, and these differences increase the likelihood of divorce. People who cohabit tend to be less religious and religious people are less likely to divorce. Those who cohabit are more likely to be African American, occupy a lower class position, hold more liberal political and sexual views, and have parents who divorced.

5. Sociological Significance of Cohabitation

    a. Sociologists have debated the meaning and significance of cohabitation for two decades. Some view it as a prelude to marriage or a new form of marriage. Others think cohabitation is more like being single.

B. Same-Sex Unions and Partnerships

1. In February 2004, the mayor of San Francisco ordered his county clerk to begin issuing marriage licenses to gay and lesbian couples.

    a. Although California laws do not allow such marriages, he argued that that these laws are discriminatory and contradict protections laid out in the state constitution.

b. Thousands of marriages took place in San Francisco, and, in a massive show of civil disobedience, spread across the country.

c. The events in California followed closely on the heels of a Massachusetts Supreme Court decision declaring the ban on gays and lesbians marrying in that state to be unconstitutional.

2. Opponents of same-sex marriage quickly began legal efforts to block issuance of marriage licenses in California and to prevent same-sex marriages in their own states.

a. Amid heavy lobbying from religious conservatives, President Bush declared support for an amendment to the federal constitution that would prevent same-sex marriages.

3. In 2001 the Netherlands became the first county in the world to legalize same-sex marriage. By 2003 Belgium had followed suit, and by 2005, Spain and Canada legalized same-sex marriages.

4. Eight other countries allow homosexuals to register their partnerships under the law in so-called *civil unions*. Civil unions recognize the partnerships as having some or all of the legal rights of marriage. These countries include Denmark, Greenland, Hungary, Norway, Sweden, France, Iceland, and Germany. In the United States, there is more opposition to registered partnerships and same-sex marriages than in these other countries.

a. By 2005, 40 states had passed laws opposing gay marriage or banning civil unions.

b. A Harris poll in 2004 showed that 53% of Americans oppose same-sex marriages and 41% oppose civil unions for same-sex couples.

5. Nevertheless, the direction of change is clear. The legal and social definition of "family" is being broadened to include cohabiting, same-sex partners in long-term relationships.

6. Research shows that most homosexuals want a long-term, intimate relationship with one other adult, like most heterosexuals. More than 2 million same-sex partners cohabit in the United States today, where there is more opposition to same-sex marriages.

7. Raising Children in Homosexual Families

a. Most people believe that children raised in homosexual families will develop a confused sexual identity, exhibit tendencies toward homosexuality themselves, and suffer discrimination.

b. There is little research in this area.

c. Much of the research is based on small, unrepresentative samples.

d. Nevertheless, the research findings consistently suggest that children from homosexual families are much like children from heterosexual families.

8. Differences between homosexual and heterosexual families:

a. Lesbian couples with children report higher satisfaction with their partnerships than lesbian couples without children; —while childless heterosexual couples report higher marital satisfaction.

b. The partners of lesbian mothers spend more time than husbands of heterosexual mothers caring for children.

c. Homosexual couples tend to be more egalitarian.

C. Single-Parent Families: Racial and Ethnic Differences

1. Single-Mother Families

    a.   Whites have the lowest incidence of single-mother families. African-Americans have the highest. Although the last few years of the 20th century witnessed a reversal in the trend toward more single-mother families in the African-American community.

    b.   Single-mother families outnumber two-parent families.

    c.   Some sources of single-parent families:
        i.   Separation, divorce, or death.
        ii.   People not getting married in the first place.

    d.   The Decline of the Two-Parent Family among African Americans
        i.   Some scholars trace the decline of the African American family back to slavery.
        ii.   Since about 1925, proportionately fewer black men have been able to help support a family, as a result of discrimination, unemployment, and the decline of manufacturing industries in the Northeast and the movement of blue-collar jobs to the suburbs in the 1970s and 1980s.
        iii.   The declining ratio of eligible black men to black women due to the disadvantaged economic and social position of the African American community, and the disproportionate number of black men imprisoned, murdered, and addicted to drugs; the disproportionately large number of black men in the armed forces; and the fact that a black man is nearly twice as likely as a black woman to marry a nonblack, and intermarriage has increased to nearly 10% of all marriages involving at least one black person.
        iv.   The relative earnings of black women and men. In recent decades, the average income of African-American women has increased, while the earning power of African-American men has fallen.

2.   Adaptations to Poverty
    a.   Poor African-American women have adapted to harsh economic realities by developing strong kinship and friendship networks.
        i.   Networks help with childcare, money, and hand-me-downs.
        ii.   However, the drawback is that if an individual experiences economic mobility out of poverty, the network is likely to be used by the network, leaving the individual still in poverty.
    b.   Similarly, Hispanics developed a family system in which the godfather (*padrino*) and godmother (*madrina*) act as coparents, providing childcare and emotional and financial support as needed.
    c.   The tendency to rely on the extended kin network declines with migration status and upward mobility.

VII. FAMILY POLICY

A.   Is the decline of the nuclear family a bad thing for society?
    1.   As suggested by the research, the answer is yes and no.

B.   Crossnational Differences: The United States and Sweden
    1.   The United States is a good example of how social problems can emerge from nuclear family decline. Sweden is a good example of how such problems can be averted.

2. *Sweden leads the United States on most indicators of nuclear family decline.*
   a. In Sweden:
      i. There is a smaller percentage of people.
      ii. People marry at a later age.
      iii. The proportion of births outside of marriage is twice as high.
      iv. A much larger proportion of Swedish women with children under age three work in the paid labor force.

3. *Sweden leads the United States on most measures of children's well being.*
   a. In Sweden:
      i. Children enjoy higher average reading test scores.
      ii. The poverty rate is only one twelfth as high.
      iii. The rate of infant abuse is one-eleventh the U.S. rate.
      iv. The rate of juvenile drug offenses is less than half as high.
      v. There is a slightly higher rate of juvenile delinquency for minor offenses.

4. Sweden has a substantial family support policy, which the United States does not. In Sweden:
   a. Parental leave.
      i. When a child is born, a parent is entitled to 360 days of parental leave at 80% of his or her salary, and an additional 90 days at a flat rate.
      ii. Fathers can take an additional 10 days of leave with pay when the baby is born.
   b. Health *care.*
      i. Parents are entitled to free "well baby clinics."
      ii. All Swedish citizens receive free health care from the state-run system.
   c. *Childcare.*
      i. Temporary parental benefits are available for parents with a sick child under the age of 12.
      ii. One parent can take up to 60 days off per sick child per year at 80% salary.
      iii. Government subsidized high-quality day care is available for all parents.
      iv. Direct cash payments based on the number of children in each family.
   d. Family Support Policies in the United States
      i. Since 1993, a parent is entitled to 12 weeks of unpaid parental leave.
      ii. About 44 million citizens have no health care coverage. Millions more experience a low standard of health care.
      iii. No system of state day care and no direct cash payments based on the number children.
      iv. The value of the dependent deduction on income tax has fallen by nearly 50% since the 1940s.

5. Three criticisms commonly raised in the United States against generous family support policies.
   a. Some say it encourages long-term dependence on welfare, illegitimate births, and the breakup of two-parent families. Research shows that:
      i. The divorce rate and the rate of births to unmarried mothers are not higher when welfare payments are higher.
      ii. Welfare dependency is not widespread in general, even among African-American teen mothers, who are thought of a particularly susceptible.

    b. Some say nonfamily childcare is bad for children under age 3. Research shows that:

       i. When comparing family care with high quality day care (as opposed to the comparison of upper-middle class family care with child care in existing facilities in the United States), studies find no negative consequences for children over age 1.

      ii. Negative effects of day care usually disappear by the time a child reaches age 5.

    iii. Day care has some benefits, like making friends, and is generally beneficial to low-income families.

    c. They are expensive and have to be paid by high taxes.

       i. This is true. It is a trade-off. Swedes have made the political decisions to pay higher taxes partly to avoid the social problems and costs that emerge when the traditional nuclear family is replaced with other family forms and no institutional support.

Study Activity: Applying The Sociological Compass

After reading Chapter 15, review the abortion debate in the section, "Reproductive Choice". Make a list of the "right-to-life" and "pro-choice" positions and social characteristics as described in the text. Consider the debate from the functionalist and feminist perspectives. Make a list of possible arguments for or against abortion from each perspective. Which perspective makes most sense to you, with respect to your own personal views, and why? As Brym and Lie ask, "Do you think your opinions are influenced by your social characteristics"? In what ways do your social characteristics influence your preference for either the functionalist or feminist perspective?

Right-To-Life

Positions and Arguments:

1.

2.

3.

Social Characteristics:

1.

2.

3.

Pro-Choice

Positions and Arguments:

1.

2.

3.

Social Characteristics:

1.

2.

3.

Functionalist Arguments

1.

2.

3.

Feminist Arguments

1.

2.

3.

Infotrac College Edition Online Exercises

For the following exercises, log on to the online library of InfoTrac College Edition at http://www.infotrac-college.com/. Make note that InfoTrac has implemented a new registration system that will allow easier access to InfoTrac through the use of a personalized username and password. Once you've created your username and password you may proceed directly to the Log On page. To create an account, register your passcode packaged with your textbook, and create a username and password, by following the online prompt. After you are logged in, click on "Infotrac College Edition." You will arrive at a screen that enables you to search topics.

Keyword: **Extended family**. Choose an article that discusses the prevalence of extended families in contemporary society. Make a list of reasons why the extended family still exists.

Keyword: **Family policy**. Search for the key words "family policy". Identify two specific policies from at least two articles, and explain its advantages and disadvantages based on what you learned in Chapter 15.

Keyword: **divorce**. What are recent divorce trends in the United States based on any relevant article(s)? Are the trends described consistent with the information in Chapter 15?

Keyword: **Stepfamily, blended family, reconstituted family**. Look up each of these key words separately. Do these terms really mean the same thing? Assess the similarities and differences between these three terms. What conclusions can you draw?

Keyword: **Daddy stress, second shift**. Look up each of these keywords separately. Many people assume these two terms are equivalent for men and women. Investigate daddy stress and distinguish it from the second shift. What are your thoughts? Support your answer with research findings from the articles.

Keyword: **Polygamous families**. Learn about family life in polygamous families. See the article on Bedouin - Arab families in The Journal of Social Psychology. Compare your life with life in polygamous families. How is it similar? Different?

Keyword: **Cohabitation**. Assess the legal consequences of cohabitation. Do you think that cohabitating couples should be recognized the same under the law as legally married couples? Do you think that same-sex couples should be recognized the same under the law as legally married couples? Write a short opinion paper stating your answers.

Internet Exercises

Go to: http://www.trinity.edu/mkearl/family.html.

There is a lot of information about marriage and family processes. Read over the different definitions of the family that are presented. How do these definitions compare with the definition of family found in Chapter 15? Come up with your own definition of the family. Compare and contrast all of the definitions. Are there similar patterns?

There are many ways to define the family. Look at the way the U.S. government defined family during the 2000 Census. Go to: http://www.census.gov/ and click on 'search' and then type in 'family'.

This will give you the top 100 documents. Look at two or three different documents and see how the government defines the family.

What are the consequences for families on welfare when they leave welfare? To learn more about these success stories, go to: http://newfederalism.urban.org/html/discussion99-02.html

Where do individuals go for advice when they are thinking of divorce? To learn more about issues of concern, go to: http://www.divorce-online.com/.

What advice would you give to one of your friends who is thinking of divorce?

Everyone wants to know what the keys to successful relationships are. Make your own list of what you think are the keys to successful relationship. Then go to:

http://www.smartmarriage.com and compare your list with those listed here.

Practice Tests

Fill-In-The-Blank

Fill in the blank with the appropriate term from the above list of "key terms."

1. A (An) _____ is a nuclear family in which the husband works outside the home for money and the wife works for free in the home.

2. _____ expands the nuclear family horizontally by adding one or more spouses to the household.

3. The _____ is the number of marriages that occur in a year for every 1,000 people in the population.

4. The _____ expands the nuclear family vertically by adding another generation to the household.

5. The _____ is the number of divorces that occur in a year for every 1,000 people in the population.

Multiple-Choice Questions

Select the response that best answers the question or completes the statement:

1. A family unit made up of parents, children and grandparents living in one household is:
   a. a nuclear family.
   b. a traditional nuclear family.
   c. an extended family.
   d. a blended family.
   e. None of the above.

2. Between 1960 and 2000, the percentage of married-couples families:
   a. fell.
   b. increased.
   c. doubled.
   d. remained relatively constant.
   e. None of the above.

3. Which form of the family was considered the most widespread and ideal in the 1940s and 1950s?
   a. the extended family
   b. the blended family
   c. the traditional nuclear family
   d. All of these are correct.
   e. None of the above.

4. Which is NOT a main function of the family?
   a. egalitarian division of labor
   b. sexual regulation
   c. economic cooperation
   d. reproduction
   e. socialization

5. In foraging societies, relative gender equality is based on:
   a. the recognition and value of child care, in which women assume most responsibility.
   b. men's ability to hunt, while women prepare the food.
   c. the communal distribution of food beyond the needs of individual nuclear families.
   d. women's production of up to 80% of the food.
   e. Shared duties between men and women involving production of necessary goods.

6. The extended family expands the nuclear family _____ by adding another generation to the household.
   a. horizontally
   b. vertically
   c. laterally
   d. All of these are correct.
   e. None of the above.

7. What trends indicate a gradual weakening of the nuclear family from the second half of the 19ᵗʰ century until the mid-1940s, according to Andrew J. Cherlin?
   a. divorce rate
   b. marriage rate
   c. total fertility rate
   d. All of these are correct.
   e. None of the above.

8. Which is true of love and marriage?
   a. The idea of love in the choice of marriage originated in the early 20ᵗʰ century, when Hollywood began promoting self-gratification and heterosexual romance.
   b. Love was always a prerequisite for marriage in most societies.
   c. In industrial societies, love alone determines mate selection.
   d. All of these are correct.
   e. None of the above.

9. Which is NOT one of the forces underlying marital dissatisfaction as highlighted in the textbook?
   a. the entry of women into the paid labor force
   b. economic factors
   c. divorce laws
   d. housework and childcare
   e. sex

10. Children of divorced parents:
    a. tend to develop behavioral problems and do less well in school.
    b. are more likely to engage in delinquent acts.
    c. often experience an emotional crisis, especially in the first 2 years following divorce.
    d. All of these are correct.
    e. None of the above.

11. According to Amato and Keith, the overall effect of divorce on children's well being is not strong and is declining over time. Which factor does NOT account for the distress among children of divorce, according to Amato and Keith?
    a. a high level of parental conflict
    b. a decline in living standards
    c. the absence of a parent
    d. the divorce itself
    e. All of the answers are factors according to Amato and Keith.

12. "Right-to-life" advocates argue that:
    a. abortion is morally indefensible.
    b. abortion is only an option.
    c. all killing is morally indefensible.
    d. life begins at birth.
    e. women have the right to live

13. The sociological and ethical issues surrounding reproductive technologies include:
    a. discrimination
    b. legal issues.
    c. social problems rendering the terms "mother" and "father" obsolete.
    d. All of these are correct.
    e. None of the above.

14. Arlie Hochschild showed that:
    a. as women's participation in the paid labor force increased, equality in the home became more common.
    b. even though income difference remains between men and women, the division of household labor becomes more equal as long as the wife strongly agrees there should be equality in the household.
    c. Even women who work full-time in the paid labor force remain primarily responsible for taking care of the household and children.
    d. All of these are correct.
    e. None of the above.

15. Severe wife abuse:
    a. is common among couples who as children witnessed their mothers being abused.
    b. is most common in lower classes, less highly education families.
    c. is most common where men are more likely to believe that male domination is justified.
    d. All of these are correct.
    e. None of the above.

16. Which country became the first to legalize same-sex marriages?
    a. Spain
    b. Canada
    c. Thailand
    d. California
    e. Netherlands

17. Which is NOT true of homosexuality based on the research cited in the textbook?
    a. Most homosexuals a want long-term, intimate relationship with one other adult.
    b. Children raised in homosexual families develop confused sexual identities and exhibit homosexual tendencies themselves.
    c. Partners of lesbian mothers spend more time than husbands of heterosexual mothers caring for children.
    d. Lesbian couples with children report higher satisfaction with their partners than lesbian couples without children.
    e. All of the answers are true.

18. Which is an adaptation to poverty by poor African-American women?
    a. the development of strong kinship and friendship networks
    b. economic mobility through interracial marriage
    c. economic mobility by marrying up in terms of social class
    d. welfare dependency
    e. None of the above.

19. From the 1940s to the 1950s in the United States, the percentage of women age 20-24 who never married:
    a. fell.
    b. increased.
    c. remained constant.
    d. All of these are correct.
    e. None of the above.

20. The movie My Big Fat Greek Wedding:
    a. celebrates ethnic intermarriage.
    b. reflects the opportunities and aspirations of only one segment of the population.
    c. says that despite obstacles, mixing cultural "apples" and "oranges" is possible.
    d. ignores two important issues regarding ethnic and racial intermarriage.
    e. All of these are correct.

21. There is rough gender equality among the !Kung-San, a foraging society in the Kalahari Desert in Botswana, which is partly due to the fact that:
    a. women play such a key economic role in providing food.
    b. women do not do much of the childcare.
    c. women do both the hunting and the gathering.
    d. women control the power and authority regarding communal "band" matters.
    e. women's ability to bear children creates a protected reverence for all females.

22. Which type of household has emerged since the 1980s?
    a. women living alone.
    b. men living alone.
    c. other types of families without children.
    d. All of these are correct.
    e. None of the above.

23. In the last three decades, the number of American couples who are unmarried and cohabiting increased:
    a. approximately 50%.
    b. between 100% and 150%.
    c. approximately 200%.
    d. more than 500%.
    e. None of the above.

True False

1. In foraging societies, men and women care for each other's children and socialization is a public matter.

    TRUE or FALSE

2. The TV portrayals of the family as a happy, smooth functioning, mother-householder, father-breadwinner in the 1950s were quite accurate depictions of postwar families.

    TRUE or FALSE

3. The entry of millions of women into higher education and the paid labor force made it easier for women to leave unhappy marriages.

    TRUE or FALSE

4. As a result of the gender revolution, the domain of housework, childcare, and senior care has changed dramatically.

    TRUE or FALSE

5. Following divorce, men often experience upward mobility, while women often experience a decline in their standard of living.

    TRUE or FALSE

6. High levels of wife assault are associated with gender inequality in the larger society.

    TRUE or FALSE

7. Spain and Canada legalized same-sex marriage in 2005.

    TRUE or FALSE

8. Two-parent families outnumber single-mother families.

    TRUE or FALSE

9. The tendency to rely on the extended kin network for social support declines with migration status and upward mobility.

    TRUE or FALSE

10. Cohabitation is a relatively unstable relationship.

    TRUE or FALSE

Short Answer

1. List and define the five main functions of the family. (p. 438-439)

    _____

    _____

2. List and define the four main reproductive technologies. Explain the two major sociological and ethical issues concerning reproductive technologies in general. (p. 452-454)

_____

_____

3. What does Arlie Hochschild mean by "the second shift"? (p. 454-455)

_____

_____

4. Briefly explain why those who cohabit before marriage are at greater risk for marital instability and divorce, than married couples who did not cohabit. (p. 457-458)

_____

_____

5. What are three main differences between homosexual and heterosexual families based on research cited in the textbook? (p. 459-460)

_____

_____

Essay Questions

1. Americans have been sharply divided on the abortion issue, especially since the 1970s. Provide an overview of the "right-to-life" versus the "pro-choice" arguments. What are the social characteristics of each group of activists? Explain the position of Randall Collins and Scott Coltrane on abortion. (p. 451-454)

2. Describe the sources and explain the reasons for the decline of the two-parent family among African Americans. (p. 460-462)

3. Sweden and the United States are described as opposite extremes on a continuum of family policy. Compare and contrast family policy in Sweden and the United States. What are the relative advantages and disadvantages of family support policies in each country? (p. 462-465)

4. Compare and contrast functionalist, Marxist, and feminist perspectives on the family. (p. 438–444)

5. Explain how love and mate selection works in the United States. Be sure to frame your discussion with an outline of the three social forces that influence whom you are likely to fall in love with and marry. (p. 444-450)

Solutions

Practice Tests

Fill-In-The-Blank

1. traditional nuclear family
2. Polygamy
3. marriage rate
4. extended family
5. divorce rate

Multiple-Choice Questions

1. C, (p. 438)
2. A, (p. 436)
3. C, (p. 436)
4. A, (p. 438-439)
5. D, (p. 440)
6. B, (p. 438)
7. D, (p. 441)
8. E, (p. 444-447)
9. A, (p. 447-449)
10. D, (p. 450-451)
11. D, (p. 450-451).
12. A, (p. 453)
13. D, (p. 452-454)
14. C, (p. 454)
15. D, (p. 456)
16. E, (p. 459)
17. B, (p. 459-460)
18. A, (p. 461)
19. A, (p. 442)
20. E, (p. 448)
21. A, (p. 439-441)
22. D, (p. 437)
23. D, (p. 457)

True False

1. T, (p. 439-441)
2. F, (p. 441-442)
3. T, (p. 447)
4. F, (p. 454-455)
5. T, (p. 450-451)
6. T, (p. 455-456)
7. T, (p. 458-459)
8. F, (p. 460-461)
9. T, (p. 462)
10. T, (p. 457-458)

RELIGION

Student Learning Objectives

After reading Chapter 16, you should be able to:

1. Compare and contrast Durkheimian, Marxist, and Weberian approaches to religion.

2. Identify and explain the general components of "religious" experiences with respect to collective conscience and social solidarity.

3. Explain the role of religion in justifying inequality and promoting conflict.

4. Explain the connection between the rise of Protestantism and capitalist development.

5. Describe the historical patterns of religious rise, decline and revival.

6. Identify and define the types of religious groups.

7. Discuss the major trends and indicators of religiosity, and the future of religion.

Key Terms

collective conscience (471)

profane (471)

sacred (471)

totems (471)

rituals (471)

civil religion (472)

secularization thesis (476)

fundamentalists (477)

revised secularization (481)

church (483)

ecclesia (484)

denominations (484)

sects (485)

cults (485)

routinization of charisma (487)

religiosity (493)

Detailed Chapter Outline

I. INTRODUCTION

    A. Psychologically, religion is a common human response and coping device to the fact that we must all die. It provides meaning and purpose.

    B. Religious response takes thousands of forms. Sociologists of religion study this variation.
        1. Why does one religion predominate here, another there?
        2. Why is religious belief more fervent at one time than another?
        3. Under what circumstances does religion act as a source of social stability and under what circumstances does it act as a source of social change?
        4. Are we becoming more or less religious?

    C. Although God is still very much alive in America, the scope of religious authority has declined in the United States and many other parts of the world. Consequently, other institutions have grown in importance as the scope of religious authority has declined.

II. CLASSICAL APPROACHES IN THE SOCIOLOGY OF RELIGION

    A. Durkheim: A Functionalist Approach
        1. Super Bowl Sunday is second only to Christmas as a religious holiday in the United States." Although there is no god of the Super Bowl, the Super Bowl meets Durkheim's definition of a religious experience.
        2. The Super Bowl generates a sense of "collective effervescence" and **collective conscience**, which is composed of common sentiments and values that people share as a result of living together.
        3. We distinguish everyday world of the **profane** from the religious transcendent world of the **sacred**. **Totems** are objects that symbolize the sacred.
            a. The game is a sacred event.
            b. The Super Bowl trophy and team logos would be considered totems.
        4. **Rituals** are public practices that connect us to the sacred.
            a. The football game itself is a public ritual.
        5. Rituals and religion reinforce *social solidarity*.

    B. Religion, Conflict Theory, and Feminist Theory
        1. Two criticisms of Durkheim's account of religion:
            a. It overemphasizes religion's role in maintaining social cohesion, when in reality religion often incites conflict.
            b. When religion does increase social cohesion, it often reinforces social inequality.
        2. Religion and Social Inequality
            a. Marx called religion "the opium of the people"—that is, how religion often tranquilizes the underprivileged into accepting their lot in life.
            b. We can draw evidence for Marx's interpretation from many time, places, and institutions. For example, all the major world religions have traditionally placed women in a subordinate position.
        3. Religion and Class Inequality

     a. Religion has traditionally supported gender inequality and class inequality.

     b. According to Robert Bellah, **civil religion** is a set of quasi-religious beliefs and practices that binds the population together and justifies our way of life.

     c. Example—The celebration of social equality and the American Dream diverts our attention from social inequalities in the United States.

  4. Religion and Social Conflict

     a. Discrimination in white churches and the separation of black and white churches in the South in the 1940s.

     b. Black churches and the Civil Rights movement in the 1950s and 1960s.

C. Weber and the Problem of Social Change: A Symbolic Interactionist Interpretation

  1. Weber stressed the way religion can contribute to social change; his book *The Protestant Ethic and the Spirit of Capitalism*.

     a. According to Weber, what prompted vigorous capitalist development in Catholic Europe and North America was a combination of:

        i. Favorable economic conditions, like Marx discussed.

        ii. The spread of moral values by the Protestant reformers of the 16th century and their followers.

     b. The *Protestant Ethic*

        i. Followers of John Calvin stressed the need to engage in intense worldly activity, to display industry, punctuality, and frugality in everyday life.

        ii. The idea that people could assure a state of grace by working diligently and living simply.

     c. Protestantism was constructed on the foundation of two relatively rational religions: Judaism and Catholicism.

  2. Two problems of Weber's argument:

     a. The correlation between the Protestant work ethic and the strength of capitalist development is weaker than Weber thought.

     b. Weber's followers have not always applied the Protestant ethnic thesis as carefully as Weber did (i.e. Confucianism and the economic growth of Taiwan, South Korea, Hong Kong, and Singapore).

III. THE RISE, DECLINE, AND PARTIAL REVIVAL OF RELIGION

A. Secularization

  1. In medieval and early modern Europe, Christianity replaced magic as the popular source of answers to mysterious, painful, and capricious events,—and became a powerful presence in religious affairs, music, art, architecture, literature, philosophy, marriage, education, morality, economic affairs, politics, etc.

  2. The Church was the center of both spiritual and worldly life. European countries proclaimed official state religions, and persecuted members of religious minorities.

  3. A few hundred years later, Weber's observations of the replacement of religious authority with forms of rationalism by the turn of the 20th century became known as the **secularization thesis**, which says that religious institutions, actions, and consciousness are on the decline worldwide.

B. Religious Revival
1. Many sociologists revised their judgments about secularization in the 1990s, because accumulated survey evidence showed that religion was not in a state of decay in many places including the United States.
2. Religious Fundamentalism in the United States
    a. The second reason many sociologists have modified their views about secularization is that an intensification of religious belief and practice has taken place among some people in recent decades. Fundamentalist religious organizations have rapidly increased their membership since the 1960s, especially among Protestants. **Fundamentalists** interpret their scriptures literally, seek to establish a direct, personal relationship with their higher being(s), are relatively intolerant of nonfundamentalists, and often support conservative social issues. For example, in 2000 only 23% of Americans who identified themselves as fundamentalist agreed with the statement that abortion is acceptable if "the women want it for any reason."
    b. Such social issues often spill over into political struggles, which is why religion in American politics is resurgent.
3. Religious Fundamentalism Worldwide
    a. Fundamentalism has spread throughout the world since the 1970s, and is typically driven by politics.
        i. Hindu nationalists in India.
        ii. Jewish fundamentalists in Israeli political life.
        iii. Muslim fundamentalism in the Middle East, Africa, and parts of Asia.
    b. Fundamentalism and Extremist Politics in the Arab World
        i. Personal Anecdote. Robert Brym at the Hebrew University of Jerusalem in 1972.
        ii. Religious fundamentalism often provides a convenient vehicle for framing political extremism, enhancing its appeal, legitimizing it, and providing a foundation of solidarity of political groups.
        iii. Many people fail to see the religious underpinnings of religious fundamentalism. They regard religious fundamentalism as an independent variable, and extremist politics as a dependent variable. In this view, some people happen to become religious fanatics, which commands them to kill their opponents – for example, George W. Bush's presidential address to Congress on September 20, 2001.
        iv. What President Bush ignored are the political sources of this fringe form of Islamic extremism. Fundamentalism, like other forms of religion, is powerfully influenced by the social context in which it emerges.

C. The Revised Secularization Thesis
1. According to the **revised secularization** thesis, in most countries, worldly institutions have broken off or "differentiated" from the institution of religion over time.
2. Education is an example of a worldly institution. Religious institutions used to run schools.

IV. THE STRUCTURE OF RELIGION IN THE UNITED STATES AND THE WORLD

A. Types of Religious Organization

1. Three types of religious groups:
    a. A **Church** is any bureaucratic religious organization that has accommodated itself to mainstream society and culture.
        i. Two main forms of churches:
        ii. (a) **Ecclesia** are state-supported churches
        iii. (b) **Denominations** are the various streams of belief and practice that some churches allow to coexist under their overarching authority.
    b. **Sects** often form by breaking away from churches due to disagreement about church doctrine.
        i. Sects are less integrated into society and less bureaucratized than churches.
        ii. Sects are often led by charismatic leaders, men and women claim to be inspired by supernatural power and whose followers believe them to be so inspired.
    c. **Cults** are small groups of people deeply committed to a religious vision that rejects mainstream culture and society.

V. WORLD RELIGIONS

A. There are five major world religions:
    1. *Judaism, Christianity, Islam, Hinduism, and Buddhism.*
    2. They are *major* in the sense that they have had a big impact on world history and, aside from Judaism, continue to have hundreds of millions of adherents. They are *world* religions in the sense that their adherents live in many countries.
    3. Some scholars consider Confucianism as a major world religion. However, Brym and Lie believe Confucianism is best seen as a worldview or a philosophy of life.
    4. Three ways in which the five major religions are similar:
        a. With the exception of Hinduism, charismatic leaders helped to turn them into world religions.
        b. With the exception of Hinduism, the world religions had egalitarian and emancipatory messages at their origins.
        c. Over time, the charismatic leadership of the world religions became routinized. The **routinization of charisma** is Weber's term for the transformation of divine enlightenment into a permanent feature of everyday life. It involves turning religious inspiration into a stable social institution with defined roles, such as interpreters of the divine message, teachers, dues-paying lay people, and so forth.
    5. Judaism
        a. Historical description and development.
        b. More than 5 million Jews live in the United States – about the same number as in Israel.
    6. Christianity
        a. Historical description and development.
        b. Christianity remains the dominant religion in the West. It can be found virtually everywhere in the world. It remains a heterogeneous religion.
    7. Islam
        a. Historical description and development.

b. Islam spread rapidly after Muhammad's death, replacing Christianity in much of the Middle East, Africa, and parts of southern Europe.

c. Only a few Islamic sects developed in the modern era.

8. Hinduism

a. Hinduism is the dominant religion of India.

b. Description of Hinduism.

c. There are wide regional and class variations in Hindu beliefs and practices; Hinduism as it is practiced bears the stamp of many other religions.

9. Buddhism

a. Historical description and development.

B. Bases of Formation of World Religions:

1. New world religions are founded by charismatic personalities in times of great trouble.

2. The founding of new religions is typically animated by the desire for freedom and equality, always in the afterlife, and often in this one.

3. The routinization of charisma typically makes religion less responsive to the needs of ordinary people, and it often supports injustices

4. New world religions could well emerge in the future.

VI. RELIGIOSITY

A. **Religiosity** refers to how important religion is to people.

1. Indicators of religiosity include:

a. Strength of belief;

b. Emotional attachment to religion;

c. Knowledge about a religion;

d. Frequency of performing rituals; and

e. Frequency of applying religious principles in daily life.

2. How often do people attend religious services?

a. Based on GSS data:

i. Older people attend religious services more frequently than younger people. Older people have more time and more need for religion and are closer to death.

ii. Frequent church attendance is more common among African Americans than whites, because the central political role the church historically played in helping African Americans combat oppression.

iii. Respondents whose mothers attended religious services frequently are more likely to do so themselves. Religiosity is a *learned* behavior.

VII. THE FUTURE OF RELIGION

A. Secularization is one of the two dominant trends influencing religion throughout the world.

1. Even as secularization absorbs many people, many others in the United States and throughout the world have been caught up by a religious revival of vast proportions.

2. The contradictory social processes of secularization and revival are likely to persist for some time, resulting in a world that is neither more religious nor more secular, but that is certainly more polarized.

Study Activity: Applying The Sociological Compass

After reading the chapter, review the historical development and social significance of the five major world religions. After reviewing the major world religions, identify the charismatic leaders of each religion (with the exception of Hinduism), take notes on the egalitarian principles of each religion, and make note of the point at which each religion became institutionalized through routinization. Finally, explain the function of each religion, and examples of inequality perpetuated by each religion, based on functionalist and conflict perspectives, respectively.

Charismatic Leaders

Egalitarian Principles and Value

Institutionalization and Routinization of Charisma

Functionalist Characteristics

Conflict Characteristics

Infotrac College Edition Online Exercises

For the following exercises, log on to the online library of InfoTrac College Edition at http://www.infotrac-college.com/. Make note that InfoTrac has implemented a new registration system that will allow easier access to InfoTrac through the use of a personalized username and password. Once you've created your username and password you may proceed directly to the Log On page. To create an account, register your passcode packaged with your textbook, and create a username and password, by following the online prompt. After you are logged in, click on "Infotrac College Edition." You will arrive at a screen that enables you to search topics.

Keyword: **religiosity**. Read at least two articles that reveal patterns of religiosity across at least two denominations under the same religious authority (i.e. Catholicism and Protestantism) or two religions (i.e. Christianity and Hinduism). What are the major findings? What are the differences and/or similarities in levels of religiosity?

Keyword: **liberation theology**. Determine the key issues according to one article.

Keyword: **prayer and school**. Search this term and see how the two ideas intersect. What are the major areas of focus in the literature on prayer and school? Do you think this debate will ever end? Write a short paper stating your findings.

Keyword: **Supreme Court and religion**. What is the emphasis of the Supreme Court cases dealing with religion and the state? Is there a specific pattern? Is so, how long has the pattern existed? Write a short paper stating your findings.

Internet Exercises

The Church of Jesus Christ of Latter-day Saints is one of the fastest growing global churches. Go to this Web site and see how this religion places an emphasis on education. Go to:

http://www.lds.org/

The official website of the American Atheists is located at: http://www.atheists.org/schoolhouse/faqs. prayer.html

They present much information about prayer in schools. Write a short paper about the interplay between the major social institutions of education and religion.

Search for "religion" using the search engine of the General Social Survey (GSS) found at: http://www.icpsr.umich.edu/GSS99/search.htm

Choose one research report and read it. List the major conclusions from the report you chose.

Practice Tests

Fill-In-The-Blank

Fill in the blank with the appropriate term from the above list of "key terms."

1. The _____ is composed of the common sentiments and values that people share as a result of living together.

2. A(An) _____ is a set of quasi-religious beliefs and practices that bind a population together and justify its way of life.

3. The _____ says that religious institutions, actions, and consciousness are on the decline worldwide.

4. _____ are state-supported churches.

5. _____ usually form by breaking away from churches due to disagreement about church doctrine.

6. _____ refers to how important religion is to people.

7. The _____ refers to the religious, transcendent world.

8. _____ are public practices designed to connect people to the sacred.

9. _____ interpret their scriptures literally, and seek to establish a direct, personal relationship with the higher being(s) they worship.

10. The _____ holds that worldly institutions break off from the institution of religion over time.

Multiple-Choice Questions

Select the response that best answers the question or completes the statement:

1. What is a collective conscience?
   a. Common sentiments and values that people share as a result of living together.
   b. A feeling of guilt that emerges in the presence of a collective.
   c. A transcendence of social influences in order to be closer to one's religion.
   d. The shaming of an individual by a social group in an attempt to sanction.
   e. None of the above.

2. _____ are public practices that connect us to the sacred.
   a. Routines
   b. Public patterns
   c. Rituals
   d. Secularization
   e. None of the above.

3. We distinguish the everyday world of the _____, from the religious transcendent world of the _____.
   a. sacred; profane
   b. profane; sacred
   c. profane; secular
   d. sacred; secular
   e. None of the above.

4. Which theorist equated religion with a drug based on the argument that religion often tranquilizes the underprivileged into accepting their lot in life?
   a. Emile Durkheim
   b. Max Weber
   c. Thomas Hobbes
   d. John Calvin
   e. Karl Marx

5. In Durkheim's terms, the Super Bowl trophy is an example of a _____.
   a. charismatic trophy
   b. secular icon
   c. votive object
   d. totem
   e. religious statuette

6. Which is a criticism of Durkheim's view of religion?
   a. Although Durkheim recognizes that religion increases social cohesion, he fails to show how rituals reinforce solidarity.
   b. Durkheim overemphasizes religions' role in maintaining social cohesion.
   c. Durkheim overemphasizes religious justification of oppression.
   d. All of these are correct.
   e. None of the above.

7. Weber made a historical connection between _____ and capitalism.
   a. the opium of the masses
   b. social solidarity
   c. secularization
   d. the Protestant ethic
   e. None of the above.

8. Protestantism was constructed on the foundation of two relatively rational religions according to Weber. Which two religions?
   a. Gnosticism and Catholicism
   b. Liberation Theology and Catholicism
   c. Judaism and Catholicism
   d. Islam and Catholicism
   e. Calvinism and Catholicism

9. Which type of religious organizations interprets their scriptures literally, and seeks to establish a direct personal relationship with their higher beings?
   a. orthodox religious organizations
   b. liberation organizations
   c. fundamentalists
   d. All of these are correct.
   e. None of the above.

10. Which is NOT an indicator of religiosity?
    a. strength of belief
    b. emotional attachment to religion
    c. knowledge about religion
    d. level of charisma
    e. frequency of performing rituals

11. Frequent church attendance is more common among _____ than _____.
    a. African Americans; whites.
    b. whites; African Americans.
    c. whites; racial minorities.
    d. All of these are correct.
    e. None of the above.

12. Which type of religious groups tends to be high on social integration, bureaucratization and longevity?
    a. sect
    b. church
    c. cult
    d. All of these are correct.
    e. None of the above.

13. What percentage of Americans who identified themselves as fundamentalist, agree with the statement that abortion is acceptable if "the woman wants it for any reason?"
    a. 72%
    b. 56%
    c. 44%
    d. 38%
    e. 23%

14. Which statement provides evidence that the American increase of fundamentalism is by no means unique?
    a. Fundamentalism has spread throughout the world since the 1970s.
    b. Fundamentalism is typically driven by politics.
    c. A revival of Muslim fundamentalism began in Iran in the 1970s.
    d. All of these are correct.
    e. None of the above.

15. Most of the five major religions are *major* in the sense that:
    a. they have had a big impact on world history.
    b. they are seen more as a worldview than a religion.
    c. they have had overwhelmingly positive effects on society.
    d. they control the lives of millions of individuals.
    e. all of these are correct.

16. Which of the five major religions is characterized by the lack of a charismatic leader who helped turn the religion into a world religion?
    a. Judaism
    b. Christianity
    c. Islam
    d. Hinduism
    e. Buddhism

17. Most of the five major religions are similar in three ways. Which is NOT among the ways in which they are similar?
    a. Charismatic leaders helped to turn them into world religions.
    b. Egalitarian and emancipatory messages are foundational beliefs.
    c. Charismatic leadership became routinized and institutionalized.
    d. The social context of the development of most of the world religions is marked by a lack of inequality and conflict.
    e. All of the answers are among the ways in which most of the major religions are similar.

18. Even in urban settings, strictly enforced rules concerning dress, diet, prayer, and intimate contact with outsiders can separate members from the larger society. This is an example of a (an):
    a. cult
    b. sect
    c. charismatic
    d. ecclesia
    e. fundamentalist

19. Which age cohort tends to display the highest attendance of religious services?
    a. 70+
    b. 50-59
    c. 40-49
    d. 30-39
    e. 18-29

20. In the 1950s and 1960s, black churches formed the breeding ground of:
    a. fundamentalism.
    b. secularization.
    c. religious revivalism.
    d. the Civil Rights movement.
    e. American Civil Liberties Union.

21. Some scholars argue that Confucianism in East Asia acted much like:
    a. Irish Catholicism in 1800s America.
    b. Spanish missionary expansion in the Philippines.
    c. Protestantism in 19th-century Europe.
    d. All of these are correct.
    e. None of the above.

22. Among the various groups that denounced the film and book, *Harry Potter and the Sorcerer's Stone* in 2001, conservative Protestants made which of the following claim(s)?
    a. The film and book glorifies witchcraft.
    b. The film and book made evil look innocent.
    c. The film and book subtly draws children into an unhealthy interest in a darker working that is occultic and dangerous to physical, psychological and spiritual life.
    d. All of these are correct.
    e. None of the above.

23. Which is an unintended, negative consequence of religious and ethnic profiling, especially in the wake of September 11, 2001?
    a. Profiling may divert scarce resources from other approaches that have a higher "hit rate" or have a higher chance of resulting in detection.
    b. Profiling may cause terrorist organizations to use more infiltrators who appear Muslim.
    c. Profiling can deter terrorist attacks.
    d. All of these are correct.
    e. None of the above.

24. Brym and Lie regard Confucianism as:
    a. one of the major world religions.
    b. a formal institutionalized religion.
    c. as a worldview or a philosophy of life.
    d. All of these are correct.
    e. None of the above.

25. Which religion is the most widespread in North and South America, Europe, and Oceania?
    a. Judaism
    b. Christianity
    c. Islam
    d. Hinduism
    e. Buddhism

True False

1. Durkheim would argue that the "psyche-up" ritual football players engage in to prepare for a game has a religious quality.

   TRUE or FALSE

2. Marx called religion the "opium of the people."

   TRUE or FALSE

3. Religiosity refers to one level of spirituality, despite the strength of belief in a particular religion.

   TRUE or FALSE

4. Based on the General Social Survey, younger people now attend religious services more frequently than older people, because of the recognition of the increased risks they endure in this day and age.

   TRUE or FALSE

5. Religiosity is a learned behavior.

   TRUE or FALSE

6. In light of the fact that a religious revival is taking place, secularization is likely to decline in the future according the Brym and Lie.

   TRUE or FALSE

7. Religion in American politics is resurgent in part because social issues regarding religion often spill over into political struggles.

   TRUE or FALSE

8. Religious fundamentalism has become a worldwide political phenomenon.

   TRUE or FALSE

9. From a Durkheimian point of view, Super Bowl Sunday can be considered an example of the routinization of charisma.

   TRUE or FALSE

Short Answer

1. What is the "secularization thesis"? Provide one historical example of it. (p. 475-476)

   _____

   _____

2. Define sacred and profane. (p. 471)

   _____

   _____

3. Define fundamentalism and provide an example of America fundamentalist attitudes toward abortion. (p. 477-481)

   _____

   _____

4. Although 60% of Americans support religious and ethnic profiling, based on the assumption that it will deter and help detect terrorist activity, explain two unintended negative consequences of profiling. (p. 482-483)

_____

_____

5. Explain the difference between a church, sect, and cult, and provide an example for each. (p. 483-486)

_____

_____

Essay Questions

1. Explain how Super Bowl Sunday is like a religious holiday according to Durkheim, and provide other similar examples and requisite explanations. (p. 470-471).

2. Compare and contrast the perspectives of Marx and Weber on religion and capitalism. (p. 472-475).

3. Explain the impact of religion and the Protestant ethic on capitalist development according to Weber, and detail the two problems with Weber's argument. Do you believe that religion affects social change? (p. 473-475)

4. Choose three major religions. For each major religion, identify the charismatic leader, describe the egalitarian principles, and provide an example of the routinization of charisma in the institutionalization of the religion. (p. 486-493)

5. Explain the rise of religious fundamentalism in the United States in recent decades. Based on your examination of this phenomenon, do you see this trend continuing to increase in the future? Use concrete examples to illustrate your points. (p. 475-483)

Solutions

Practice Tests

Fill-In-The-Blank

1. collective conscience
2. civil religion
3. secularization thesis
4. Ecclesia

5. Sects
6. Religiosity
7. sacred
8. Rituals

9. Fundamentalists
10. revised secularization thesis

Multiple-Choice Questions

1. A, (p. 471)
2. C, (p. 471)
3. B, (p. 471)
4. E, (p. 472)
5. D, (p. 471)
6. B, (p. 472)
7. D, (p. 473-475)
8. C, (p. 474)
9. C, (p. 477)

10. D, (p. 493-494)
11. A, (p. 493-494)
12. B, (p. 483)
13. E, (p. 477)
14. D, (p. 477-481)
15. A, (p. 486)
16. D, (p. 491-492)
17. D, (p. 487)
18. B, (p. 485)

19. A, (p. 494)
20. D, (p. 438)
21. C, (p. 474)
22. D, (p. 478)
23. A, (p. 482)
24. C, (p. 487)
25. B, (p. 486)

True False

1. T, (p. 470-471)
2. T, (p. 472)
3. F, (p. 493)

4. F, (p. 493-494)
5. T, (p. 494)
6. F, (p. 475-477, 481)

7. T, (p. 478)
8. T, (p. 479)
9. F, (p. 470-471)

EDUCATION

Student Learning Objectives

After reading Chapter 17, you should be able to:

1. Engage the debate and controversy over affirmative action.

2. Identify and describe the manifest and latent functions of schools.

3. Explain how education reproduces the existing social stratification system in terms of standardized tests, tracking, self-fulfilling prophecies, cultural capital, and educational institutions.

4. Describe the emergence and spread of mass schooling.

5. Explain the contemporary dynamics of credential inflation and

professionalization.

1. Critically discuss pressing educational issues in terms of schools standards, educational attainment, and inequality.

2. Discuss proposed solutions to school crisis and their limitations.

Key Terms

meritocracy (501)

educational attainment (502)

educational achievement (502)

manifest functions (502)

latent functions (502)

tracking (504)

stereotype threat (512)

cultural capital (513)

credential inflation (516)

professionalization (517)

Detailed Chapter Outline

I. AFFIRMATIVE ACTION AND CLASS PRIVILEGE

   A. Legacy considerations affect acceptance rates.
      1. President George W. Bush was accepted at Yale with an SAT score of 1206, which was below average at Yale for his admission year.

   B. Affirmative Action
      1. Much controversy surrounds the awarding of admission points based on cash gifts, family ties, and minority status. Consider the arguments for and against special treatment:
         a. *Advocates of affirmative action.*
         b. *Opponents of affirmative action.*
         c. *Advocates of special treatment for the well to do.*
         d. *Advocates of meritocracy.* **Meritocracy** is a stratification system in which equality of opportunity allows people to rise or fall to a position that matches their talent and effort.

II. MACRO-SOCIOLOGICAL PROCESSES

   A. The Functions of Education
      1. Many Americans believe that there is equal access to basic schooling, and think schools identify and sort students based on merit and effort. The school system is the American Dream in action. **Educational attainment** is seen as largely an outcome of individual talent and hard work. Educational attainment, which refers to the number of years of school completed, should not be confused with **educational achievement**, which refers to how much students actually learn.
      2. The view that the American educational system *sorts* students based on talent and effort is central to the *functional theory* of education, which also stresses the *training* role of schools. A third function involves the *socialization* of the young. Finally, schools *transmit culture* generationally, fostering a common identity and social cohesion in the process.
      3. Sorting, training, socializing and transmitting culture are **manifest** functions or positive goals that schools accomplish intentionally. Schools also perform **latent** or unintentional functions, such as encouraging the development of a separate youth culture, serving the "marriage market," custodial service, and encouraging critical thinking and dissent.

   B. The Effect of Economic Inequality from the Conflict Perspective
      1. From the *conflict perspective*, the problem with the functionalist view is that it exaggerates the degree to which schools sort students by ability to ensure that the most talented students get the most rewarding jobs. Conflict theorists argue that schools distribute the benefits of education unequally, thus reproducing social stratification generation after generation, rather than functioning as meritocracy.
      2. Schools reproduce stratification partly because they vary in quality.
      3. In the 1950s and 1960s, U.S. schools were homogeneous in resources and educational outcomes. Since the 1970s, big gaps have opened up the quality of education.
      4. For example, Jonathan Kozol's (1991) research in Chicago city schools, compared average spending in an upper-middle-class, suburban school.

5. Wide variations in the wealth of communities and a system of school funding based mainly on local property taxes ensure that most children from poor families learn inadequately in ill-equipped schools, and most well-to-do children learn well in better-equipped schools.

6. Some schools, mainly those in poor neighborhoods, have a lot of students from disadvantaged homes, dropouts, and disciplinary problems. They pay less well and thus tend to have weaker teachers.

C. Standardized Tests

1. Standardized tests are a second feature of schools that perpetuates stratification. In a society supposedly based on equality of opportunity, one's ability must be judged by objective standardized testing.

2. Schools sort students into high-ability, middle-ability, and low-ability classes based on IQ and other tests (i.e. ACT and SAT), which is called **tracking**.

3. How Do IQ and Social Status Influence Academic and Economic Success?

    a. IQ by itself contributes to academic success and to economic success later in life, which is what *functionalists* would predict.

    b. However, home environment background factors such as family income, parents' education, encouragement of creativity and studying, number of siblings, and so forth, influence success. This is in accordance with *conflict* predictions.

    c. Home environment, community environment, and schooling experience affect IQ test results. IQ is partly genetic and partly social in origin.

    d. Two cases where changed social circumstances resulted in improved IQ scores.

    e. The example of 300 black and Latino students at the Hostos-Lincoln Academy of Science in the South Bronx.

    f. IQ tests measure cognitive ability and social status, and reflect genetic endowment and underlying social stratification. Hence, when schools use IQ tests to determine entrance into a high-ability track, they are not increasing the opportunities of only the most talented, but rather for some less-talented students with advantaged backgrounds. In turn, this decreases opportunities for some talented students from disadvantaged backgrounds.

4. Are SAT and ACT Tests Biased?

    a. SAT and ACT tests sort students for college entrance.

    b. Culturally biased items can be found in the SAT and ACT tests, in which equally skilled member of different groups have significantly different rates of correct response to test items.

    c. More affluent students can afford coaching that boosts scores.

    d. According to Brym and Lie, the biggest problem with the SAT and ACT tests lies in the background factors that determine who gets to take the test in the first place and how well prepared different groups of test-takers are.

D. Case Study: Functionalist versus Conflict Theories of the American Community College

1. Between 1900-2000 the number of community colleges grew from zero to more than 1400 in the United States.

2. About 5.5 million students are enrolled in community colleges today.

3. Two social forces gave rise to the community college system.
   a. The country needed skilled workers.
   b. The belief that higher education would contribute to upward mobility and greater equality in American society, the accuracy of which has become a point of contention among sociologists.
4. *Functionalists* examined the social composition of the student bodies.
   a. Found a disproportionate number of students from lower socioeconomic strata and minority ethnic groups.
   b. Many students live at home while studying, and have lower tuition cost than four-year colleges.
   c. Graduates are usually able to find relatively good jobs and steady employment.
5. *Conflict theorists* deny that community colleges increase upward mobility and equality.
   a. They believe that the entire stratification system is upwardly mobile; that is, the quality of nearly all jobs improves but the *relative* position of community college graduates versus four-year college graduates remains the same.
   b. They argue that community colleges *reinforce* prevailing patterns of social and class inequality by directing students from disadvantaged backgrounds away from four-year colleges, thus *decreasing* the probability that they will earn a four-year degree and a high-status position in society.
6. Functionalists and Conflict theorists both have a point.
   a. Community colleges *do* create opportunities for individual upward mobility that some students would not otherwise have.
   b. Community colleges *do not* change the overall pattern of inequality in American society.
7. Brym and Lie conclude that functionalists paint a somewhat idealized picture of education.
   a. While identifying latent and manifest functions of education, functionalists fail to emphasize sufficiently the far-reaching effects of stratified home, community, and school environments on student achievement and placement.
   b. A similar conclusion is reached about the effects of gender on education.

E. Gender and Education: The Feminist Contribution
   1. In some respects, women are doing better than men in the American education system.
      a. Overall, women in college have higher GPAs and complete degrees faster.
      b. The number of women enrolled as college undergraduates has exceeded the number of men since 1978; in graduate school, since 1984.
         i. The enrollment gap is growing internationally, also.
   2. There is, then, considerable improvement over time in the position of women in the educational system. Yet feminists have established that women are still at a disadvantage.
      a. A disproportionately large number of men earn Ph.D.s and professional degrees in fields requiring a strong math and science background, and that result in high paying jobs.
      b. A disproportionately large number of women earn Ph.D.s and professional degrees in fields requiring little background in math and science, and that result in low paying jobs.
   3. Parents and teachers are partially responsible for these choices.

       a. They tend to direct boys and girls toward what they regard as masculine and feminine fields of study.

       b. Sex segregation in the labor market also influences choice of field of study.

       c. College students know that women are more likely to jobs in certain fields and make career choices accordingly.

    4. Gender, like class and race, structures the educational system and its consequences.

III. MICRO-SOCIOLOGICAL PROCESSES

  A. The Stereotype Threat: A Symbolic Interactionist Perspective

    1. Sociologists have contributed much about our understanding of face-to-face interaction processes that influence the educational process.

       a. Sociological researchers examining the self-fulfilling prophecy concluded that teachers' expectations influenced students' performance.

    2. Low expectations often encourage low achievement. A minority student is often under suspicion of intellectual inferiority and feels rejected. This **stereotype threat** has a negative impact on the school performance of disadvantaged groups.

    3. As a result of resentment and defiance of authority, many minority students reject academic achievement as a goal, because they see it as a goal of the dominant culture, hence can lead to discipline problems, apathy, and disruptive and illegal behavior.

    4. Challenging minority students, giving them emotional support and encouragement, giving greater recognition in the curriculum to the accomplishments of their group, creating an environment where they can relax and achieve- can explode self-fulfilling prophecies and improve academic performance.

  B. Cultural Capital

    1. **Cultural capital** accounts for some differences in school performance.

    2. Although the independent effect of cultural capital is sometimes exaggerated, the most often cited American study on the subject suggests that cultural capital may be as important as measured ability in determining grades.

    3. The original research on cultural capital was conducted in France, and it showed that possession of cultural capital is linked to being born in high-status families. American research shows a weaker and fluid link between one's family status and acquiring cultural capital. Having parents with or without education does not guarantee the acquisition of cultural capital, or lack thereof.

IV. HISTORICAL AND COMPARATIVE PERSPECTIVES

  A. The Rise of Mass Schooling

    1. In Europe 300 years ago, the nobility and the wealthy hired personal tutors to teach their children reading, writing, history, geography, foreign languages, etiquette and demeanor. Few people went to college. The majority of Europeans were illiterate.

    2. Today, compulsory mass education has become a universal feature of European life by the early 320th century, and nearly universal literacy was achieved by the middle of the 20th century. Every country has a system of mass schooling. Four factors account for the spread of mass schooling: the Protestant Reformation, the democratic revolution, the rise of the modern state, and globalization.

a. The Protestant Reformation
   i. The Catholic Church, which dominated pre-industrial Europe, did not want mass literacy because one of the sources of power was the monopoly of priests over reading and writing to interpret the bible.
   ii. In the early 16th century, Martin Luther, a German monk, criticized the Catholic Church, and Protestantism grew out of his criticisms.
b. The Democratic Revolution
   i. By the late 18th century, the populations of France, the United States, and other new democracies demanded access to centers of learning, and the right of public education.
   ii. In the United States, educational opportunities expanded in the 19th and 20th centuries.
c. The Modern State
   i. The modern state encouraged compulsory mass education, because education promoted loyalty and social order in an era when the Church was weakening. Industrialization also encouraged the state to promote compulsory mass education.
d. Globalization
   i. Many of the conditions that contributed to mass education in the West now exist in the world's less developed countries.

B. Credential Inflation and Professionalization
1. According the Randall Collins, **credential inflation** refers to the fact that it takes ever more certificates and diplomas to qualify for a given job. Why?
   a. The increasing technical requirements of many jobs.
   b. However, on-the-job training often gives people the needed skills. Credentialism is also a convenient sorting device for employers to restrict high status positions to certain people.
   c. Credential inflation is fueled by professionalization.
2. **Professionalization** occurs when members of an occupation insist that people earn certain credentials in order to enter the occupation.
   a. It ensures standards are maintained.
   b. It keeps earnings high, by limiting the number of people entering a profession, which would drive down the cost of services provided.
      i. Example. General earnings between physicians and lawyers on the one hand, and professors on the other hand. The former have powerful organizations (The American Medical Association and American Bar Association) that regulate and effectively limit entry.
   c. Because professionalization ensures high standards and high earnings, it has spread widely. Example: the World Clown Association (WCA) and professional clowns.

V. CONTESTED TERRAIN: CRISIS AND REFORM IN U.S. SCHOOLS

A. School Standards
1. Many Americans believe that the public school system has turned soft if not rotten. In comparison to Japan and South Korea, it is argued that American students spend fewer hours in school and study more nonbasic subjects that have little practical value.

2. Unsettling evidence apparently proving the inferiority of the American school system was made public by the United States Department of Education in 1997 and 1998.
   a. The Trends in International Math and Science Study (TIMSS) found that although American fourth graders did respectably well (placed eighth in combined math and science scores among 25 other countries), the relative standing of American students dropped as they progressed through the school system (by their last year in school, Americans placed 18th out of 21 countries.
   b. Criticisms of the TIMSS test.
      i. Most of the participating countries did not follow sampling guidelines.
      ii. (a) For political reasons, many of them excluded groups of students whom education administrators thought would do poorly, hence inflating their scores.
      iii. Different countries have different kinds of secondary school systems.
      iv. (a) Some have 14-year school systems in comparison to the U.S. 12-year system.
      v. (b) Some countries have higher dropout rates than the United States, but siphon off poor academic performers to trade schools and job-training before they graduate high school—while the United States tried to graduate as many students as possible because it enhances democracy and social cohesion in the context of U.S. diversity.
   c. Nonetheless, this does not excuse the academic flabbiness of many U.S. schools. The real crisis in American education can be found in schools that contain many disadvantaged minority students, most in inner cities.
B. Solutions to the School Crisis
   1. Much research underscores the crisis in the American school system.
   2. Statistics on completion and dropout rates suggest that the crisis of the American school system is strongly related to minority status, which is strongly related to social class.
      a. In 2002, more than 21% of non-Hispanic whites older than 24 had a B.A., while only 17% of African Americans and 11% of Hispanic Americans older than 24 had a B.A.
      b. Eighty nine percent of non-Hispanic whites over age 25 had a high school diploma, compared to 79% of African Americans and 57% of Hispanic Americans.
      c. The high school dropout rate for white non-Hispanics was less than 7%, compared to 13% for black students and 28% for Latinos.
   3. For 40 years, the main strategy to improve educational attainment of disadvantage minorities has been school desegregation by busing to make schools more racially and ethnically integrated. Things did not work out as hoped.
      a. Many whites moved to all-white suburbs or enrolled their children in private schools.
      b. In integrated public schools, gains among minority students were limited by racial tensions, competition among academically mismatched students with different family backgrounds, and related factors.
      c. Many think a wiser course of action to improve the quality of underfinanced minority schools is needed, while many educators fear that ignoring integration will deny American students of different races and ethnic groups the opportunity to learn to work and live together.
   4. Local Initiatives

a. In light of self-fulfilling prophecies, challenge minority students, give them emotional support and encouragement, prepare a curriculum that gives more recognition to the accomplishments of their group, and create a relaxing and achievement oriented environment.

b. The mentoring movement involving community members volunteering to work with disadvantaged students in schools, church basements, community centers, and housing projects (i.e. "computer clubhouses").

5. Redistributing and Increasing School Budgets

a. The federal budget surplus is one potential source of funds.

b. It would be helpful if the federal and state governments collected school taxes and tied them in part to people's ability to pay.

6. Economic Reform and Comprehensive Preschools

a. Limitations with the first two proposed solutions.

i. With James Coleman's survey of American schools, sociologists discovered in the 1960s that schools were limited in what they could do on their own to encourage upward mobility and end poverty. Coleman found that differences in the quality of schools accounted for at the most a third of the variation in student's academic performance, while two thirds was due to inequalities imposed on children by their homes, neighborhoods, and peers.

ii. Little research contradicts his findings today. Follow-up studies of Head Start found gains in cognitive and socioemotional functioning displayed by Head Start graduates within 2 years of leaving the program. A 1997 Department of Education study of Title I of the Elementary and Secondary Education Act found no academic differences between students who received assistance and those who did not.

b. Programs aimed at increasing school budgets and encouraging local reform initiatives need to be augmented by policies that improve the social environment of young, disadvantaged children *before and outside* school.

i. Children from disadvantaged homes do better when their parents create a healthy, supportive, and academically enriching environment at home and if peers do not lead children to a life of drugs, crime, and disdain for academic achievement.

ii. Policies aimed at creating these conditions would be far-reaching (i.e. job training for parents, comprehensive child and family assistance programs that start when a child is born).

c. A few model comprehensive child and family assistance programs exist.

i. The federally funded Abecedarian Project in North Carolina.

ii. A state funded preschool program in Vineland, New Jersey.

C. Realistically, many people are likely to oppose such reforms because they are costly in an era of school budget cuts in the inner cities, and an era when many parents prefer private schools or move to neighborhoods in suburbs with excellent public schools. On the other hand, 72% of Americans said too little money is being spent on the nation's schools based on 2000 GSS data.

D. This suggests that most Americans still want the education system to live up to its ideals and serve as a path to upward mobility.

Study Activity: Applying The Sociological Compass

Consider the various functions and problems at the high school you graduated from. Did your high school efficiently perform the functions as described in Chapter 17? If so, in what ways were these functions performed efficiently? Make a list of these functions and provide specific examples of their efficiency or lack thereof. Are the problems you remember consistent with the school problems discussed in Chapter 17? Make a list of common problems you recall from your high school. Attempt to explain the sources of these problems from the functionalist and conflict perspective. Make a list of possible solutions.

Functions of Your High School:

1.

2.

3.

4.

Examples of Each Function:

1.

2.

3.

4.

School Problems:

1.

2.

3.

4.

Sources of School Problems

Functionalist explanations:

Conflict explanations:

Possible Solutions:

1.

2.

3.

4.

Infotrac College Edition Online Exercises

For the following exercises, log on to the online library of InfoTrac College Edition at http://www.infotrac-college.com/. Make note that InfoTrac has implemented a new registration system that will allow easier access to InfoTrac through the use of a personalized username and password. Once you've created your username and password you may proceed directly to the Log On page. To create an account, register your passcode packaged with your textbook, and create a username and password, by following the online prompt. After you are logged in, click on "Infotrac College Edition." You will arrive at a screen that enables you to search topics.

In the textbook, Brym and Lie caution us not to confuse "educational achievement" with "educational attainment." Explore the research on educational achievement using Infotrac. Read several articles that highlight racial, ethnic, and class differences in educational achievement of students. Are the findings on educational achievement consistent with the research on educational attainment cited in Chapter 17? In what ways are the findings consistent or inconsistent?

Keyword: **school vouchers**. Examine the different viewpoints on school vouchers from two articles. What are the arguments for and against the use of school vouchers?

Keyword: **distance education**. Distance education, through the use of telecourses and online courses, is increasing in popularity. What are the arguments for and against the use of distance education in higher education? What sociological issues lie behind this debate?

Keyword: **education** and **child poverty**. Look for three articles that address educational concerns and children who live in poverty. What suggestions does the research give to help children from poverty homes get an "equal" education?

Internet Exercises

Look at the homepage of the National Education Association (NEA) at: http://www.nea.org/

Click on *issues*. This list provides a good basis for contemporary education issues. Write a "pro" and "con" paper on a specific issue facing the American education system.

Distance education has become a hot topic in the area of education. Recently, people being home-schooled have increasingly been using distance education as another medium of learning. Go to this site and see what this method of learning and teaching has to offer our society: http://www.internethomeschool.com/

Do you agree with the use of distance education in grades 1-12? Why or why not? What are the strengths and weakness of such an approach to education?

In recent years there has been less funding available for special education. Programs are being cut at both the federal and state level. Funding has also been cut in the area of disability concerns.

To examine these issues in more detail go to: http://www.cec.sped.org/

Do you agree or disagree that funding should be cut in special education programs in the U.S. educational system?

Search for "education" using the search engine of the General Social Survey (GSS) found at: http://www.icpsr.umich.edu/GSS99/search.htm.

Choose one research report and read it. List the major conclusions from the report you chose.

Practice Tests

Fill-In-The-Blank

Fill in the blank with the appropriate term from the above list of "key terms."

1. _____ involves sorting students into high-ability, middle-ability, and low-ability classes based on the results of IQ and other tests.

2. _____ takes place when members of an occupation insist that people earn certain credentials to enter the occupation.

3. _____ refers to how much students actually learn.

4. A(An) _____ is a stratification system in which equality of opportunity allows people to rise or fall to a position that matches their talent and effort.

5. _____ refers to widely shared, high-status cultural signals used for social and cultural exclusion.

Multiple-Choice Questions

Select the response that best answers the question or completes the statement:

1. The fact that employers have been increasing the requirements to qualify for professional positions to the extent that far more people are earning college degrees than ever before, and the value of degrees has decline, is an example of:
   a. educational screening
   b. credential inflation
   c. tracking
   d. All of these are correct.
   e. None of the above.

2. Which factor accounts for the spread of mass schooling?
   a. the Protestant transformation
   b. democratic forces that resulted from the French and American Revolutions
   c. industrialization
   d. All of these are correct.
   e. None of the above.

3. _____ refers to the number of years of school completed.
   a. educational attainment
   b. educational achievement
   c. educational potential
   d. educational tracking
   e. educational enhancement

4. Which is a social mechanism that plays out in the school system to reproduce inequality based on the conflict perspective?
   a. unequal funding
   b. testing
   c. tracking
   d. the self-fulfilling prophecy
   e. All of these are correct

5. Schools commonly teach students to view their nation with pride. This is considered a _____
   function of schools.
   a. manifest
   b. latent
   c. unintended
   d. hidden
   e. None of the above.

6. The Trends in International Math and Science Study (TIMSS) provides evidence that the public
   educational system in the United States is in crisis. However, their findings need to be considered with
   caution. Which is a major criticism of the TIMSS test?
   a. The test ignores the fact that there are so many different cultural learning styles globally, that it is
      difficult to compare academic achievement across countries.
   b. Most participating countries did not follow sampling guidelines, some for political reasons in
      order to inflate their scores.
   c. Although secondary school systems are standardized globally, many countries fail to follow
      the world standards.
   d. All of these are correct.
   e. None of the above.

7. As a result of 40 years of school desegregation, educational attainment between racial minorities
   and whites:
   a. have approached equity.
   b. remain sharply different.
   c. have improved dramatically.
   d. All of these are correct.
   e. None of the above.

8. A historically patterned response to school desegregation by whites has been:
   a. an overwhelming acceptance as indicated by the number of integrated neighborhoods.
   b. a practically formal return to Jim Crow legislation.
   c. the movement of whites to all-white suburbs or enrollment of their children in private schools.
   d. All of these are correct.
   e. None of the above.

9. Which statement(s) regarding the relationship between schools and youth culture is (are) accurate
   according to Chapter 17 of the textbook?
   a. Schools discourage the development of youth culture.
   b. Schools encourage students to become like their parents for the most part.
   c. Schools encourage the development of a separate youth culture that often conflicts with parents'
      values.
   d. All of these are correct.
   e. None of the above.

10. About what fraction of schools are in need of immediate repair in New York City?
    a. three-fourths
    b. one-half
    c. one-fifth
    d. one-eighth
    e. one-sixteenth

11. According to a 2002 nationwide poll, what percentage of 'Americans with an opinion on the subject of national expenditure on schools' indicated that we are not spending enough on education?
    a. 74%
    b. 65%
    c. 50%
    d. 42%
    e. 35%

12. Which of the following factors directly influences social inequality according to Figure 17.1 in Chapter 17 of the textbook?
    a. parental home environment
    b. schooling
    c. adolescent community environment
    d. All of these are correct.
    e. None of the above.

13. Which racial or ethnic group makes up the lowest percentage of eighth-grade students in high-ability classes in U.S. public schools?
    a. Asian
    b. White
    c. Hispanic
    d. Black
    e. Native American

14. Which racial or ethnic group had the lowest SAT score among college-bound seniors in 2001?
    a. Native American
    b. Asian American
    c. African American
    d. Mexican American
    e. Puerto Rican

15. The movie *Stand and Deliver* is a story of:
    a. the overwhelming effect of urban decay on educational failure.
    b. the virtually inescapable impact of self-fulfilling prophecies on minority students.
    c. how students' school performances can be improved if they are encouraged to think highly of themselves.
    d. All of these are correct.
    e. None of the above.

16. In 2003, which country had both the lowest math scores and lowest science scores among the 51 countries indicated on Table 17.4 in Chapter 17 of the textbook?
    a. Singapore
    b. Taiwan
    c. Australia
    d. Iran
    e. South Africa

17. Which of the following is NOT a latent function of educational institutions?
    a. transmission of culture
    b. bringing potential mates together
    c. custodial service
    d. encouragement of critical thinking and dissent
    e. serving the marriage market

18. The negative impact of negative stereotypes on school performance of disadvantaged groups is referred to as:
    a. tracking
    b. stereotype threat
    c. cultural capital
    d. stereotyped inflation
    e. meritocracy

19. The fact that it takes ever more certificates and diplomas to qualify for a given job is referred to as:
    a. meritocracy.
    b. professionalization.
    c. credential inflation.
    d. self-fulfilling prophecy.
    e. tracking.

20. In 1997 46 percent of ethnic minority students in American higher education were enrolled in community colleges, whereas community colleges accounted for ___% of the total enrollment in American higher education?
    a. 52%
    b. 38%
    c. 23%
    d. 11%
    e. 5%

21. Which of the following statements is consistent with the conflict perspective on education?
    a. Schools identify and sort students based on merit and effort.
    b. The school system is the American Dream in action.
    c. Schools distribute the benefits of educational unequally.
    d. All of these are correct.
    e. None of the above.

22. What is Jonathan Kozol's research on Chicago city schools known for?
    a. Comparing the disparity in average spending in Chicago city schools with spending in an upper-middle-class suburban Chicago school.
    b. Documenting the upward mobility of minority students.
    c. Documenting the underrepresentation of racial minorities in Ivy league universities.
    d. All of these are correct.
    e. None of the above.

23. What is the biggest problem with the SAT and ACT scored tests according to Brym and Lie?
    a. More affluent students can afford to be coached, hence raise their scores.
    b. These tests are culturally biased.
    c. Background factors determine who gets to take the test in the first place and how well prepared different groups of test-takers are.
    d. IQ tests are used to track students.
    e. None of the above.

24. Which of the following professional degrees has 60-69.9%+ female overrepresentation?
    a. home economics
    b. education
    c. visual and performing arts
    d. architecture and related programs
    e. all of these are correct.

25. Which of the following factors influenced the spread of mass schooling?
    a. the Protestant Reformation
    b. the democratic revolution
    c. the modern state
    d. globalization
    e. All of these are correct.

True False

1. Educational attainment and educational achievement are synonymous.

   TRUE or FALSE

2. The function of schools in providing custodial care is a latent function.

   TRUE or FALSE

3. The majority of Americans believe that the public school system has increased the level of school standards to prepare ever more students for college.

   TRUE or FALSE

4. While other countries siphon off poor academic performers to trade schools before they graduate, the United States tries to graduate as many students as possible because it enhances democracy and social cohesion in the context of U.S. diversity.

   TRUE or FALSE

5. William Tyndale, whose translation formed the basis of the King James version of the Bible, was burned at the stake for translating and publishing the Bible in English.

   TRUE or FALSE

6. American research shows a much stronger relationship between the status of one's family and the acquisition of cultural capital, compared to earlier research conducted in France.

   TRUE or FALSE

7. Conflict theorists argue that community colleges reinforce prevailing patterns of social and class inequality.

   TRUE or FALSE

Short Answer

1. Explain the difference between educational attainment and educational achievement. Why is it important to make a distinction in the context of inequality in education? (p. 502)

   _____

   _____

2. Explain Randall Collins' concept of credential inflation. Provide any specific examples you have observed. (p. 516-517)

   _____

   _____

3. Define manifest function and latent function. List four manifest and four latent functions of schools. (p. 502)

   _____

   _____

4. Provide a conflict criticism of the functionalist view on education. Be sure to provide one concrete example from the textbook to support the criticism. (p. 502-504)

   _____

   _____

5. Define cultural capital and give concrete examples. (p. 513)

_____

_____

Essay Questions

1. Many argue that the public school system in the United States is in crisis. What is the evidence for this argument? Provide an overview of the three main solutions to school crisis, and also discuss the *No Child Left Behind* act. Be sure to explain their merits, drawbacks and likelihood of implementation. What do you think is the best path to solving the school crisis? (p. 519-525)

2. Explain the functions of education from a functionalist perspective. Also, explain educational inequality from a conflict perspective. In your answer, be sure to discuss manifest functions, latent functions, and standardized tests, and make reference to the discussion of *Theories of the American Community College*. (p. 502-510)

3. Historically, describe and explain the rise of mass schooling in terms of the Protestant Reformation, the democratic revolution, the modern state, and globalization. How do you see things changing in the future? (p. 514-515, 519-525)

4. Describe how schools reproduce existing stratification systems. Has this always been the case? Examine your own experience in an educational system, and use real examples to illustrate your points. (p. 499-511)

5. Compare and contrast the school system in the United States with those of other countries. How are we doing internationally? Where do our problems lie? Do you have any suggestions in the debate over school reform that you think could help solve the school crisis? (p. 517-525)

Solutions

Practice Tests

Fill-In-The-Blank

1. Tracking
2. Professionalization
3. Educational achievement
4. meritocracy
5. Cultural capital

Multiple-Choice Questions

1. B, (p. 516)
2. D, (p. 514-515)
3. A, (p. 502)
4. E, (p. 502-511)
5. A, (p. 502)
6. B, (p. 518)
7. B, (p. 520-522)
8. C, (p. 520-522)
9. C, (p. 502)
10. C, (p. 503)
11. A, (p. 504)
12. D, (p. 505)
13. E, (p. 506)
14. E, (p. 507)
15. C, (p. 513)
16. E, (p. 519)
17. A, (p. 502)
18. B, (p. 512)
19. C, (p. 516)
20. B, (p. 509)
21. C, (p. 502)
22. A, (p. 503-504)
23. C, (p. 507-508)
24. B, (p. 511)
25. E, (p. 514-515)

True False

1. F, (p. 502)
2. T, (p. 502)
3. F, (p. 517-519)
4. T, (p. 517-519)
5. T, (p. 515)
6. F, (p. 514)
7. T, (p. 510)

THE MASS MEDIA

Student Learning Objectives

After reading Chapter 18, you should be able to:

1. Define mass media, identify its various forms, and describe its historical emergence.

2. Explain the historical causes of media growth.

3. Compare and contrast functionalist, conflict, and interpretive and feminist approaches to the effects of media on society.

4. Identify the functions of mass media.

5. Explain ways in which the mass media favors and benefits dominant social classes and political groups.

6. Discuss issues of diversity and the mass media in terms of numerical representation and biased portrayals by generation, social class, gender, race and ethnicity.

7. Explain the link between persuasive media messages and actual behavior.

8. Discuss contemporary and future implications of domination and resistance on the Internet, in terms of access, content, and interactive television.

Key Terms

mass media (531)

two-step flow of communication (542)

cultural studies (542)

media imperialism (547)

media convergence (548)

Detailed Chapter Outline

I. THE SIGNIFICANCE OF THE MASS MEDIA

   A. Illusion Becomes Reality

1. At the turn of the 21st century, there were many movies about the blurred line separating reality from fantasy (*The Truman Show, The Matrix, Pleasantville, EdTV, Nurse Betty,* and *American Psycho*).
   a. In different ways, these movies suggest that the fantasy worlds created by the mass media are increasingly the only realities we know, and are as pervasive as religion was 500 or 600 years ago.
2. Much of our reality is indeed media generated.
   a. Of the 8,760 hours in a year, the average American spends 3,649 hours (42%) interacting with the mass media.
   b. We spend more time interacting with the mass media than we do sleeping, working, or going to school.

B. What are the Mass Media?
   1. **Mass media** refers to print, radio, television, and other communication technologies.
      a. The word "mass" implies that the media reach many people. The word "media" signifies that communication does not take place directly through face-to-face interaction. Technology intervenes in transmitting messages from senders to receivers. Communication via the mass media is usually one way, or at least one sided.
      b. Members of the audience cannot exert much influence on the mass media.

C. The Rise of Mass Media
   1. Although it may be difficult to imagine a world without the mass media, most of the mass media are recent inventions.
      a. The first developed system of writing appeared only about 5,500 years ago in Egypt and Mesopotamia.
      b. The print media became mass in the 19th century.
      c. The daily newspaper first appeared in the United States in the 1830s.
      d. Long-distance communication required physical transportation using a horse or a railroad in the 1830s.
      e. The newspaper was the dominant mass medium as late as 1950.
      f. In 1876, a nationwide system of telegraphic communication was established, and long-distance communication no longer required physical transportation since then.
      g. Most electronic media are products of the 20th century.
         i. The first television signal was transmitted in 1928.
         ii. Network TV began in the United States in 1948.
         iii. In 1969, the U.S. Department of Defense established ARPANET (a system of communication between computers that would automatically find alternative transmission routes if one or more nodes in the network broke down).
         iv. By around 1991, the World Wide Web and the Internet were established, and grew out of ARPANET.
         v. By 2005, some 817 million people worldwide used the Web routinely.

D. Causes of Media Growth
   1. *The Protestant Reformation.*

a. In the 16th-century Catholic Church, people relied on priests to tell them what is in the Bible.

b. In 1517, Martin Luther protested and wanted people to develop a more personal relationship with the Bible, which formed Protestantism 40 years later.

c. People were encouraged to read and the Bible became the first mass media product in the West.

d. Technological improvements in papermaking and printing, especially Johannes Gutenberg's invention of the printing press, made the diffusion of the Bible and other books far-reaching.

   i. This contributed to the Renaissance and the rise of modern science.

   ii. A remarkable feature of the book is its durability.

2. *Democratic movements.*

a. With the achievement of representation in government in France since the 18th century, the United States, and other countries, ordinary citizens wanted to be literate and gain access to centers of learning.

b. Democratic governments depended on an informed citizenry, and encouraged popular literacy and the growth of the free press.

c. Today, the mass media, especially television, mold our outlook on politics.

d. Analysts criticize television for oversimplifying and reducing politics to a series of catchy slogans.

3. *Capitalist industrialization.*

a. Modern industries required a literate and numerate workforce, and rapid means of communication.

b. Mass media in itself became a major source of profit.

II. THEORIES OF MEDIA EFFECTS

A. Functionalism

1. As societies become larger and more complex, face-to-face interaction becomes less viable as a means of communication. The mass media do an important job in meeting the increasing needs of the various parts of society.

2. Functions of the mass media:

a. The mass media performs an important function of *coordinating* the operation of industrial and postindustrial societies.

b. The mass media are important agents of *socialization*.

c. The mass media helps insure conformity through *social control*.

d. The mass media provides *entertainment*.

B. Conflict Theory

1. Conflict theorists contend that some people (dominant classes and political groups) benefit from the mass media disproportionately more than others. The mass media are a source of inequality.

2. Two ways in which dominant classes and political groups benefit disproportionately from the mass media:

a. The mass media broadcasts beliefs, values, and ideas that create widespread acceptance of the basic structure of society.

    b. Ownership of the mass media is highly concentrated in the hands of a small number of people and is highly profitable for them.

3. Media Ownership

    a. The *degree* of media concentration has increased over time.

        i. In 1984, around 50 corporations controlled half of the media organizations in the United States. By 1993, about 20 corporations maintained control.

        ii. Between 1992 and 1996, the proportion of U.S. television stations owned by the 10 biggest owners quadrupled.

        iii. U.S. book production, film production, newspaper publishing, and cable TV are each dominated by only six firms.

        iv. Five firms dominate the U.S. music industry.

    b. The *form* of media concentration began to shift in the 1990s.

        i. Prior to the 1990s, media concentration involved mainly "horizontal integration,"—that is a small number of firms controlled as much production as possible in their particular fields.

        ii. In the 1990s, "vertical integration" (media firms sought control of production and distribution in many fields) became more widespread. They became media "conglomerates."

        iii. AOL/Time Warner is the biggest media conglomerate.

    c. In the interest of maintaining diversity and competing view points in a vibrant democracy, the FCC (Federal Communications Commission – the federal watchdog over interstate and international communications) imposed ownership restrictions on radio, television, newspaper, and cable companies from its founding in 1934 until 1996. In 1996, it removed its restriction on radio stations, resulting in concentrated ownership in the radio industry.

    d. Aside from AOL/Time Warner, the biggest media conglomerates in the United States include Disney, Viacom, and News Corporation. Others include Bertelsmann (Germany), and Sony (Japan),

4. Media Bias

    a. As Edward Hermann and Noam Chomsky argue, several mechanisms help to bias the news in favor of powerful corporate interests and political groups.

        i. *Advertising.* Corporations routinely seek to influence the news so it will reflect well on them.

        ii. *Sourcing.* Sources such as press releases, news conferences, and interviews organized by large corporations and government agencies, routinely slant information to reflect favorably on their policies and preferences.

        iii. *Flak.* Governments and big corporations routinely attack journalists who depart from official and corporate points of view.

    b. The mass media virtually unanimously supports core values, democracy, capitalism, and consumerism.

    c. It is only when mass media deal with issues that touch on less central values that one may witness diversity of media opinion. Thus despite media ownership, the mass media are diverse and often contentious on specific issues.

C. Interpretive Approaches
   1. Many people view the mass media as powerfully influential to a passive public, and believe that violence on TV causes violence in real life, pornography leads to immoral sexual behavior, and celebrities lighting up leads adolescents to smoke.
   2. The degree to which TV violence encourages behavior is unclear. The sociological consensus seems to be that TV violence has a weak effect on a small percentage of viewers.
      a. Research for a half century has shown that people do not change their attitudes and behaviors just because media tell them to do so. Rather the link between persuasive media messages and actual behavior is indirect, and takes place in a **two-step flow of communication** that involves:
         i. Respected people of high status and independent judgment evaluating messages; and
         ii. Other members of the community being influenced to varying degrees by these opinion leaders.
   3. Interpretive sociologists such as symbolic interactionists and interdisciplinary **cultural studies** experts provide another argument that questions the effects of the mass media. Cultural studies focus not just on cultural meanings producers try to transmit but also on the way audiences filter and interpret mass media messages in the context of their own interests, experiences, and values.
      a. British sociologist, Stuart Hall, emphasizes that:
         i. People are not empty vessels, but rather take an active role in consuming the products of mass media;
         ii. People filter and interpret mass media messages in the context of their own interests, experiences, and values;
         iii. An adequate analysis of the mass media needs to consider both the production and consumption of media products;
         iv. We need to study the meanings intended by the producers, and then how audiences consume media products, because intended and received messages may diverge.
      b. Television viewing in terms of age and social class, turn out to be complex when analyzed through the interpretive approach.
         i. Young children clearly distinguish "make-believe" media violence from real-life violence.
         ii. Working-class women tend to evaluate TV programs more skeptically than middle-class women.
         iii. Elderly tend to be selective and focused in their television viewing.

D. Feminist Approaches
   1. In the 1970s, feminist researchers focused on the representation and the misrepresentation of women in the mass media.
   2. Most of the early feminist research assumed that audiences are passive. Much influenced by cultural studies, in the 1980s and 1990s, feminist researchers criticized this simple formula, and realized that audience members selectively interpret media messages and even contest them.
   3. Andrea Press and Elizabeth Cole's (1999) study of audience reaction to abortion as portrayed on TV shows, found complex, ambivalent, and contradictory attitudes toward abortion among audience members.

a. *"Pro-life" women from all social classes* form the most homogeneous group, think abortion is never justified, and reject the mass media's justification for abortion.

b. *Pro-choice working-class women who think of themselves as member of the working class* adopt a pro-choice stand as a survival strategy, not on principle.

c. *Pro-choice middle-class women* believe that only an individual woman's feelings can determine whether abortion is right or wrong in her own case.

4. One of the most striking aspects of Press and Cole's findings is that, for different reasons, three or four categories of audience members (categories 1, 2, and 3) are highly skeptical of TV portrayals of the abortion issue.

5. In recent years, some feminists have focused on the capacity of the mass media to reproduce and change the system of racial inequality in American society. The twin issues of female misrepresentation and active audience interpretation reappear, with a racial twist. On the one hand, African American women, for example, often appear in the role of the welfare mother, the highly sexualized Jezebel, and the mammy. On the other hand, they recognize that some mass media have enabled women of color to challenge these stereotypes.

6. In some ways the situation is improving. Moreover, the portrayal of women, racial minorities, the poor, and people with disabilities still tends to reinforce traditional, mainstream, negative stereotypes.

E. Summing Up
1. Brym and Lie conclude that each of the theoretical approaches reviewed contributes to our understanding of how the mass media influence us.
   a. Functionalism
   b. Conflict theory
   c. Interpretive approaches
   d. Feminist approaches

III. DOMINATION AND RESISTANCE ON THE INTERNET

A. The Internet has the potential to make the mass media more democratic-at least for those who can afford access.

B. Access
1. Because the Internet requires an expensive infrastructure, it has to be paid for by individual users, thus access is not open to everyone.
2. In the United States, college-educated whites with above-average incomes are most likely to enjoy Internet access.
3. Internet is not evenly distributed globally, and mirrors global inequalities overall.
   a. In 2005, the United States was the overwhelming leader in Internet connection.

C. Content
1. Roughly two thirds of the servers that provide content on the Internet are in the United States, indicating U.S. dominance.
2. Some refer to American domination of the Web as an example of **media imperialism**, which is the control of a mass medium by a single national culture and the undermining of other national cultures.

3. Some media analysts contend that the Internet not only restricts access and promotes American content, but also increases the power of media conglomerates, which is most evident in the realm of **media convergence** (the blending of the World Wide Web, television, telephone, and other communications media into new, hybrid media forms. Interactive TV involves the reception of digital signals via cable, satellite dish, or fiber-optic telephone line; and is connected to the Web through a built-in computer. It enables those with interactive TV to program and record a blend of programs on a long menu of specialty channels; feature-length movies, use of e-mail, and videoconferences.

4. Interactive TV seems poised to repeat the pattern of media concentration more quickly.

5. However, because consumers can *interact* with new media, millions of people help create it. Thus, although media conglomerates may be able to carve out a new lucrative niche, they can never fully dominate the Internet.

Study Activity: Applying The Sociological Compass

After reading Chapter 18, re-read the introductory section, "Illusion becomes Reality". Also, review the functionalist, conflict, interpretive, and feminist approaches to the mass media. Watch one of the movies mentioned, in particular *The Truman Show*. Whether you have already seen the movie or not, view the movie with a sociological eye. As you watch the movie, take notes by listing the functionalist, conflict, interpretive, and feminist implications of each scene as the plot of the movie unfolds.

Functionalist Implications

Conflict Implications

Interpretive Implications

Feminist Implications

Infotrac College Edition Online Exercises

For the following exercises, log on to the online library of InfoTrac College Edition at http://www.infotrac-college.com/. Make note that InfoTrac has implemented a new registration system that will allow easier access to InfoTrac through the use of a personalized username and password. Once you've created your username and password you may proceed directly to the Log On page. To create an account, register your passcode packaged with your textbook, and create a username and password, by following the online prompt. After you are logged in, click on "Infotrac College Edition." You will arrive at a screen that enables you to search topics.

Keyword: **cyberspace violence**. Pay particular attention to the article called <u>Antidotes to Pop Culture Poison</u> by Kristine Napier. What proposals does the author present for cyberspace violence?

Keyword: **mass media violence**. Research whether or not the mass media perpetuates aggressive behavior in children. List five reasons why researchers say the mass media causes aggressive behavior in children. List five reasons why researchers say the mass media does not cause aggressive behavior in children. What is your personal opinion?

Keyword: **media bias**. What issues or concerns are presented about bias in the mass media? What solutions are proposed to curtail the problem of media bias? Do you think the mass media has to be biased to a certain point, or is there a way they can really ever be unbiased?

Keyword: **television violence**. Read three articles that deal with television violence.

Make a list of the policies and programs that are being recommended to reduce television violence?

Internet Exercises

From any Web browser or search engine (Yahoo, America Online, Google, Internet Explorer, etc.) accomplish the following tasks and answer the following questions.

Search any five topics of your interest. The topic could be related to a hobby, sports, shopping, academic interests, and so forth.

For each topic you search, click on at least ten Web sites.

Make a list of the location in which each Web site is housed. In other words, what geographical location of the world or country was the web site created and is maintained.

How many of the Web sites originated in the United States?

Are your findings consistent with the discussion in the textbook on "media imperialism"? Provide an explanation.

Go to this site to read an article about the digital divide:

http://news.com.com/2100-1023_3-213840.html

Answer the following questions. What is the digital divide? What suggestions does the author give for lessening the digital divide? What suggestions do you have for weakening the digital divide? Do you think the digital divide is a positive thing for your society?

For an interesting site for information on corporate owners in the mass media check out this web site: http://www.fair.org/media-woes/corporate.html Write a short paper applying this information to sociological theories explaining the distribution of power.

If you love reality television programs *WebcamSearch.com* is a meta-directory with over 11,000 links to camera broadcasting to the worldwide Web. Go to: http://www.webcamsearch.com/.

Pick one and watch 30 minutes of "reality". Do you think that this is a healthy extension of the mass media? Why or why not?

To find out information about the extent of censorship in our society, go to *project censored* at http://projectcensored.org/. Write a brief report about censorship. Do you agree or disagree with the overall concept of censorship?

Practice Tests

Fill-In-The-Blank

Fill in the blank with the appropriate term from the above list of "key terms."

1. The _____ occurs between mass media and audience members.

2. The _____ are print, radio, television, and other communication technologies.

3. _____ is the blending of the World Wide Web, television, and other communications media as new, hybrid media forms.

4. _____ is an increasingly popular interdisciplinary area of media research.

5. _____ is the domination of a mass medium by a single national culture and the undermining of other national cultures.

Multiple-Choice Questions

Select the response that best answers the question or completes the statement:

1. The newspaper was the dominant mass medium as late as:
   a. 1970
   b. 1950
   c. 1986
   d. the 1830s
   e. 1910

2. Which is NOT a cause of media growth?
   a. the Protestant Reformation
   b. democratic movements
   c. capitalist industrialization
   d. socialist revolutions
   e. All of the answers are causes of media growth.

3. Which is NOT a function of the mass media according to functionalism?
   a. The mass media is filtered and interpreted by people who exercise human agency.
   b. The mass media are important agents of socialization.
   c. The mass media provides entertainment.
   d. The mass media helps ensure conformity.
   e. All of the answers are functions of mass media according to functionalism.

4. Conflict theorists argue that dominant classes and political groups benefit disproportionately from mass media. Which is a way that makes this possible according to conflict theorists?
   a. The mass media broadcasts a diversity of beliefs, values, and ideas that represents diversity in society.
   b. Ownership of the mass media is highly concentrated in the hands of a small number of people.
   c. Access of mass media is highly profitable, but evenly distributed.
   d. All of these are answers.
   e. None of the above.

5. Between 1992 and 1996, the proportion of U.S. television stations owned by the 10 largest owners quadrupled. This is an example of:
   a. the degree of media consumerism.
   b. the form of media monopolization.
   c. the degree of media concentration.
   d. the form of screening and interpretation.
   e. None of the above.

6. In the 1990s, media firms sought to control the production and distribution of many fields, as opposed to limiting their scope to their traditionally specific fields. This is referred to as:
   a. vertical integration.
   b. horizontal integration.
   c. lateral integration.
   d. total integration.
   e. cultural integration.

7. Disney, Viacom, and News Corporation are examples of:
   a. popular culture experts.
   b. media cultural producers.
   c. media conglomerates.
   d. media transnationals.
   e. None of the above.

8. In the film *The Fog of War* former secretary of defense Robert McNamara says that you win the most and lose the least if you empathize with your enemy. Some analysts argue that the mass media contribute to the fog of war today:
   a. by utilizing new technologies, thereby bringing the war home to everybody.
   b. by creating a dizzying barrage of conflicting reports and viewpoints, numbing people.
   c. by allowing easy access to dissenting or subversive positions, destroying empathy.
   d. through manipulative images, creating a false sense of empathy to suffering and death.
   e. by obstructing the development of empathy with the enemies of the United States.

9. Which mechanism helps to bias the news in favor of powerful corporate interests?
   a. advertising
   b. sourcing
   c. flak
   d. All of these are correct.
   e. None of the above.

10. Based on the conflict perspective, the mass media virtually unanimously supports:
    a. core values.
    b. democracy.
    c. capitalism.
    d. consumerism.
    e. All of these are correct.

11. It is only when the mass media deal with issues and news stories that deal with non-core values that one may witness a _____.
    a. diversified audience
    b. dissent among media constituents
    c. contentious, non-productive debate
    d. diversity of media opinion
    e. None of the above

12. From an interpretive approach, when viewing media violence, young children:
    a. become quite violent in real-life.
    b. almost always imitate the violence in real-life.
    c. clearly distinguish "make-believe" media violence from real-life violence.
    d. All of these are correct.
    e. None of the above.

13. In Chapter 18 of the textbook, natives in traditional dress on Enaotai Island in West Papua New Guinea are photographed observing the movie *Grease* on television. This is an example of the global influence and control of television by the United States. Which term best describes this phenomenon?
    a. media convergence
    b. media integration
    c. media imperialism
    d. media functioning
    e. None of the above.

14. Because consumers can interact with new media, in the future it is likely that:
    a. consumers will continue to be passive observers.
    b. media conglomerates will not be able to carve out a new lucrative media niche.
    c. media conglomerates will never fully dominate the Internet.
    d. All of these are correct.
    e. None of the above.

15. In the 1970s, feminists:
    a. focused on misrepresentation of women in the mass media.
    b. realized that audience members selectively interpret media messages.
    c. emphasized that audience members contest media messages.
    d. All of these are correct.
    e. None of the above.

16. Which is a finding(s) of Andrea Press and Elizabeth Cole's study of audience reaction to abortion as portrayed on TV shows?
    a. Pro-life women from all classes form the most homogeneous group.
    b. Pro-choice working-class women who think of themselves as members of the working class adopt a pro-choice stand as a survival strategy, not in principle.
    c. Pro-choice working-class women who aspire to middle-class status distance themselves from the "reckless" members of their own class who sought abortions on the TV shows.
    d. Pro-choice middle-class women believe that only an individual woman's feelings can determine whether abortion is right or wrong in her own case.
    e. All of these are correct.

17. Which is true of racial and ethnic minority representation in the mass media?
    a. The representation of Asian American remained the same between 1991-1992 and 2001-2002.
    b. The representation of Hispanic Americans more than tripled between 1991-1992 and 2001-2002.
    c. Comparing 1991-1992 and 2001-2002, the representation of African Americans is now substantially greater than their representation in the U.S. population.
    d. The representation of Native Americans between 1991-1992 and 2001-2002 has nearly doubled.
    e. All of these are correct.

18. Approximately what percentage of waking hours does the average American use the mass media?
    a. 91%
    b. 76%
    c. 63%
    d. 42%
    e. 21%

19. Invented in 1455 by Johannes Gutenberg, the printing press and the printed book:
    a. allowed ideas to be mass marketed, thereby minimizing content.
    b. contributed to the colonization of the new world.
    c. created a new forum for propaganda.
    d. allowed religious ideas to spread unchecked.
    e. enabled the widespread diffusion and exchange of ideas.

20. What was the main media development in 1975?
    a. VCR invented.
    b. First cable television.
    c. First microcomputer marketed.
    d. Cell phone invented.
    e. Development of the World Wide Web.

21. What was the main media development in 1895?
    a. Alexander Graham Bell sends the first telephone message.
    b. The first phonograph.
    c. Motion pictures are invented.
    d. First radio voice transmission.
    e. First commercial TV broadcast.

22. When did the AOL-Time Warner merger take place?
    a. in 2003
    b. in 2000.
    c. in 1998.
    d. in 1992.
    e. in 1986.

23. Which is the largest media conglomerate?
    a. Disney.
    b. Verizon.
    c. General Electric.
    d. AOL/Time Warner.
    e. None of the above.

24. Critics of concentrated ownership of mass media:
    a. cite examples of corporation influencing the flow of information.
    b. argue that freedom of speech is protected.
    c. suggest that corporate ownership does not have a major impact on the televised flow of information.
    d. All of these are correct.
    e. None of the above.

25. Considering the relationship between the centrality of values and diversity of media opinion, core values tend to:
    a. spark high levels of diversity of media opinion.
    b. be moderately diverse in media opinion.
    c. be low on diversity of media opinion.
    d. All of these are correct.
    e. None of the above.

26. Which is true of hip-hop artist Missy "Misdemeanor" Elliot?
    a. She rarely writes and produces her own music.
    b. Her work is a striking example of the ongoing mammy stereotypes of African-American women in American media.
    c. Her work is an example of how some mass media have enabled women of color challenge stereotypes, and break down established roles and images of African American women in America.
    d. All of these are correct.
    e. None of the above.

True False

1. The word "media" in "mass media" signifies that communication does not take place directly through face-to-face interaction.

   TRUE or FALSE

2. The average American spends more time interacting with mass media than sleeping, working, or going to school.

   TRUE or FALSE

3. All electronic media are products of the 20th century.

   TRUE or FALSE

4. Both functionalists and conflict theorists emphasize how mass media reinforces core values.

   TRUE or FALSE

5. Media conglomerates tend to be domestic in their operation.

   TRUE or FALSE

6. Because mass media are concentrated in the hands of a few media firms, there is virtually no reflection of social diversity in mass media.

   TRUE or FALSE

7. Functionalists and conflict theorists understate the degree to which audience members interpret media messages in different ways.

   TRUE or FALSE

Short Answer

1. Define and describe the mass media using clear examples. (p. 531)

   _____

   _____

2. List and explain the three causes of media growth. (p. 533-535)

   _____

   _____

3. List and explain three mechanisms of news bias that favor powerful corporate interests, according to Edward Hermann and Noam Chomsky. (p. 538-541)

   _____

   _____

4. Utilize the "two-step flow of communication" to explain the connection between persuasive media messages and actual behavior. (p. 542)

   _____

   _____

5. Some analysts refer to United States' domination of the Web as an example of "media imperialism". Explain this concept. Do you agree with this assertion? (p. 547-550)

   _____

   _____

Essay Questions

1. Compare and contrast functionalist and conflict theories of media effects. (p. 535-541).

2. Despite the mass media's support of core values, the mass media also reflects social diversity. Which theoretical approach best captures this fact, and why? (p. 535–546)

3. Explain theories of the effects of the mass media by Stuart Hall and other interpretive sociologists, such as symbolic interactionists and cultural studies experts. (p. 541-543)

4. Feminist scholars have highlighted two contradictory issues regarding women and racial minorities and the mass media. Describe these issues and give examples, and explain how women and racial minorities have successfully challenged the characterization of race and gender by the mass media. (p. 543-546)

5. Describe media conglomerates and explain how domination and resistance on the Internet provides opportunities for wealth accumulation by media conglomerates, and simultaneously, the potential to make the mass media more democratic. (p. 538, 546-550)

Solutions

Practice Tests

Fill-In-The-Blank

1. two-step flow of communication
2. mass media
3. Media convergence
4. Cultural studies
5. Media imperialism

Multiple-Choice Questions

1. B, (p. 532)
2. D, (p. 533-535)
3. A, (p. 535-536)
4. B, (p. 536-541)
5. C, (p. 537-538)
6. A, (p. 537)
7. C, (p. 537-538)
8. E, (p. 539)
9. D, (p. 538-541)
10. E, (p. 536-541)
11. D, (p. 541)
12. C, (p. 541-543)
13. C, (p. 547-550)
14. C, (p. 548)
15. A, (p. 543)
16. E, (p. 543)
17. C, (p. 544-545)
18. C, (p. 530, 534)
19. E, (p. 534)
20. C, (p. 533)
21. C, (p. 533)
22. B, (p. 537)
23. D, (p. 537)
24. A, (p. 538)
25. C, (p. 541)
26. C, (p. 544)

True False

1. T, (p. 531)
2. T, (p. 530-531)
3. F, (p. 532-533)
4. T, (p. 541)
5. F, (p. 537-541)
6. F, (p. 537-543)
7. T, (p. 541-543)

CHAPTER 19

HEALTH AND MEDICINE

Student Learning Objectives

After reading Chapter 19, you should be able to:

1. Describe how and explain why health problems change over time.

2. Explain the contradiction of medical advances and new problems.

3. Describe how health is measured.

4. Discuss the social causes of illness and health in terms of public health and health care systems.

5. Describe variations in health risk, public health, and health care systems in the context of global inequality.

6. Identify indicators of health care inequality.

7. Explain the American anomaly of health in terms of social class, race, and gender.

8. Explain the connections among health, politics, inequality, and privatization from a conflict perspective.

9. Explain the medicalization of deviance.

10. Describe the political controversies regarding the labeling of mental disorders.

11. Describe the emergent dominance of scientific medicine and professionalization.

12. Discuss patient activism, alternative medicine, and holistic medicine as challenges to traditional scientific medicine within the larger social, political, and cultural contexts.

Key Terms

life expectancy (554)

health (555)

maximum average human life span (555)

public health system (557)

health-care system (557)

infant mortality (559)

medicine (565)

medicalization of deviance (566)

sick role (572)   holistic medicine (576)

placebo effect (575)

Detailed Chapter Outline

I. HEALTH

    A. The Black Death

        1. In the 1300s, the Black Death killed a third of Europe's population, and ranks as the most devastating catastrophe in human history. Today we know the cause was a bacillus that spread from lice to rats to people, that spread so efficiently because people lived close together in unsanitary conditions.

        2. The only people who had luck avoiding the plague were the well to do, who could afford to flee densely populated cities for the countryside, and the Jews, whose religion required that they wash their hands before meals, bathe once a week, and conduct burials soon after death.

    B. Sociological Issues of Health and Medicine

        1. The sociology of health, aging, and medicine are embedded or implied in the story of the Black Death.

        2. Health risks are unevenly distributed across gender, social class, race, and ethnicity.

        3. Health problems change over time.

            a. Today we know that sanitation and hygiene prevent the spread of disease.

            b. We are able to treat many infectious diseases with antibiotics.

            c. People live longer; that is, there is greater **life expectancy** (the average age of death of the members of a population).

        4. In contrast, we cannot help but to be struck by the superstition and ignorance surrounding the treatment of the ill in medieval times.

        5. However, medieval doctors stressed the importance of prevention, exercise, a balanced diet, and a congenial environment in maintaining good health. One of the great shortcomings of modern medicine is its emphasis on high-tech cures rather than preventive and environmental measures.

II. HEALTH AND INEQUALITY

    A. Defining and Measuring Health

        1. **Health**, according to the World Health Organization, is "the ability of an individual to achieve his [or her] potential and to respond positively to the challenges of the environment."

        2. Sociologists typically examine negative aspects such as rates of illness and death, when *measuring* the health of the population.

        3. The **maximum human life span** is the longest an individual can live under current conditions (currently about 122 years), which may well increase in the 21st century.

        4. The **maximum average human life span**, the average age of death for a population under *ideal* conditions (currently at about 87 years), is also likely to increase in the 21st century.

5. Conditions are nowhere ideal, and vary across countries. Life expectancy tends to be lower in poor countries.

6. Sociologists of health account for the difference between the maximum average human life span and life expectancy, and the impact of social causes on illness and death.

B. The Social Causes of Illness and Death

1. Three types of social causes of illness and death
   a. Lifestyle factors.
   b. Human-environmental factors
   c. The public health and health care systems.
      i. The **public health system** is composed of government-run programs that ensure access to clean drinking water, basic sewage and sanitation services, and inoculation against infectious diseases.
      ii. The **health care system** is composed of a nation's clinics, hospitals, and other facilities for ensuring health and treating illness. The absence of a public health system and minimum standard health care is associated with high rates of disease and low life expectancy.

C. Global Health Inequalities

1. AIDS is the leading cause of death in urban Haiti and in the poverty-stricken part of Africa south of the Sahara desert. The absence of adequate health care and medical facilities makes the epidemic's impact more devastating.
   a. AIDS/HIV is nearly 12 times more common in sub-Saharan Africa than in North America and nearly 25 times more common than in Western Europe. However, spending on research and treatment is concentrated in the rich countries of North America and Western Europe.
   b. Global inequality influences the exposure of people to different health risks.
   c. Biomedical advances that increase with prosperity partly increases life expectancy.
   d. A sound public health system is more important.
   e. While industrialized countries were able to develop their public health systems to help create a healthier labor force and citizenry in the 19th century, it is not possible for many developing countries to do so today.
   f. Indicators of health inequality.
      i. Access to sanitary water supply
      ii. **Infant mortality** (the annual number of deaths before the age of one for every 1,000 live births)
      iii. Health expenditures
      iv. Malnutrition
      v. Number of Physicians
      vi. Number of nurses and midwives
      vii. Immunization of children against measles
   g. Efficient health care systems achieve high outputs given resources invested; inefficient health care systems achieve low outputs given resources invested.

i.   The American health care system is only a little above average among 191 countries, putting it in the same league as health care systems in South America, Eastern Europe, and India.

2.   Class Inequalities in Health Care
   a.   An "American anomaly". Why does the United States spend far more health care than any other country in the world yet on average, is less healthy than the population of other rich countries, such as Japan, Sweden, Canada, and France?
   b.   The higher the level of inequality in a country, the more unhealthy its population tends to be.
      i.   The gap between the rich and the poor is greater in the United States.
      ii. The United States contains a higher percentage of poor people than other countries contain.
   c.   Manifestations of health inequality.
      i.   Infant mortality rates.
      ii. Life expectancy.
   d.   Reasons for disparities in health inequality.
      i.   The poor are more likely to be exposed to violence, high-risk behavior, and environmental hazards.
      ii. Because poverty is more common among African Americans, they have higher mortality rates than whites.
      iii. The poor cannot afford adequate, and in some cases even minimal, health care, and often live in areas where medical treatment facilities are inadequate.
      iv. Poor diets; healthy food is expensive or there is no access.
      v.   The poor tend to have less knowledge about healthy lifestyles as a result of less access to doctors and hospitals.

3.   Racial Inequalities in Health Care
   a.   Economic differences between racial groups. Studies show that blacks and whites *at the same income level* have similar health statuses.
      i.   *Racism.* The health status of African Americans is somewhat lower than the health status of European Americans even within the same income group. Three ways racism affects health: Income and other rewards do not have the same value across racial groups. For instance, due to discrimination, each year of education completed by African Americans results in smaller income gains than for white Americans, which consequently affects health. Also, racism affects access to health services, because African Americans at all income levels tend to live in racially segregated neighborhoods with fewer health-related facilities. Finally, the experience of racism induces psychological distress that negatively affects health.
   b.   Increases in income have a greater positive health impact on below-median income earners than on above-median income earners.
   c.   Beyond the issue of *access* to medical resources, even among people with the *same access*, people of higher rank tend to live healthier and longer lives. Why?
      i.   People of high rank experience less stress because they are more in control of their lives (i.e. autonomy, creativity, and freedoms).

      ii.   Subordinates have little control of their work environments, and experience a continuous sense of vulnerably that lead to high-levels stress, which results in reduced immune function, hardening of the arteries, increased chance of heart attack, and other ailments.

    4. Gender Inequalities in Health Care: The Feminist Contribution

      a. Feminist scholars have brought gender inequalities to the attention of the sociological community. Conclusions:

        i.   Gender bias exists in medical research.

        ii. Gender bias exists in medical treatment (i.e. fewer kidney transplants and various cardiac procedures).

        iii. Although women experience greater lifetime risk of functional disability and chronic illness, and greater need for long-term care because they live longer, more is spent on men's health care in the United States.

        iv. There are 40% more poor women than poor men in the United States.

      b. Although women live longer than men, gender inequalities have a negative impact on women's health.

III. HEALTH AND POLITICS: THE UNITED STATES FROM CONFLICT AND FUNCTIONAL PERSPECTIVES

  A. Another reason for the American anomaly lies in the nature of the American health care system.

  B. From a conflict perspective, the United States health care system is seen as a system of privilege for some and disadvantage for others.

    1. About 80% of medical costs are paid by governments or taxes in other rich postindustrial countries, while about 40 million Americans have no health insurance, and another 40 million are inadequately covered. Only the elderly, the poor, and veterans receive Medicaid and Medicare.

    2. Private insurance programs run by employers and unions cover the vast majority of Americans.

      a. About 85% of employees buy their health coverage from HMO's.

      b. HMOs are private organizations that collect regular payments from employers and employees, and administer medical treatment.

      c. HMOs are corporations that pursue profit.

  C. Problems with Private Health Insurance and Health Maintenance Organizations

    1. Some HMOs avoid covering sick people and people who are likely to get sick. An HMO can show you had a preexisting medical condition.

    2. HMOs try to minimize the cost of treating sick people they can't avoid covering. Doctor-compensation formulas for withholding unprofitable treatments exist.

    3. Allegations that HMOs routinely inflate diagnoses to maximize reimbursements.

    4. HMOs keep overhead charges high.

  D. Advantages of Private and For-Profit Health-Care Institutions

    1. From a functionalist point of view, one big advantage of running for-profit organizations is that they can invest enormous sums in research, development, latest diagnostic equipment, and high

salaries to attract the best medical researchers and practitioners. Thus the United States has the best health-care system in the world for those who can afford it.

    2. Although HMOs and the American Medical Association attempt to convince Americans that the private-system serves the public better than a state-run system, they have been only partly successful.

        a. Based on GSS data, 41% of Americans disagree or strongly disagree that HMOs improve the quality of medical care.

  E. A National Health-Care System for the United States?

    1. Attempts to create a national system of health care in which everyone is covered have failed. Congress rejected President Clinton's 1993–94 Health Security proposal, because media campaigns were largely funded by health insurance companies.

    2. Summing up, Brym and Lie say that apparently *natural* processes of health and illness are in fact deeply *social* processes.

    3. A similar argument can be made about medicine. **Medicine** is a social institution devoted to prolonging life by fighting disease and promoting health.

IV. MEDICINE

  A. The Medicalization of Deviance

    1. The **medicalization of deviance** refers to the tendency for medical definitions of deviant behavior to become more prevalent over time.

    2. What used to be regarded as willful deviance and "badness" is now considered involuntary deviance and "sickness". For example:

        a. A person prone to drinking sprees is likely to be declared an alcoholic and treated in a detoxification center.

        b. A person prone to violent rages is likely to be medicated.

        c. A person inclined to overeating is likely to seek therapy, and even surgery in extreme cases.

        d. A heroin addict is likely to seek a methadone program.

  B. The Political Sociology of Mental Illness

    1. Political debates among American psychiatrists in the 1970s and 1980s.

        a. The American Psychological Association (APA), the Diagnostic and Statistical Manual of Mental Disorders (DSM), and homosexuality.

        b. DSM task force and the term "neurosis."

        c. The APA, Vietnam War veterans, and "post traumatic stress disorder."

        d. Feminists, DSM, and "self-defeating personality disorder."

    2. Although some mental disorders have obvious organic causes that can be identified, the definition of a host of mental disorders depends on social values and political compromise.

    3. Four reasons for the numerical expansion of conditions labeled "mental disorder."

        a. Mental health problems are more widespread because Americans are experiencing more stress and depression, while traditional institutions, like the church and the family, are less able to cope with them.

b. Powerful organizations demand inflation in a number of mental disorders, because classification of mental disorders is useful. HMOs demand precise diagnostic codes to pay for psychiatric care.

c. Culturally, Americans are inclined to turn their problems into medical and psychological issues, sometimes without considering the disadvantages of doing so. For instance, children diagnosed with attention deficit disorder (ADD) (a label for hyperactive and inattentive schoolchildren, mainly boys, who are believed to suffer chemical imbalances that affect the functioning of the brain to regulate behavior) may have no organic disorder, and their behaviors may be a response to deprivation at home or school system failure to capture children's imagination.

d. Professional organizations have promoted it. For example the trivialization of post-traumatic stress disorder.

C. The Professionalization of Medicine

1. In 1850, the practice of medicine was chaotic, which involved competition among herbalists, faith healers, midwives, druggists, and medical doctors. A century later, medical science was victorious following its first series of breakthroughs involving the identification of bacteria and viruses, and effective procedures and vaccines to combat them.

2. Sociologically, scientific medicine also became dominant because doctors were able to professionalize—that is, the process by which people gain control and authority over their occupation and their clients.

3. The American Medical Association (AMA)

   a. Founded in 1847, the AMA quickly broadcasted the successes of medical science, and criticized alternative approaches to health.

   b. By the early years of the 20th century, the AMA convinced state licensing boards to certify only doctors who had been trained in programs recognized by the AMA.

   c. Historically, doctors did not earn much until it was possible to lay virtually exclusive claim to health.

4. The Rise of Modern Hospitals

   a. The modern hospital is the institutional manifestation of the medical doctor's professional dominance.

D. The Social Limits of Modern Medicine

1. The cases of severe acute respiratory syndrome, or SARS (a pneumonia-like illness for which there is no vaccine and no cure).

2. Hospitals are ideal environments for the spread of germs. The hospital system in the United States is perhaps the most dangerous in this respect.

3. Cutting costs and catering to paying customers are among the chief means of keeping shareholders happy in a health-care system by profitability. This results in disproportionate investment in expensive, high-tech, cutting-edge diagnostic equipment and treatment for those who can afford it; and scrimping on simple, laboratory intensive, time-consuming hygiene for those who cannot.

4. From the 1860s to the 1940s, American hospital staffs were obsessed with cleanliness, because infection often meant death. In the 1950s, the prevention of infections in hospitals became less of a priority because penicillin and antibiotics became widely available. It was less expensive to wait until a patient got sick and respond to symptoms by prescribing drugs than preventing the sickness in the first place. Doctors and nurses have grown lax about hygiene over the past half-century.

5. The use of penicillin and antibiotics indiscriminately has its own costs. With the use of a lot of penicillin and antibiotics, "super germs" that are resistant to these drugs multiply.

6. The epidemic of infectious disease caused by slack hospital hygiene and the overuse of antibiotics suggests that social circumstances constrain the success of modern medicine.

E. Recent Challenges to Traditional Medical Science
   1. Patient Activism
      a. According to Talcott Parsons, playing the **sick role** involves the nondeliberate suspension of routine responsibilities, wanting to be well, seeking competent help, and cooperating with health care practitioners at all times. By definition, a patient cannot reasonably question doctor's orders, no matter how well educated the patient and how debatable the effect of the prescribed treatment.
      b. The sick role probably sounds authoritarian and foreign to most young people because things have changed. Patients want to play a more active role in their own care.
         i. The American public is more highly educated.
         ii. Patients are taught to perform simple, routine medical procedures themselves.
         iii. People now seek the Internet for medical advice.
         iv. Patients are increasingly uncomfortable with doctors that act like patriarchal fathers.
         v. Doctors seek patient's informed consent.
         vi. Most hospitals have established ethics committees.
      c. Organized and political challenges to the authority of medical science.
         i. The challenge of AIDS as a "gay disease" by AIDS activists.
         ii. Feminist support of reintroducing midwifery, and raised attention to women's bodies and diseases.
         iii. Movements that empower people with disabilities, who have been traditionally treated like incompetent children.
      d. Recent challenges to the pharmaceutical industry:
         i. Major drug companies fund most of the research that government panels use to decide whether new drugs should go on the market.
         ii. Some observers worry about bias in the approval process or whether drug companies rush new drugs to the market without proper safeguards.
         iii. Given such close ties between industry and government, many observers are calling for a more independent approval process that will consider only the public interest in reaching its decisions, avoiding another Vioxx type problem.
   2. Alternative Medicine
      a. Alternative medicine is a less organized and less political challenge to the authority of medical science.

b. The most frequent types of alternative medicine include chiropractic, acupuncture, massage therapy, and various relaxation techniques.

c. Alternative medicine is most frequently use by highly educated, upper-income white Americans in the 25- to 49-year age group, and has grown in popularity.

d. Many medical doctors were hostile to alternative medicine, and dismissed them as unscientific, until recently.

e. By the 1990s, a more tolerant attitude was evident.

    i. For instance, a 1998 editorial in *New England Journal of Medicine* admitted that beneficial effect of chiropractic on low back pain is "no longer in dispute." This was due in part to new scientific evidence.

f. The uneasy relationship between scientific and alternative medicine is likely to continue in the coming decades.

3. Holistic Medicine

a. Research shows that strong belief in the effectiveness of a cure can by itself improve the condition of about a third of people suffering from chronic pain or fatigue, which is known as the **placebo effect**.

b. Traditional scientific medicine often responds to illness by treating disease symptoms as physical and individual problems, with drugs and high-tech machinery.

c. In contrast, traditional Indian medicine sees individuals in terms of the flow of vital fluids or "humors" and their health in the context of their environment. Traditional Chinese medicine seeks to restore individuals' internal balance, as well as their relationship to the outside world.

d. The third challenge to traditional scientific medicine is contemporary **holistic medicine**, which is like these "ethnomedical" traditions, and argues that good health requires maintaining a balance between mind and body, and between individual and the environment.

e. Most holistic practitioners:

    i. Do not reject scientific medicine, but emphasize disease prevention;

    ii. Seek to establish close ties with their patients and treat them in their homes or other relaxed settings;

    iii. Expect patients to take an active role in maintaining their good health; and

    iv. Recognize that industrial pollution, poverty, racial and gender inequality, and other social factors contribute heavily to disease.

Study Activity: Applying The Sociological Compass

After reading Chapter 19, re-read Box 19.3, Social Policy: What Do You Think?, "The High Cost of Prescription Drugs", which describes the issue, possible policy solutions, and arguments on each side of the issue. Make a list of explanations on each side of the issue from functionalist and conflict perspectives. Which perspective and corresponding positions seems most supported by what you learned in Chapter 19?

Key Arguments:

1.

2.

3.

4.

Functionalist Explanations:

1.

2.

3.

4.

Conflict Explanations:

1.

2.

3.

4.

Infotrac College Edition Online Exercises

For the following exercises, log on to the online library of InfoTrac College Edition at http://www.infotrac-college.com/. Make note that InfoTrac has implemented a new registration system that will allow easier access to InfoTrac through the use of a personalized username and password. Once you've created your username and password you may proceed directly to the Log On page. To create an account, register your passcode packaged with your textbook, and create a username and password, by following the online prompt. After you are logged in, click on "Infotrac College Edition." You will arrive at a screen that enables you to search topics.

Keyword: **alternative medicine**. Pick two articles that discuss the problems of alternative medicine. What are the main problems that are discussed in your articles about alternative medicine? How do these concerns compare to the criticisms of alternative medicine discussed in Chapter 19?

Keyword: **Alzheimer's disease**. Look for articles that stress social support programs for Alzheimer's. What characteristics do they have in common? How do these programs relate to the sociological discipline?

Keyword: **malpractice insurance**. Find three articles that discuss the malpractice insurance crises. What are the reasons behind the increase in malpractice insurance? What solutions do they proposed to end this crisis? Do you believe that the tremendous hike in malpractice insurance is justified? Why or why not? Write a short opinion paper on the subject.

Keyword: **euthanasia**. Arguments for and against euthanasia have become more vocal in our society. Look for an article that discusses any policies in other countries about euthanasia and physician assisted suicide? How do their views compare with those in the United States?

Keyword: **health maintenance organizations (HMOs)**. HMOs focus on preventive health care for a fixed fee. Some HMOs are under the gun for not providing adequate care for their patients. What are some of the growing concerns of unhappy HMO subscribers? What solutions do they call for?

Internet Exercises

The Center for Disease Control and Prevention web site located at http://www.cdc.gov/ has an immense supply of information and statistics on a wide spectrum of health related issues. Look up *anthrax*. What does it say are some critical tips if you come across anthrax? Write up a short report with your conclusions.

Go to this newspaper article from Newday: http://tinyurl.com/8myxe. Read about the changing face of AIDS/HIV. How has the social distribution of AIDS changed recently? Which groups are currently at "high risk?" After reading the article and answering the two questions, propose several solutions to the AIDS/HIV problem.

An Initiative of the Robert Wood Johnson Foundation sponsors the following Internet site: http://www.statecoverage.net/who.htm.

Go to this page and read about those in the United States who are the uninsured. Compile a list of characteristics of those who don't have insurance. Why do you think so many people in the United States are not insured? Write a brief opinion paper about your point of view.

The World Health Organization web site at http://www.who.int/ is an excellent source of information on health issues. Type in "obesity." What type of subjects on obesity does your search yield? Make a list of different problems and subjects associated with obesity. What does this tell you about the widespread problem of obesity in the United States?

Practice Tests

Fill-In-The-Blank

Fill in the blank with the appropriate term from the above list of "key terms."

1. The _____ is composed of a nation's clinics, hospitals, and other facilities for ensuring health and treating illness.

2. The _____ is the positive influence of a strong belief in the effectiveness of a cure on healing.

3. The _____ is the average age of death for a population under ideal conditions.

4. The _____ is the tendency for medical definitions of deviant behavior to become more prevalent over time.

5. Playing the _____ involves the nondeliberate suspension of routine responsibilities, wanting to be well, seeking competent help, and cooperating with health-care practitioners at all times.

6. According to the World Health Organization, _____ is the ability of an individual to achieve his or her potential and to respond positively to the challenges of the environment.

Multiple-Choice Questions

Select the response that best answers the question or completes the statement:

1. What is life expectancy?
   a. the average age of death of the members of a population.
   b. the maximum age of death of eldest members of a population.
   c. the minimum age of death of among members of an elderly population.
   d. All of these are correct.
   e. None of the above.

2. The average age of death for a population under ideal conditions is referred to as:
   a. the maximum average human life span.
   b. the maximum life expectancy.
   c. the maximum average human life span.
   d. All of these are correct.
   e. None of the above.

3. The _____ is composed of government-run programs that ensure access to clean drinking water, basic sewage and sanitation services, and inoculation against infectious diseases.
   a. health care system
   b. public health system
   c. public works system
   d. Medicare system
   e. All of these are correct.

4. Which is a type of social cause of illness and death?
   a. human-environmental factors
   b. lifestyle factors
   c. factors related to public health systems.
   d. factors related to health care systems.
   e. All of these are correct.

5. AIDS/HIV is nearly _____ more common in sub-Saharan Africa than in North America.
   a. 2 times
   b. 5 times
   c. 12 times
   d. 50 times
   e. 80 times

6. Which is NOT an indicator of health inequality?
   a. infant mortality
   b. health expenditures
   c. access to sanitary water supply
   d. number of nurses and midwives
   e. All of the answers are indicators of health inequality.

7. Which is considered a sociological reason for disparities in health inequality?
   a. The poor are more likely exposed to violence, and high-risk behavior.
   b. The poor are more likely exposed to environmental hazards.
   c. The poor cannot afford adequate health care.
   d. The poor tend to have less access to knowledge about healthy lifestyles.
   e. All of these are correct.

8. Which is NOT a reason for racial disparities in health status?
    a. Economic differences between racial groups exist.
    b. Blacks and whites at the same income level have similar health statuses.
    c. Income and other rewards do not have the same value across racial groups.
    d. The experience of racism induces psychological distress that negatively affects health.
    e. All of the answers are reasons for racial disparities in health status.

9. Which is true regarding health inequality based on gender?
    a. Gender bias exists in medical research.
    b. Gender bias exists in medical treatment.
    c. Socially, more money is spent on men's health care in the United States.
    d. There are 40% more poor women than poor men in the United States, thus the impact of poverty on health disproportionately affects women.
    e. All of these are correct.

10. Which is NOT a strategy used by HMOs to keep shareholders happy?
    a. HMOs avoid covering sick people and people who are likely to get sick.
    b. HMOs try to minimize the cost of treating sick people.
    c. HMOs keep overhead charges high.
    d. HMOs work hand in hand with government subsidized health care agencies.
    e. All of the answers are strategies used by HMOs.

11. Which statement about health care organizations is most consistent with a functionalist point of view?
    a. An advantage of for-profit health organizations is that they invest enormous sums in research, development, latest diagnostic equipment, and have high salaries to attract the best medical researchers and practitioners.
    b. For-profit health organizations are largely responsible the United States leading in quality health care in the world, however is only accessible for those who can afford it.
    c. Social meanings are attached to health care and medicine in the context of cultural systems.
    d. All of these are correct.
    e. None of above.

12. Which is an example of the "medicalization of deviance"?
    a. A person prone to drinking sprees is likely to be treated in a detoxification center.
    b. The prescription of medication to a person prone to violent rages.
    c. A person inclined to overeating who seeks therapy.
    d. All of these are correct.
    e. None of the above.

13. Which is NOT a sociological reason for the numerical expansion of conditions labeled "mental disorder"?
    a. Mental health problems are more widespread because Americans are experiencing more stress and depression in society.
    b. Scientific medical breakthroughs have identified a host of new organic causes of mental disorders.
    c. Culturally, Americans are inclined to turn their problems into medical and psychological issues.
    d. Powerful organizations demand an inflation in the number of mental disorders.
    e. All of the answers are sociological reasons.

14. _____ argues that good health requires maintaining a balance between mind and body, and between individual and the environment.
    a. Traditional scientific medicine
    b. The sick role
    c. Holistic medicine
    d. All of these are correct.
    e. None of the above.

15. The hospital system in the United States is the _____, with respect to hospitals being ideal environments for the spread of germs.
    a. most dangerous
    b. safest
    c. model system of hygiene and preventive health care
    d. All of these are correct.
    e. None of the above.

16. Which statement is true regarding SARS?
    a. SARS stands for sever acute respiratory syndrome.
    b. SARS is a pneumonia-like illness for which there is no vaccine or cure.
    c. SARS originated in Guangdong Province in South China.
    d. All of these are correct.
    e. None of the above.

17. In 2002, the Chicago Tribune published a major investigative report on the problem of the spread of disease in U.S. hospitals, and found that about _____ of 35 million Americans admitted to a hospital contacted a hospital-acquired infection.
    a. 45%
    b. 23%
    c. 13%
    d. 6%
    e. 1%

18. What was the leading cause of death in the United States in 2001?

    a. pneumonia/influenza
    b. diabetes
    c. chronic lung disease
    d. lung inflammation
    e. heart disease

19. Pneumonia/influenza ranked as the _____ leading cause of death in the United States in 2001.

    a. 3rd
    b. 5th
    c. 7th
    d. 10th
    e. 15th

20. The record for the maximum human life span is the age of:

    a. 101.
    b. 122.
    c. 138.
    d. 145.
    e. 200.

21. Which country had the highest life expectancy in 2004, among the following countries?

    a. India
    b. the United States
    c. China
    d. Japan
    e. All of these are correct.

22. Approximately how many people were there with HIV/AIDS in the world in 2004?

    a. 55 million
    b. 39.4 million
    c. 30 million
    d. 23 million
    e. 14 million

23. Which region of the world has the greatest number of people with HIV/AIDS?

    a. North America
    b. Latin America
    c. Western Europe
    d. Eastern Europe
    e. Sub-Saharan Africa

24. What percentage of sub-Saharan Africans were living with HIV/AIDS in December 2004?
    a. 12%
    b. 7.4%
    c. 7.2%
    d. 4.9%
    e. 3%

25. Which country had the greatest health expenditure per capita from 1999-2002?
    a. United States
    b. Japan
    c. Canada
    d. Mexico
    e. Zambia

26. Which country had the highest infant mortality rate per 1000 live births from 1999-to 2002?
    a. the United States
    b. Japan
    c. Canada
    d. Mexico
    e. Zambia

True False

1. Over time, life expectancy has declined because of the depletion of our environment.

   TRUE or FALSE

2. AIDS is the leading cause of death in the poverty stricken part of Africa south of the Sahara desert.

   TRUE or FALSE

3. The higher the level of inequality in a country, the more unhealthy its population tends to be.

   TRUE or FALSE

4. Upward mobility in income has a greater positive health impact on above-median income earners than below-median income earners.

   TRUE or FALSE

5. Aside from the issue of access to medical resources, people of higher rank tend to live healthier lives even in comparison to people who have the same access but lower rank.

   TRUE or FALSE

6. The overwhelming majority of Americans agree that HMOs improve the quality of medical care.

   TRUE or FALSE

7. Most recently discovered mental disorders are almost exclusively attributable to organic causes.

   TRUE or FALSE

8. Most holistic practitioners reject scientific medicine.

   TRUE or FALSE

9. Pneumonia/Influenza was the leading cause of death in the United States in 1900.

   TRUE or FALSE

Short Answer

1. Briefly explain the medicalization of deviance, and provide one example. (p. 566)

2. What do Brym and Lie mean by an "American anomaly" of health in the United States? (p. 559–562)

3. Explain Talcott Parson's concept the "sick role". (p. 572-575)

4. What is a "placebo effect"? Provide an example in terms of health care. (p. 575)

5. Explain the contradiction of U.S. hospitals being both a place of high-tech health care, and as an ideal environment for the spread of germs. (p. 570-572)

Essay Questions

1. Explain the social benefits and consequences of for-profit health organizations from a functionalist and a conflict perspective. Which perspective makes most sense to you, and why? (p. 562-565)

2. From a feminist perspective, explain why the issues of health are impacted by gender. Discuss also gender inequality in health care and use concrete examples of why this situation exists and how it is made manifest. (p. 562)

3. Explain two ways global inequality influences the exposure of people to different health risks. (p. 557-559)

4. Explain how apparently natural processes of health and illness are in fact deeply social processes. Use clear examples. (p. 553-565)

5. Define medicine and explain how society shapes medical practice as much as it influences health and illness. Use clear examples. (p. 565-576)

Solutions

Practice Tests

Fill-In-The-Blank

1. health-care system
2. placebo effect
3. maximum average human life span
4. medicalization of deviance
5. sick role
6. health

Multiple-Choice Questions

1. A, (p. 554)
2. C, (p. 555)
3. B, (p. 557)
4. E, (p. 556-557)
5. C, (p. 557)
6. E, (p. 557-561)
7. E, (p. 557-562)
8. B, (p. 561)
9. E, (p. 562)
10. D, (p. 563)
11. A, (p. 563-564)
12. D, (p. 566)
13. B, (p. 567-569)
14. C, (p. 575-576)
15. A, (p. 570)
16. D, (p. 570)
17. D, (p. 570)
18. E, (p. 555)
19. C, (p. 555)
20. B, (p. 556)
21. D, (p. 556)
22. B, (p. 558)
23. E, (p. 558)
24. B, (p. 558)
25. A, (p. 559, 562-565)
26. E, (p. 559)

True False

1. F, (p. 554)
2. T, (p. 558)
3. T, (p. 559)
4. F, (p. 559-561)
5. T, (p. 559-561)
6. F, (p. 564)
7. F, (p. 567-569)
8. F, (p. 575-576)
9. T, (p. 555)

POPULATION AND URBANIZATION

Student Learning Objectives

After reading Chapter 20, you should be able to:

1. Describe historical trends and explanations of population growth.

2. Explain and critique the Malthusian perspective on the cycle of population growth.

3. Understand and apply demographic transition theory.

4. Explain the relationship between population growth and social inequality.

5. Describe the emergence of urbanization in the context of industrialization.

6. Explain the sociological contribution of the Chicago School to the understanding of urbanization and industrialization.

7. Contrast the theories of human ecology and urban sociology.

8. Discuss the urbanization of rural regions of the United States.

9. Describe, compare and contrast the corporate city and the postmodern city.

Key Terms

demographers (580)

Malthusian trap (583)

demographic transition theory (584)

crude death rate (584)

crude birth rate (584)

replacement level (585)

immigration (585)

in-migration (585)

emigration (585)

out-migration (585)

sex ratio (588)

The Chicago School (591)

human ecology (591)

differentiation (591)

Detailed Chapter Outline

I. POPULATION

    A. The City of God

        1. Rio de Janeiro, Brazil, is one of the world's most beautiful cities. The inner city of Rio is a place of great wealth and beauty, devoted to commerce and the pursuit of leisure.

        2. Rio is a large city, and is the 18th biggest metropolitan area in the world. Brazil has more inequality of wealth than any other country in the world.

        3. Some of Rio's slums began as government housing projects designed to segregate the poor from the rich.

        4. Population growth and urbanization are more serious now than ever. Brazil is now more urbanized than the United States, with more than three quarters of its population living in urban areas.

    B. The Population "Explosion"

        1. In 1804 the number of humans reached 1 billion, and in 1987 5 billion. On July 1, 2006, there were an estimated 6.52 billion people in the world.

        2. Where one person stood 12,000 year ago, there are now 1,050 people, and by 2100 it is projected that there will be nearly 1,700 people.

        3. Fewer than 250 of those 1,700 will be standing in rich countries.

        4. Some population analysts contend that we're in the midst of a population "explosion," an imagery that connotes horrific events, and widespread severe damage.

        5. **Demographers**, or population analysts, wish to convey this image.

        6. The concern of the population "bomb" is as old as the social sciences.

        7. In 1798 Robert Malthus proposed an influential theory of human population.

II. THEORIES OF POPULATION GROWTH

    A. The Malthusian Trap

        1. Malthus's theory rests on two facts and a questionable assumption.

            a. The facts: people must eat, and are driven by a strong sexual urge.

            b. The assumption: while food supply increases slowly and arithmetically, population size grows quickly and geometrically.

        2. "The superior power of population cannot be checked without producing misery and vice."

3. Two forces that can hold population growth in check:
    a. "Preventive" measures such as abortion, infanticide, and prostitution, which Malthus called "vices."
    b. "Positive checks" such as war, pestilence, and famine.
4. The **Malthusian trap** is a cycle of population growth followed by an outbreak of war, pestilence, or famine that keeps population growth in check.

B. A Critique of Malthus
1. Events that have cast doubt on several of Malthus's ideas and overstated pessimism:
    a. Technological advances allowing rapid growth in food production for each person on the planet is the opposite of the slow growth Malthus predicted.
    b. It is unclear what that natural upper limit to population growth is.
    c. Population growth does not always produce misery.
    d. Helping the poor does not generally result in the poor having more children.
    e. Although the human sexual urge is strong, people have developed contraceptive devices and techniques to control the consequences of their sexual activity. There is no necessary connection between sexual activity and childbirth.

C. Demographic Transition Theory
1. According to **demographic transition theory**, the main factors underlying population dynamics are industrialization and the growth of modern cultural values. The theory is based on the observation that the European population developed in four distinct stages.
2. The Preindustrial Period
    a. The **crude death rate**, which is the annual number of deaths per 1,000 people in a population, was high, due to poor nutrition, hygiene, and uncontrollable disease.
    b. The **crude birth rate**, which is the annual number of live births per 1,000 people in a population, was high.
3. The Early industrial Period.
    a. Increased economic growth and improved nutrition and hygiene.
    b. The crude death rate dropped.
    c. Life expectancy increased.
4. The Mature industrial Period
    a. The crude death rate continued to fall.
    b. The crude birth rate fell more dramatically because economic growth eventually changed people's traditional beliefs about the value of having many children.
    c. The crude birth rate took longer to decline than the crude death rate did, because people's values often change more slowly than their technologies.
    d. Population stabilized. "Economic development is the best contraceptive."
5. The Postindustrial Period
    a. In the last decade of the 20th century, the total fertility rate, which is the annual number of live births per 1,000 women in a population, continued to fall.
    b. The total fertility rate even fell below the **replacement level** of some countries—that is, the number of children each women must have on average for population size to

remain stable. The replacement level is 2:1, ignoring any inflow of settlers from other countries (**immigration** or **in-migration**) and any outflow of other countries (**emigration** or **out-migration**).

6. Due to the proliferation of low fertility of societies, some scholars suggest that we have entered a fourth, *postmodern* stage of population development, in which the number of deaths per year exceeds the number of births.

7. A Critique of Demographic Transition Theory
   a. An adequate theory of population growth must pay more attention to social factors other than industrialization and in particular to the role of social forces.

III. POPULATION AND SOCIAL INEQUALITY

A. Karl Marx

1. As one of Malthus's staunchest intellectual opponents, Karl Marx argued that the problem of overpopulation is specific to capitalism—not a problem of too many people, but a problem of too much poverty. If a society is rich enough to eliminate poverty, then its population is not too large. Thus exploitation of workers needs to be done away with for poverty to disappear.

2. Although Marx was wrong since overpopulation is not a serious problem in rich capitalist countries and more of a problem where capitalism is weakly developed, social inequality is the main cause of overpopulation.

B. Gender Inequality and Overpopulation

1. The case of Kerala:
   a. A state in India with more than 30 million people.
   b. Total fertility rate of 1.8 in 1991, half India's national rate and below the replacement level of 2:1.
   c. Not highly industrialized, and among the poorer Indian states.
   d. The decision to have children remains strictly a private matter, with no government intervention or prevention.
   e. How?
      i. Their government raised the status of women over a period of decades.
      ii. The government helped create a realistic alternative to a life of continuous childbearing and childrearing.
      iii. The government organized successful campaigns and programs to educate women, increase labor force participation, and make family planning available.
      iv. Keralan women enjoy the highest literacy rate, labor force participation, and political participation in India.
      v. Keralan women want small families, so they use contraception to prevent unwanted births.

2. Impact of the sex ratio: How can we find 100 million missing women?
   a. The **sex ratio** is the ratio of women to men in a geographical area.

C. Class Inequality and Overpopulation

1. The South Korean case:

      a. South Korea had a total fertility rate of 6.0 in 1960. Yet by 1989, the total fertility rate fell to 1.6.

      b. Why?

          i. Land reform not industrialization: the government took land from big landowners and gave it to small farmers.

          ii. The standard of living of small farmers increased, which eliminated the major reason for high fertility,—child labor and support for elderly.

      c. Increasing social inequality can lead to overpopulation, war, and famine.

          i. For example, in the 1960s the governments of El Salvador and Honduras encouraged the expansion of commercial agriculture and the acquisition of large farms by wealthy landowners.

          ii. Peasants were driven off the land and migrated to the cities, where they found squalor, unemployment, and disease.

          iii. These two countries with a combined population size of less than 5 million suddenly had a big "overpopulation" problem.

          iv. Increased competition for land contributed to rising tensions that eventually led to the war between El Salvador and Honduras in 1969.

      d. Economic inequality helps to create famines.

          i. Amartya Sen notes, "[f]amine is the characteristic of some people not having enough food to eat. It is not the characteristic of there not being enough food to eat." The source of famine is not underproduction or overpopulation, but inequality of access to food.

          ii. In his analysis of several famines, Sen found available food to be depleted or withheld.

          iii. Suppliers and speculators took advantage of short supply, as they hoarded grain and increased prices beyond the means of most people.

          iv. Food was withheld for political reasons to bring a population to its knees, or because many were not considered entitled.

   D. Summing Up

     1. Population growth is influenced by a variety of social causes, chiefly including social inequality, not from natural causes and industrialization and modernization alone.

     2. Well-intentioned Western analysts continue to insist that people in developing countries should be forced to stop multiplying, and some even suggest diverting scarce resources from education, health, and industrialization into forms of birth control including forced sterilization.

     3. They fail to see how measures that lower social inequality help to control overpopulation and its consequences.

IV. URBANIZATION

   A. Overpopulation is largely an urban problem. Most of the fastest growing cities are in semi-industrialized countries where the factory system is not highly developed. Urbanization is also taking place in the world's rich countries, but is expected to increase faster in semi-industrialized countries.

   B. From the Preindustrial to the Industrial City

     1. To a degree, urbanization results from industrialization.

2. However, industrialization is not the whole story.
    a. The connection between industrialization and urbanization is weak in the world's less developed countries.
    b. Cities emerged in Syria, Mesopotamia, and Egypt long before the growth of the modern factory.
    c. It was not industry, but trade in precious goods that stimulated the growth of cities in preindustrial Europe and the Middle East.
3. Preindustrial cities differed from industrial cities.
    a. Preindustrial cities were typically smaller, less densely populated, built within protective walls, and organized around a central square and places of worship.
    b. The industrial cities that emerged at the end of the 18th century were more dynamic and complex.

C. The Chicago School and the Industrial City
    1. The **Chicago School** of sociology was distinguished by their vividly detailed descriptions and analyses of urban life.
    2. Three of its leading members, Robert Park, Ernest Burgess, and Roderick McKenzie, proposed a theory of **human ecology** to illuminate the process of urbanization, which highlights the links between the physical and social dimensions of cities and identifies the dynamics and patterns of urban growth.
    3. The Concentric Zone Model
        a. The theory of human ecology holds that cities grow in ever-expanding concentric circles, sometimes called the "concentric zone model." Three social processes animate this growth.
            i. **Differentiation** refers to the process by which urban populations and their activities become more complex and heterogeneous over time.
            ii. **Competition** is an ongoing struggle by different groups to inhabit optimal locations.
            iii. **Ecological succession** takes place when a distinct group of people moves from one area to another, and another group moves into the old area to replace the first.
    4. Zonal patterns of differentiation, competition, and ecological succession in Chicago in the 1920s:
        a. Zone 1 was the central business district.
        b. Zone 2, the "zone in transition," was the area of most intense competition between residential and commercial interests.
        c. Zone 3 is the "zone of working-class homes."
        d. Zone 4 is the "residential zone" that contains small, middle-class, detached homes.
        e. Zone 5 is the "commuter zone," where middle-, upper-middle, and upper-class families live in more expensive detached homes, and commute to work in the city.
    5. Urbanism: A Way of Life
        a. For members of the Chicago school, the city involved a way of life, "a state of mind, a body of customs, ...traditions, ...attitudes and sentiments" called **urbanism**. Wirth's "urban way of life."

D. After Chicago: A Critique

1. Three major criticisms of the Chicago school approach to urban sociology over the years:
2. One criticism focuses on Wirth's characterization of the "urban way of life."
    a. Research shows that social isolation, emotional withdrawal, stress, and other problems may be just as common in rural as in urban areas.
    b. Research shows that urban life is less impersonal, anomic, and devoid of community than the Chicago sociologists made it appear.
    c. Even in the largest cities, most residents form social networks and subcultures that serve functions similar to those of the small community.
3. A second criticism focuses on the concentric zone model.
    a. The specific patterns discovered by the Chicago school are most applicable to industrial cities in the first quarter of the 20th century.
    b. With the automobile, some cities expanded in wedge-shaped sectors along natural boundaries and transportation routes.
    c. Other cities formed around many nuclei.
    d. The human ecology approach presents urban growth as an almost natural process, slighting its historical, political, and economic foundations in capitalist industrialization.

E. The Conflict View and New Urban Sociology
    1. The **new urban sociology**, heavily influenced by conflict theory, attempted to correct the problems of the Chicago School, and stressed that city growth is a process rooted in power relations toward profit.
        a. Urban space is seen as a set of *commodified* social relations.
        b. Political interests and conflicts shape the growth pattern of cities.
        c. John Logan and Harvey Molotch portray cities as machines fueled by a "growth coalition," which tries to obtain government subsidies and tax breaks to attract investment dollars.
    2. Although the growth coalition present redevelopment as a public good that benefits everyone, the benefits are often unevenly distributed.
        a. Most redevelopments are "pockets of revitalization surrounded by areas of extreme poverty."
        b. Local residents rarely enjoy any direct benefits from redevelopment.
    3. Community activism often targets local governments and corporations that seek unrestricted growth.

F. The Corporate City
    1. The **corporate city** refers to the growing post-World War II perception and organization of the North American city as a vehicle for capital accumulation.
    2. The Growth of Suburbs
        a. **Suburbanism** is a way of life organized mainly around the needs of children and involving higher levels of conformity and sociability than life in the central city.
        b. Initially restricted to the well to do, the suburban life became in reach of middle-class Americans following World War II, due to economic growth and government assistance. By 1970, more Americans lived in suburbs than in urban core areas, which is the case today.
    3. Gated Communities, Exurbs, and Edge Cities

a. **Metropolitan areas** include downtown city cores, surrounding suburbs, **gated communities, exurbs** (rural residential areas within commuting distance to the city, and **edge cities** (exurban clusters of malls, offices, and entertainment complexes that arise at the convergence point of major highways).

4. Factors that have stimulated the growth of exurban residential areas and edge cities:
   a. Costs of operating businesses in city cores;
   b. The growth of new telecommunication technologies that allow businesses to operate in the exurbs.

5. Some sociologists, urban and regional planners, and others lump all these developments together as indicators of **urban sprawl**.

6. City cores continued to decline as the middle class fled to suburbs and exurbs. Tax revenues fell. Despite urban renewal in the 1950s and 1960s, many cities remained in a state of decay.

7. Urban Renewal
   a. In a spate of urban renewal in the 1950s and 1960s, many low-income and minority homes were torn down and replaced with high-rise apartment buildings and office towers in the city core. In the 1970s and 1980s, some middle-class people moved into rundown areas and restored them in a process called **gentrification**.

G. The Urbanization of Rural America
   1. The case of Napa Valley.
   2. Napa Valley is an extreme case of what has happened throughout rural America since World War II. Agriculture has been industrialized and rural areas have been partly urbanized.
   3. Many small towns, especially those within an hour's drive of big cities, have also been partly urbanized.
   4. Newcomers to small-town America tend to be younger, more highly educated, more ethnically diverse, more urbane, and less committed to their adopted communities than old-time residents.

H. The Postmodern City
   1. The **postmodern city** is a new phenomenon that has emerged alongside of the legacy of old urban forms since about 1970. Three main features of the postmodern city:
      a. *Privatized*—exclusive organization of access to space;
      b. *Fragmented*—lacks a single way of life;
      c. *Globalized*—centers of financial decision-making.
   2. The global city has come to reflect the priorities of the global entertainment industry.
   3. New high-tech, clean, controlled, predictable and safe forms of entertainment do little to:
      a. Increase the economic well being of the communities beyond creating some low-level, dead-end jobs.
      b. Provide ways of meeting new people, seeing old friends and neighbors, and improving urban sociability.
      c. Enable cities and neighborhoods to retain and enhance their distinct traditions, architectural styles, and ambiance.

Study Activity: Applying The Sociological Compass

After reading Chapter 20, review the concentric zone model and Figure 20.5. Consider your place of residence and the larger metropolitan area in which you live. Based on the descriptions of zonal patterns described and illustrated on pages 591-593 in the textbook, what zone does your neighborhood fall in? What zones do the surrounding areas fall into? Categorize your neighborhood, the surrounding communities and businesses throughout the metropolitan area in which you live. Describe these areas, and explain why you categorized them in the way you did.

Zone 1:

Zone 2:

Zone 3:

Zone 4:

Zone 5:

Infotrac College Edition Online Exercises

For the following exercises, log on to the online library of InfoTrac College Edition at http://www.infotrac-college.com/. Make note that InfoTrac has implemented a new registration system that will allow easier access to InfoTrac through the use of a personalized username and password. Once you've created your username and password you may proceed directly to the Log On page. To create an account, register your passcode packaged with your textbook, and create a username and password, by following the online prompt. After you are logged in, click on "Infotrac College Edition." You will arrive at a screen that enables you to search topics.

Keyword: **urbanization**. Look for articles that deal with urbanization in low-income countries. How does urbanization in low-income countries differ from urbanization in higher-income countries? Write a short paper discussing the effect of urbanization in low-income and higher-income countries. How does your Infotrac research compare with what is said in Chapter 20?

Keyword: **zero population**. Search for four articles on zero population. Do most articles support the idea of zero population growth? In what context is zero population most likely to be discussed? What sociological theory is zero population related to?

Keyword: **edge city**. Look for the article by Brenda Case Scheer called, "Edge City, Morphology: A Comparison of Commercial Centers". How does this discussion of edge cities differ from the one presented in Chapter 20? Is there an edge city in your hometown? If yes, where is its location? If no, why not? Write a short paper with your findings.

Internet Exercises

Review the fertility rates of the countries listed in Table 20.1. Choose at least two countries with the lowest fertility rates, and two countries with the highest fertility rates. Conduct additional research on these countries using the Internet. Simply explore and see what general information you dig up regarding each country. Take notes on each country.

1. Notes on Countries

2. Country One:

3. Country Two:

4. Country Three:

5. Country Four:

1. Based on any descriptions or statistics on each country, which theory of population growth and global inequality makes the most sense to you, and why?

2. Theory of Population Growth:

3. Theory of Global Inequality:

Go to the following URL for international demographic data: http://www.census.gov/ftp/pub/ipc/www/idbsum.html

Write a short paragraph and answer these questions. What was the population in 1990 for any two countries? What was the projected population for 2000 for the same countries? What factors do you think accounted for the population changes?

The World PopClock from the U.S. Bureau of the Census gives the complete and total population of the world and its projected population to the day, hour, minute and second. Go to this URL: http://www.census.gov/cgi-bin/ipc/popclockw

Indicate what time and date you looked up this URL. What is the total population of the world? Tomorrow, or in two days, go back to the website and look again. What time and date is it? What is the total population of the world now? Did the population increase or decrease?

Practice Tests

Fill-In-The-Blank

Fill in the blank with the appropriate term from the above list of "key terms."

1. _____ is the spread of cities into ever-larger expanses of the surrounding countryside.

2. The _____ refers to a cycle of population growth followed by an outbreak of war, pestilence, or famine that keeps population growth in check.

3. _____ refers to the struggle by different groups for optimal locations in which to reside and set up their business.

4. The _____ is the annual number of live births per 1000 women in a population.

5. _____ is a way of life that involves increased tolerance but also emotional withdrawal and specialized, impersonal, and self-interested interaction.

6. The _____ is the number of children that each woman must have on average for population size to remain stable.

7. The _____ is the ratio of women to men in a geographical area.

8. The _____ refers to the growing post-World War II perception and organization of the North American city as a vehicle for capital accumulation.

9. _____ refers to the process by which urban populations and their activities become more complex and heterogeneous over time.

10. _____ is the process of middle-class people moving into rundown areas of the inner city and restoring them.

Multiple-Choice Questions

Select the response that best answers the question or completes the statement:

1. Which is a contention about Malthus's theory?
   a. People must eat.
   b. People are driven by a strong sexual urge.
   c. While food supply increases slowly and arithmetically, size grows quickly and geometrically.
   d. All of these are correct.
   e. None of the above.

2. The Malthusian trap refers to:
    a. a cycle of population growth followed by an outbreak of war, pestilence, or famine that keeps population growth in check.
    b. the annual number of deaths due to poor nutrition, hygiene, and uncontrollable disease.
    c. the process by which urban populations and their activities become more complex and heterogeneous over time.
    d. All of these are correct.
    e. None of the above.

3. Which is NOT a stage of European population development according to demographic transition theory?
    a. preindustrial stage
    b. early industrial stage
    c. mature industrial stage
    d. postindustrial stage
    e. All of the answers are stages according to demographic transition theory.

4. Which stage of demographic transition theory involved a high crude death rate and a high crude birth rate?
    a. preindustrial stage
    b. early industrial stage
    c. mature industrial stage
    d. postindustrial stage
    e. All of these are correct.

5. "Immigration" is synonymous with _____, and "emigration" is synonymous with _____.
    a. out-migration; in-migration
    b. in-migration; lateral migration
    c. out-migration; vertical migration
    d. lateral migration; vertical migration
    e. in-migration; out-migration

6. Kerala, one of the poorer Indian states, was marked by low fertility rates in 1991, in spite of not being highly industrialized. Why does the case of Karela contradict the common assumption that impoverished regions display high fertility rates and population growth?
    a. Their government raised the status of women.
    b. Their government prevented women from having too many children.
    c. Infanticide was a widespread practice during this time period.
    d. All of these are correct.
    e. None of the above.

7. In comparison to rich countries, urbanization is expected to _____ in semi-industrialized countries.
   a. decrease
   b. increase faster
   c. be consistent
   d. be equal as
   e. None of the above.

8. Human ecology theory:
   a. stresses that city growth is a process rooted in power relations toward profit.
   b. holds that cities grow in ever-expanding concentric circles.
   c. economic underdevelopment results from poor countries lacking Western attributes.
   d. All of these are correct.
   e. None of the above.

9. The _____ city refers to the growing post-World War II perception and organization of the North American city as a vehicle for capital accumulation.
   a. postmodern
   b. urban
   c. metropolitan
   d. corporate
   e. None of the above.

10. Which is a main feature of the postmodern city?
    a. privatized
    b. fragmented
    c. globalized
    d. All of these are correct.
    e. None of the above.

11. According to Brym and Lie, the postmodern city:
    a. does little to increase the economic well being of the community beyond creating some low-level, dead-end jobs.
    b. improves urban sociability.
    c. enables cities and neighborhoods to retain and enhance distinct traditions.
    d. All of these are correct.
    e. None of the above.

12. Which statement is true of the city Rio De Janeiro, Brazil?
    a. It is one of the world's most beautiful cities.
    b. The inner city of Rio is a place of great wealth and beauty.
    c. Rio is a larger metropolitan area than Chicago.
    d. All of these are correct.
    e. None of the above.

13. What is the projected world population for 2013?
    a. 10 billion
    b. 9 billion
    c. 8 billion
    d. 7 billion
    e. 6 billion

14. The main purpose of demography is to:
    a. figure out why the size, geographical distribution, and social composition of human populations change over time.
    b. figure out the social and geographical patterns of social inequality.
    c. figure out root social and economic causes of urban crime.
    d. All of these are correct.
    e. None of the above.

15. Among the world's 10 largest cities in 2015, only one will be in a highly industrialized country. All the others will be in developing countries. Which city will this be?
    a. Rio de Janeiro
    b. New York
    c. London
    d. Mexico City
    e. Tokyo

16. Which country has the highest fertility rate?
    a. Niger
    b. Liberia
    c. Afghanistan
    d. Taiwan
    e. Mexico

17. The sex ratio tends to be low where women have:
    a. greater access to health care than do men.
    b. about equal access to medicine men do.
    c. less adequate nutrition than men do.
    d. All of these are correct.
    e. None of the above.

18. What was the largest metropolitan area in 1900?
    a. London, England
    b. Chicago, United States
    c. Tokyo, Japan
    d. Beijing, China
    e. Mexico City, Mexico

19. Which city is the largest in the United States?
    a. Boston, MA
    b. Memphis, TN
    c. San Antonio, TX
    d. Philadelphia, PA
    e. New York, NY

20. Napa Valley, California is an extreme case of what has happened throughout America since World War II in:
    a. corporate cities.
    b. gated communities.
    c. rural areas.
    d. metropolitan areas.
    e. None of the above.

21. The movie *8 Mile* depicts the separation of the rich and the poor in the most racially segregated cities in the United States. Which city is portrayed?
    a. Chicago
    b. Los Angeles
    c. Portland
    d. Detroit
    e. San Diego

22. A decrease in the percentage of people living in high-poverty neighborhoods were biggest in the:
    a. Midwest and the South.
    b. Southwest.
    c. East.
    d. All of these are correct.
    e. None of the above.

23. Who argued that the problem of overpopulation is specific to capitalism?
    a. Weber
    b. Marx
    c. Engels
    d. Malthus
    e. Durer

24. Which region is one of the most prosperous regions of the world, despite its rapid population growth over the past 200 years?
    a. Western Europe
    b. Latin America
    c. the Caribbean
    d. the Western Pacific
    e. South America

25. Which theory especially suggests that the main factors underlying population dynamics are industrialization and the growth of modern cultural values?
    a. human ecology theory
    b. demographic transition theory
    c. Marxist theory
    d. Malthusian theory
    e. None of the above.

True False

1. The concern of population explosion is as old as the social sciences.

   TRUE or FALSE

2. According to demographic transition theory, the crude birth rate took longer than the crude death rate did in the mature industrial period.

   TRUE or FALSE

3. Although their respective theories differ on many points, Malthus and Marx were generally in agreement about the problem of overpopulation.

   TRUE or FALSE

4. Sociologically, population growth is often influenced by patterns of natural causes.

   TRUE or FALSE

5. Initially restricted to the well to do, suburban life became reachable for middle-class Americans following World War II.

   TRUE or FALSE

6. The postmodern city provides ways of meeting new people, and seeing old friends and neighbors.

   TRUE or FALSE

7. Brazil is now more urbanized than the United States.

   TRUE or FALSE

8. Hong Kong is one of the most densely populated places in the world.

   TRUE or FALSE

Short Answer

1. Explain democratic transition theory. (p. 584)

   _____

   _____

2. Explain urbanism and illustrate with concrete examples. (p. 592-593)

   _____

   _____

3. List and define the three social processes of the concentric zone model. (p. 591-592)

   _____

   _____

4. Briefly describe the contradictory characteristics of Rio de Janeiro, Brazil, as both a beautiful city, and at the same time a classic case of social inequality and urban decay. (p. 579-580)

   _____

   _____

5. Briefly explain the Malthusian trap, and provide an example. (p. 583)

   _____

   _____

Essay Questions

1. Define the terms "crude death rate", "crude birth rate", and "replacement level". What is the relationship between these terms and the four stages of European population development? (p. 584-586)

2. Explain the impact of inequality on overpopulation in terms of gender and social class, respectively. In doing so, describe the cases of Kerala, in India, and South Korea. (p. 586-589)

3. Explain the general approach of the new urban sociology, in light of the three main critiques of the Chicago School. (p. 589-594)

4. Explain the criticisms of Malthus' approach to population growth. Also, explain Marx's position in relation to Malthus. Where do you stand in relation to these ideas? (p. 583-589)

5. Compare and contrast the *corporate city* with the *postmodern city*. Where do you want to live in the future, and why? (p. 594-601)

Solutions

Practice Tests

Fill-In-The-Blank

1. Urban sprawl
2. Malthusian trap
3. Competition
4. crude birth rate

5. Urbanism
6. replacement level
7. sex ratio
8. corporate city

9. Differentiation
10. Gentrification

Multiple-Choice Questions

1. D, (p. 583)
2. A, (p. 583)
3. D, (p. 584-585)
4. A, (p. 584)
5. E, (p. 585)
6. A, (p. 586-587)
7. B, (p. 589-590)
8. B, (p. 591)
9. D, (p. 594)

10. D, (p. 597-601)
11. A, (p. 597-601)
12. D, (p. 579)
13. D, (p. 581)
14. A, (p. 581)
15. E, (p. 589)
16. A, (p. 586)
17. C, (p. 588)
18. A, (p. 589)

19. E, (p. 595)
20. C, (p. 596-597)
21. D, (p. 598)
22. A, (p. 599)
23. B, (p. 586)
24. A, (p. 583)
25. B, (p. 584-586)

True False

1. T, (p. 582)
2. T, (p. 584-585)
3. F, (p. 583-584, 586)

4. F, (p. 579-589)
5. T, (p. 591-601)
6. F, (p. 597-601)

7. T, (p. 580)
8. T, (p. 582)

COLLECTIVE ACTION AND SOCIAL MOVEMENTS

Student Learning Objectives

After reading Chapter 21, you should be able to:

1. Identify the social conditions under which people act collectively to change, or resist change, to society.

2. Explain the formation of the lynch mob phenomena, historically, by applying breakdown theory. You should also be able to critique breakdown theory.

3. Compare and contrast breakdown theory, solidarity theory, and frame alignment theory as explanations of collective action and social movements.

4. Describe the historical trends of unionization and its recent decline, in the United States.

5. Describe the history of social movements by identifying historical events.

6. Identify and define the types of citizenship that have been expanded and defended by social movements.

7. Contrast "old social movements" and "new social movements."

8. Discuss the future of social movements and the prospects for postmodern revolutions.

Key Terms

collective action (606)

social movements (607)

breakdown theory (609)

absolute deprivation (609)

relative deprivation (609)

contagion (609)

strain (610)

rumors (612)

solidarity theory (614)

resource mobilization (614)

political opportunities (615)

social control (615)

union density (618)

frame alignment (621)

civil citizenship (625)

political citizenship (625)

Detailed Chapter Outline

I. HOW TO SPARK A RIOT

    A. Personal Anecdote

        1. Robert Brym almost sparked a riot in 1968.

        2. Historical events of 1968

            a. Student riots in France.

            b. Violent suppression of students in Mexico.

            c. Free speech and anti-war student movement at Berkeley and Michigan.

            d. Civil rights for American blacks and women.

    B. Under what social conditions do people act in unison to change, or resist change, to society?

    C. The Study of Collective Action and Social Movements

        1. **Collective action** occurs when people act in unison to bring about or resist social, political, and economic change. Some collective actions are routine. Others are nonroutine. Routine collective actions are typically nonviolent and follow established patterns of behavior in existing types of social structures. Nonroutine collective actions take place when usual conventions cease to guide social actions and people transcend, by-pass, or subvert established institutional patterns and structures.

        2. **Social movements** are enduring collective attempts to change part or all of the social order by means of rioting, petitioning, striking, demonstrating, and establishing lobbies, unions, and political parties.

        3. The final section makes some observations about the changing character of social movements, in the context of historical attempts by underprivileged groups.

II. NONROUTINE COLLECTIVE ACTION: THE LYNCH MOB

    A. The Lynching of Claude Neal

        1. On October 27, 1934, a 23-year-old black man named Claude Neal was lynched near Greenwood. He was accused of raping and murdering 19-year-old Lola Cannidy, a pretty white woman, based on evidence that was not totally convincing.

        2. A lynch mob formed and brutally tortured and murdered Neal. The nude and mutilated body of Claude Neal was strung up on a tree in the lawn of the Jackson County courthouse, to symbolize justice.

    B. Breakdown Theory

        1. **Breakdown theory** suggests that social movements emerge when traditional norms and patterns of social organization are disrupted or breakdown. Breakdown theory may be seen as a functionalist theory, for it regards collective action as a form of social imbalance that results from various institutions functioning improperly.

C. Deprivation, Crowds, and the Breakdown of Norms
1. Following Charles Tilly and his associates, three factors may be grouped under breakdown theory. Most pre—1970s sociologists would have said that the lynching of Neal was caused by one or more of these factors.
   a. *A background of economic deprivation experienced by impoverished and marginal members of the community.* **Relative deprivation**, the growth of an intolerable gap between the social rewards people expect to receive and those they actually receive, generates collective action, not **absolute deprivation** or extreme poverty.
   b. *The inherent irrationality of crowd behavior.* Gustave Le Bon's **contagion** theory of crowd behavior argues that the transformation of individuals into "barbarians" occurs because people lose their individuality and will power when they join a crowd. People in crowds are often able to perform extraordinary and outrageous acts.
   c. *The serious violation of norms.* Intimate contact between blacks and whites in the South was strictly forbidden. Therefore, black-on-white rape and murder were the deepest violations of the region's norms, and bound to evoke a strong reaction by the dominant race. Pre—1970s sociologists highlighted the breakdowns in traditional norms that preceded group interest, and referred to them as indicators of **strain**.

D. Assessing Breakdown Theory
1. Since 1970, sociologists have uncovered flaws in all three elements of breakdown theory.
2. Deprivation
   a. Research shows no clear association between fluctuations in economic well-being and the number of lynchings between the 1880s and the 1930s.
   b. In most cases of collective action, leaders and early joiners are well-integrated members of their communities, not deprived socially marginal individuals. This was the case involving the Neal lynching.
3. Contagion: As the Neal lynching and research show, albeit wild, nonroutine collective action is usually structured by:
   a. Ideas and norms that emerge in the crowd itself;
   b. The predispositions that unite crowd members and predate their collective action;
   c. The degree to which different types of participants adhere to emergent preexisting norms;
   d. Preexisting social relationships among participants.
4. Strain
   a. Although the alleged rape did violate the deepest norms of the Old South in a pattern that was often repeated, lynching had deeper roots than the mere violation of norms governing black-white relations. It was a chief means by which black farm workers were disciplined and kept tied to the southern cotton industry after the abolition of slavery threatened to disrupt the industry's traditional, captive labor supply.
   b. Lynching was a two-sided phenomenon.
      i. Breakdown theory explains one side—that lynching was partly a reaction to the violation of norms that threatened to *disorganize* traditional social life in the South.
      ii. Lynching also grew out of, and was intended to maintain, the traditional *organization* of the South's cotton industry.

E. Rumors and Riots
    1. The study of rumor reinforces the idea that social organization underlies all collective action. **Rumors** are claims about the world that are not supported by authenticated information.
    2. While rumor transmission is a form of collective action in its own right, it intensifies just before and during riots.
    3. In *Henry IV* Shakespeare wrote that "Rumor is a pipe blown by surmises, jealousies, conjectures." More recent analyses link rumors to the primary emotions of hope, fear, and anger.
    4. Rumors are often false and unverifiable. They seem credible to insiders but preposterous to outsiders.
    5. Sociological understanding of rumors reveal the distribution of hope, fear, and anger in society and the structural flaws that lie beneath these emotions.
    6. Brym and Lie's discussion of lynching leads to the conclusion that collection action is rarely a short-term reaction to disorganization and deprivation. It is usually part of a long-term attempt to correct perceived injustice that requires a sound social-organizational basis.

III. SOCIAL MOVEMENTS

    A. Solidarity Theory
        1. Social movements emerge from collective action only when the discontented succeed in building up an organizational base.
        2. **Solidarity theory** is a variant of conflict theory, and suggests that social movements are social organizations that emerge when potential members can mobilize resources, take advantage of new political opportunities, and avoid high levels of social control by authorities.
        3. Resource Mobilization
            a. Most collective action is part of a power struggle. The struggle intensifies as disadvantaged groups become more powerful as a result of **resource mobilization**, or the process by which groups engage in more collective action as their power increases due to their growing size and increasing organizational, material, and other resources.
            b. Solidarity theory links the timing of collective action and social movement formation to the emergence of new **political opportunities**, which occurs when influential allies offer insurgents support, when ruling political alignments become unstable, and when elite groups become divided and weaker. Collective action takes place and social movements crystallize under these conditions.
            c. Government reactions to protest influence subsequent protest. Attempts by governments to diffuse a protest through **social control** measures, such as concessions and co-opting protest leaders, may encourage protesters to press further, in response to beliefs that the government is weak and indecisive.

IV. CASE STUDY: STRIKES AND THE UNION MOVEMENT IN AMERICA

    A. Workers have engaged in collective action in three ways: unionization, creation of labor parties, and striking.
    B. The example of 1934, as one of the bloodiest year of collective violence in American history.

C. In 1945, unionization reached its historical peak in the United States. **Union density** (union members as a percentage of non-farm workers) remained above 30% until the early 1960s.

D. The United States has the lowest union density for any rich industrialized country today. What accounts for the post—1945 drop?

E. Strikes and Resource mobilization.
  1. The industrial working class has shrunk and become weaker, yet is the precise source of strength of unionism.
  2. The industrial working class has been weakened by globalization and employer hostility to unions. For example, employers could now close American factories and relocate them in other countries like Mexico and China.

F. Strikes and Political opportunities.
  1. Government action has limited opportunities for union growth since the end of World War II.
  2. In 1947, Congress passed the Taft-Hartley Act in reaction to a massive post-World War II strike wave.
    a. Unions could no longer force employees to become members or require union membership as a condition of being hired.
    b. The act allowed employers to replace striking workers.
  3. Based on resource mobilization theory, less social organization means little protest.
  4. Comparing historical periods, we see low union density has helped to virtually extinguish the strike as a form of collective action.
  5. Over the short term, strikes tend to be more frequent during economic "booms" and less frequent during economic "busts."

G. Recent Developments in the Union Movement
  1. Since the mid-1990s, the American Federation of Labor-Congress of Industrial Organizations (AFL-CIO), the largest union umbrella organization in the United States, has sought to reverse the trends in the decline of unionization, by organizing immigrants, introducing feminist issues, and developing new forms of employee organization and representation more appropriate to a postindustrial society.

V. FRAMIING DISCONTENT

  A. **Frame alignment** is the process by which individual interests, beliefs, and values either become congruent with the activities, ideas, and goals of the movement or fail to do so. Frame alignment has recently become the subject of sustained sociological investigation by scholars in the symbolic interactionist tradition.
    1. Frame alignment can be encouraged in several ways. Social movement leaders can reach out to other organizations that contain people who may be sympathetic to their movement's cause.
    2. Movement activists can stress popular values that have so far not featured prominently in the thinking of potential recruits. They can also elevate the importance of positive beliefs and values about the movement.
    3. Social movements can stretch their objectives and activities to win recruits who are not initially sympathetic to the movement's original aims, which may water down the movement

ideals. Alternatively, movement leaders may decide to take calculated action to appeal to nonsympathizers on grounds that have little or nothing to do with the movement's purpose. For example, when rock, punk, or reggae bands play at nuclear disarmament rallies.

B. An Application of Frame Alignment Theory: Back to 1968
1. Frame alignment theory stresses face-to-face interaction strategies employed to recruit nonmembers.
2. Resource mobilization theory focuses on the broad social-structural conditions that facilitate the emergence of social movements.
3. The two theories help clarify Robert Brym's experience in 1968.
    a. He lived in a poor and relatively unindustrialized region of Canada where people had few resources to mobilize.
    b. Many of Robert's classmates did not share his sense of injustice, and thus received an unsympathetic hearing.

VI. SOCIAL MOVEMENTS FROM THE 18TH CENTURY TO THE 21ST CENTURY

A. Determinants of Collective Action and Social Movement Formation
1. Altogether, breakdown theory, solidarity theory, and frame alignment theory provide a comprehensive picture of the why, how, when, and who of collective action and social movements.
2. Breakdown theory partly answers the question of *shy* discontent is sometimes expressed collectively and in nonroutine ways.
3. Solidarity theory focuses on *how* these social changes may eventually facilitate the emergence of social movements,—and speaks to the question of *when* collective action erupts and social movements emerge.
4. Frame alignment theory directs our attention to the question of *who* is recruited to social movements.

B. The History of Social Movements
1. In 1700, social movements were typically small, localized, and violent.
2. As the state grew, social movements changed in three ways.
    a. They became national in scope, and typically directed against central governments.
    b. Their membership grew partly because potential recruits were now literate, and big new social settings could serve as recruitment bases.
    c. They became less violent, larger, more organized, bureaucratized, stabilized, and sufficiently powerful without resorting to violence.
3. In Britain, social movements often used their power to expand the rights of citizens, which may be categorized into four stages:
    a. In 18th—century Britain, rich property owners fought against the king for **civil citizenship**, which is the right to free speech, freedom of religion, and justice before the law.
    b. In the 19th century, the male middle class and the more prosperous strata of the working class fought against rich property owners for **political citizenship**, which is the right to vote and run for office. In early 20th—century Britain, women and poorer workers achieved these same rights despite the opposition of well to do men.

    c. During the remainder of the 20th century, blue- and white-collar workers fought against the well to do for **social citizenship**, which is the right to a certain level of economic security and full participation in social life with the help of the modern welfare state.

  4. In the United States, the timing of the struggle for citizenship rights was different.

    a. Universal suffrage for white males was won earlier in the 19th century than in Europe.

    b. Although the 15th Amendment to the Constitution gave African Americans the right to vote in 1870, most of them were unable to exercise that right, at least in the South, from the late 19th century until the 1960s, due to restrictions on voter registration, including poll taxes and literacy tests.

    c. The civil rights movement of the 1960s was in part a struggle over the voting rights of African Americans.

C. The Future of Social Movements

  1. Inspired by the civil rights movements of the 1960s, so-called **new social movements** emerged in the 1970s, and are distinguishable for their breadth of goals, the kinds of people they attract, and their potential for globalization.

  2. Goals

    a. Some new social movements promote the rights of humanity as a whole to peace, security, and a clean environment, such as the peace movement, environmental movement, and the human rights movement.

    b. Other new social movements promote the rights of particular groups excluded from full social participation, such as the women's movement and the gay rights movement.

    c. The emergence of the peace, environmental, human rights, gay rights, and women's movements marked the beginning of the fourth stage in the history of social movements, which involves **universal citizenship**, or the extension of citizenship right to all adult members of society and society as a whole.

  3. Membership

    a. New social movements attract a disproportionately large number of highly educated, relatively well-to-do people from the social, educational, and cultural fields, such as teachers, college professors, journalists, social workers, artists, actors, writers, and students.

    b. Their higher education exposes them to appealing radical ideas.

    c. They tend to possess jobs outside of the business community, which often opposes their values.

    d. They often get personally involved in the problems of their clients and audiences, and sometime become advocates.

  4. Globalization Potential

    a. New social movements have more potential for globalization than did old social movements.

    b. Social movements were typically *national* in scope up until the 1960s. Many new social movements since the 1970s increased the scope of protest beyond the national level.

      i. Members of peace and environmental movements pressed for *international* agreements binding all countries to protect the environment and stop the spread of nuclear weapons.

    c. Inexpensive international travel and communication facilitated the globalization of social movements.

5. An Environmental Social Movement
   a. The case of Greenpeace.
   b. In 1953, 110 international social movement organizations spanned the globe. By 1993, there were 631.
   c. "Old" social movements can go global due to changes in the technology of mobilizing supporters. For example, the case of peasant uprising in Chiapas, Mexico.
      i. Oppressed by Europeans and their descendants for nearly 500 years, the poor, indigenous people of southern Mexico were facing a government edict preventing them from gaining access to formerly communal farmland for subsistence agriculture.
      ii. The government wanted the land in the hands of large Hispanic ranchers and farmers, who could earn foreign revenue by exporting goods to the United States and Canada under the new North American Free Trade Agreement.
      iii. The peasants seized a large number of ranches and farms.
      iv. "Subcomandante Marcos", leader of the Zapatista National Liberation Army, effectively used the Internet and the international mass media against the Mexican government.
      v. The New *York Times* called this "the first postmodern revolution," combining uprising with the World Wide Web, short-wave radio, and photo spreads in *Marie Claire*.

Study Activity: Applying The Sociological Compass

After reading Chapter 21, re-read the set of questions posed by Brym and Lie on page 624. In addition to engaging the exercise described in the text, carry out the activity from a slightly different angle. As Brym and Lie suggest, "Try applying solidarity and frame alignment theories". Being a student on a college campus, you may have been part of an audience of a political rally, presentation, or discussion. If not, seek out an opportunity on your campus or community to participate in such an event that deals with some political issue, such as a faculty rally, student protest of tuition, an anti-war rally, a meeting of minority advocates, an AFL-CIO sponsored event, and so forth. Make a list of the issues discussed. Make note of personal and emotional response to the event. Did you find yourself to be emotionally charged or bored with the activity? Why? Describe what you did or did not do about it. Explain your response or lack thereof, in terms of solidarity theory and frame alignment theory.

List of Issues:

Your Personal and Emotional Responses:

Explanations of Your Personal Responses and Action or Inaction:

Solidarity theory:

Frame alignment theory:

Infotrac College Edition Online Exercises

For the following exercises, log on to the online library of InfoTrac College Edition at http://www.infotrac-college.com/. Make note that InfoTrac has implemented a new registration system that will allow easier access to InfoTrac through the use of a personalized username and password. Once you've created your username and password you may proceed directly to the Log On page. To create an account, register your passcode packaged with your textbook, and create a username and password, by following the online prompt. After you are logged in, click on "Infotrac College Edition." You will arrive at a screen that enables you to search topics.

Keyword: **Land mines**. Those who advocate the elimination of land mines may be considered part of a new global social movement. Look for articles that deal with the elimination of land mines. How widespread is this social movement? Do you think there is a new global social movement in this area?

Keyword: **Human cloning**. Look for articles that deal with human cloning and collective action. What specific aspects of collective action are portrayed in the articles? Do you think that there will be a human cloning social movement in the future?

Keyword: **bioterrorism.** Bioterrorism is a type of terrorism that strikes fear in people's hearts. Look for articles that discuss how a society can prepare for this type of terrorism. What type of bioterrorism is discussed most frequently?

Keyword: **Million Mom March**. This social movement supports gun control. What type of social movement is this?

Keyword: **Promise Keepers**. Research this social movement and classify it according to the characteristics of a social movement in Chapter 21.

Internet Exercises

For this exercise, go to: http://www.aflcio.org/, the Web site of the AFL-CIO. Navigate through the Web site to learn as much as possible about their organizational goals, activities, scope, resources, and so forth. Based on the information you acquire, identify the characteristics of the AFL-CIO that coincide with solidarity theory and frame alignment theory. Make a list of organizational goals, activities, scope, resources, and so forth. Categorize the organizational characteristics of the AFL-CIO in terms of solidarity theory and frame alignment theory. Do you have any suggestions on strategies for the AFL-CIO to consider in light of what you learned in Chapter 21?

September 11, 2001, is a tragic day to remember in American history. International terrorist groups changed American's lives forever. To learn about international terrorist groups go to this site: http://www.specialoperations.com/

Research two different groups. How do these two groups differ from the Ku Klux Klan in the U.S.? Write a short paper noting your comparisons.

The Hate Directory lists hate groups and has many links of websites. It is located at: http://www.bcpl.net/~rfrankli/hatedir.pdf

This is a helpful site for research and discussion purposes. Research two groups on this list? Does your research support the idea that your two groups could be classified as social movements or collective action? Compare the ideas of social movements and collective action in Chapter 21 with your two groups in a short paper.

Recent media attention has been given to the gay and lesbian movement. The homepage of The Gay and Lesbian Activists Alliance is located at:

http://www.glaa.org/

This organization is the oldest continuously active gay and lesbian civil rights organization. It was founded in 1971 to advance the equal rights of gay men and lesbians in Washington, DC. Look at the main issues of the alliance and discuss how they fit in with the progressive steps of collective behavior discussed in the text.

Practice Tests

Fill-In-The-Blank

Fill in the blank with the appropriate term from the above list of "key terms."

1. _____ recognizes the right of marginal groups to full citizenship and the rights of humanity as a whole.

2. _____ is the number of union members in a given location and time as a percentage of nonfarm workers.

3. _____ occurs when people act in unison to bring about or resist social, political, and economic change.

4. _____ recognizes the right to run for office and vote.

5. _____ is a condition of extreme poverty.

6. _____ is the process by which exteme passions supposedly spread rapidly through a crowd like a contagious disease.

7. _____ is the process by which individual interests, beliefs, and values become congruent and complementary with the activities, goals, and ideology of a social movement.

8. _____ are claims about the world that are not supported by authenticated information.

9. _____ refers to the process by which social movements crystallize due to increasing organizational, material, and other resources of movement members.

10. _____ recognizes the right to free speech, freedom of religion, and justice before the law.

11. _____ refers to the means by which authorities seek to contain collective action, including co-optation, concessions, and coercion.

Multiple-Choice Questions

Select the response that best answers the question or completes the statement:

1. _____ are enduring collective attempts to change part or all of the social order.
   a. Routine collective actions
   b. Nonroutine collective actions
   c. Race riots
   d. Social movements
   e. None of the above

2. Which theory of social movements is most consistent with functionalism?
   a. breakdown theory
   b. solidarity theory
   c. frame alignment theory
   d. resource mobilization theory
   e. None of the above.

3. _____ is an intolerable gap between social rewards people expect and those they actually receive.
   a. Contagious deprivation
   b. Absolute deprivation
   c. Relative deprivation
   d. Deprivation density
   e. None of the above.

4. Which is a factor that caused the lynching of Claude Neal in 1934 according to breakdown theory?
   a. A background of economic deprivation experienced by impoverished and marginal members of the community.
   b. The inherent irrationality of crowd behavior.
   c. The serious violation of norms.
   d. All of these are correct.
   e. None of the above.

5. Which theory is most closely a variant of conflict theory?
   a. breakdown theory
   b. solidarity theory
   c. frame alignment theory
   d. All of these are correct.
   e. None of the above.

6. _____ is a process by which groups engage in more collective action as their power increases due to their growing size and increasing organizational, material, and other resources.
   a. Resource mobilization
   b. Political opportunities
   c. Social control
   d. All of these are correct.
   e. None of the above.

7. Which is NOT a way workers have engaged in collective action?
   a. unionization
   b. creation of labor parties
   c. striking
   d. social control
   e. All of the answers are ways workers have engaged in collective action.

8. Union density refers to:
   a. the success rate of labor unions since the 1980s.
   b. union members as a percentage of non-farm workers.
   c. the level of resource mobilization of unions.
   d. All of these are correct.
   e. None of the above.

9. Based on resource mobilization theory, less social organization means:
   a. more protest.
   b. violent protest.
   c. little protest.
   d. All of these are correct.
   e. None of the above.

10. Frame alignment can be encouraged when:
    a. social movement leaders limit recruitment only to people with the same social class position.
    b. social movements constrain their activities strictly to the relevance of the mission and goals.
    c. movement activists stress popular values that have so far not featured prominently in the thinking of potential recruits.
    d. All of these are correct.
    e. None of the above.

11. Which theory focuses on *why* discontent is sometimes expressed collectively and in nonroutine ways?
    a. breakdown theory
    b. solidarity theory
    c. frame alignment theory
    d. All of these are correct.
    e. None of the above.

12. _____ theory directs our attention to the question of *who* is recruited to social movements.
    a. Breakdown
    b. Solidarity
    c. Frame alignment
    d. All of these are correct.
    e. None of the above.

13. Which theory focuses on *how* social changes may facilitate social movements?
    a. breakdown theory
    b. solidarity theory
    c. frame alignment theory
    d. All of these are correct.
    e. None of the above.

14. In what way did social movements change as the state grew?
    a. They became national in scope.
    b. Their membership grew.
    c. They became more organized and less violent.
    d. All of these are correct.
    e. None of the above.

15. Which type of citizenship refers to the right to free speech, freedom of religion, and justice before the law?
    a. civil citizenship
    b. political citizenship
    c. social citizenship
    d. universal citizenship
    e. None of the above.

16. Which is NOT a characteristic of new social movements that distinguish them from old social movements?
    a. goals
    b. membership
    c. globalization
    d. organization
    e. All of the answers are characteristics of new social movements that distinguish them from old social movements.

17. The integration of a peasant uprising with the World Wide Web, by the people of Chiapas, Mexico, may be considered:
    a. the first postmodern revolution.
    b. an example of new social movements.
    c. an example of the globalization potential.
    d. All of these are correct.
    e. None of the above.

18. Which of the following is an indicator of the organizational power of unions?
    a. resource mobilization
    b. union density
    c. organizational inertia
    d. collective action
    e. None of the above.

19. The National Association for the Advancement of Colored People (NAACP) took out a full-page ad in the *New York Times* in 1922 to:
    a. encourage people to support passage of the Dyer anti-lynching bill in Congress.
    b. celebrate the passage of the Dyer anti-lynching bill in Congress.
    c. protest the fact that the Dyer anti-lynching bill was never passed in Congress.
    d. criticize the media for not reporting on lynching.
    e. encourage people to fight back against oppressive treatment and lynching.

20. Which state had the highest frequency of lynchings of black victims from 1882 to 1964?
    a. Mississippi
    b. Louisiana
    c. Alabama
    d. Florida
    e. Tennessee

21. Around what year did the frequency of lynching of blacks peak between 1882 and 1932?
    a. 1932
    b. 1922
    c. 1912
    d. 1907
    e. 1892

22. Which of the following is NOT one of the social determinants of rumors in Figure 21.2?
    a. characteristics of the speaker
    b. rumor content
    c. characteristics of the audience
    d. social structure
    e. rumor authentication

23. According to Table 21.1, "Correlates of Collective Violence, France, 1830-1960", the correlation between major crimes and the rate of collective violence is:
    a. positive
    b. negative
    c. perfect
    d. spurious
    e. antecedent

24. Which is (are) among the statements made in Chapter 21 of the textbook regarding social movements, in the wake of the destruction of the twin towers of the World Trade Center on September 11, 2001?
    a. Increased government surveillance of social movements is entirely unnecessary.
    b. Giving the government a free hand to increase surveillance of social movements could endanger democracy.
    c. Americans have very clear and easy choices to make about which movements are dangerous since September 11, 2001.
    d. All of these are correct.
    e. None of the above.

25. Between 1925 and 2004, union density peaked around:
    a. 2000
    b. 1985
    c. 1965
    d. 1945
    e. 1935

True False

1. In support of breakdown theory, research shows a clear association between fluctuations in economic well-being and the number of lynchings between the 1880s and the 1930s.

   TRUE or FALSE

2. Although lynching was barbaric and wild, nonroutine collective action such as lynching is usually structured.

   TRUE or FALSE

3. The United States has the lowest union density for any rich industrialized country.

   TRUE or FALSE

4. The Taft-Hartley Act sparked the proliferation of contemporary labor unions.

   TRUE or FALSE

5. In 1700, social movements were typically small, localized, and violent.

   TRUE or FALSE

6. The environmental movement is considered a "new social movement."

   TRUE or FALSE

7. New social movements tend to attract a disproportionately large number of disaffected, undereducated youth.

   TRUE or FALSE

8. International travel and communication facilitated the globalization of social movements, since these resources remain inaccessible to most members of social movements due to their continued social disadvantage and lack of education.

   TRUE or FALSE

Short Answer

1. Define collective action and give examples. (p. 606)

   _____

   _____

2. Define social movements and give examples. (p. 606-607)

   _____

   _____

3. Explain how government's attempts to diffuse a protest by offering concessions can backfire and further encourage protesters to press further. (p. 613-615)

    _____

    _____

4. Explain the impact of the Taft-Hartley Act on unionization. (p. 615-621)

    _____

    _____

5. List and define three types of citizenship that movements have historically fought for in Europe and the United States. (p. 625-627)

    _____

    _____

Essay Questions

1. Explain how lynching is a two-sided phenomenon in terms of "disorganization" and "organization". (p. 611-612)

2. Draw Figure 21.6, which illustrates the determinants of collective action and social movement formation by integrating breakdown theory, solidarity theory, and frame alignment theory. Explain the terms and how they fit together. (p. 625)

3. Compare and contrast breakdown theory and solidarity theory. (p. 608-612, 614-615)

4. Explain frame alignment theory and give examples. Try applying solidarity or frame alignment theory to a time when you personally felt a sense of injustice against a social institution. (p. 613-624)

5. Explain the major differences between new social movements and old social movements. (p. 624-629)

Solutions

Practice Tests

Fill-In-The-Blank

1. Universal citizenship
2. Union density
3. Collective action
4. Political citizenship
5. Absolute deprivation
6. Contagion
7. Frame alignment
8. Rumors
9. Resource mobilization
10. Civil citizenship
11. Social control

Multiple-Choice Questions

1. D, (p. 607)
2. A, (p. 608-609)
3. C, (p. 609)
4. D, (p. 608-612)
5. B, (p. 614)
6. A, (p. 614)
7. D, (p. 615)
8. B, (p. 618)
9. C, (p. 619-620)
10. C, (p. 621-622)
11. A, (p. 608-612)
12. C, (p. 621-622)
13. B, (p. 614-615)
14. D, (p. 624-627)
15. A, (p. 625)
16. D, (p. 627-629)
17. D, (p. 628-629)
18. B, (p. 618)
19. A, (p. 608)
20. A, (p. 611)
21. E, (p.611)
22. E, (p. 613)
23. B, (p. 614)
24. B, (p. 616)
25. D, (p. 619)

True False

1. F, (p. 608-610)
2. T, (p. 607-608)
3. T, (p. 618-619)
4. F, (p. 619)
5. T, (p. 625)
6. T, (p. 627-629)
7. F, (p. 627)
8. F, (p. 627-629)

CHAPTER 22

TECHNOLOGY AND THE GLOBAL ENVIRONMENT

Student Learning Objectives

After reading Chapter 22, you should be able to:

1. Describe the impact of Hiroshima on American perceptions of the benefits and pitfalls of technology.

2. Discuss the notion of the United States as a "risk society" and a "technopoly."

3. Explain the idea that technology and people make history, by describing major historical technological breakthroughs in the context of social conditions.

4. Describe the transformation of high-tech since the 19th century.

5. Identify and describe the major forms and sources of environmental degradation.

6. Explain environmental problems from a social constructionist perspective.

7. Discuss the social distribution of environmental risk in the context social class inequality and racism.

8. Identify the major controversies and engage in the debate of environmental degradation and the future.

9. Discuss the larger implications of evolution and sociology in terms of the challenges, obstacles, and hopes of dealing with environmental crisis.

Key Terms

technology (633)

normal accidents (635)

risk society (636)

technopoly (637)

technological determinism (638)

greenhouse effect (643)

global warming (643)

acid rain (644)

ozone layer (644)

biodiversity (645)

genetic pollution (646)

recombinant DNA (646)

environmental racism (649)

Detailed Chapter Outline

I. TECHNOLOGY: SAVIOR OR FRANKENSTEIN?

A. Hiroshima divided the 20th century into two distinct periods.

  1. The period before Hiroshima may be called the era of naïve optimism.

    a. **Technology** was widely defined as the application of scientific principles to the *improvement* of human life.

  2. With Hiroshima, and in fact three weeks before, pessimism grew.

    a. J. Robert Oppenheimer, who organized the largest and most sophisticated technological project up to that time, is considered the "father" of the atom bomb.

    b. In response to the first nuclear bomb explosion at the Alamagordo Bombing Range in New Mexico on July 16, 1945, Oppenheimer quoted from Hindu scripture: "I am become Death, the shatterer of worlds."

    c. Oppenheimer's misgivings continued after the war, and he wanted the United States and the Soviet Union to halt thermonuclear research and refuse to develop the hydrogen bomb.

  3. The United States is by far the world leader in scientific research, publications, and elite achievements. In 1998, 59% of Americans agreed that science and technology do more good than harm.

  4. In the postwar years, a growing number of people have come to share Oppenheimer's doubts. Increasingly, ordinary citizens and leading scientists are beginning to think of technology as a Frankenstein rather than a savior.

B. The Environmental Awakening

  1. In the 1970s, a series of disasters woke people up to the fact that technological advance is not always beneficial. Examples of the most infamous technological disasters of the 1970s and 1980s:

    a. An outbreak of "Legionnaires Disease" in a Philadelphia hotel in 1976.

    b. High levels of toxic chemicals in Love Canal in 1977.

    c. The partial meltdown of the reactor core at the Three Mile Island nuclear facility in Pennsylvania in 1979.

    d. A gas leak at Union Carbide pesticide plant in Bhopal, India in 1984.

    e. The explosion of the No. 4 reactor at Chernobyl, Ukraine in 1986.

    f. The spilling of 11 million gallons of crude oil by Exxon Valdez in Prince William sound, Alaska in 1989.

C. Normal Accidents and Risk Society

  1. By the mid-1980s, sociologist Charles Perrow was referring to such disastrous events as **normal accidents**, a term that recognizes that the very complexity of modern technologies ensures they will *inevitably* fail, though in predictable ways.

  2. Sociologist Ulrich Beck also coined a pertinent term when he said we live in a **risk society**, or a society in which technology distributes danger among all categories of the population. However, some categories are more exposed to technological disasters than others.

    a. Increased risk is due to mounting *environmental* threats, which are more widespread, chronic, and ambiguous than technological accidents. For example, new terms have entered our vocabulary, such as "greenhouse effect," "global warming," "acid rain," "ozone depletion," and "endangered species."

3. Neil Postman refers to the United States as the first **technopoly**, or a country in which technology has taken control of culture.

4. The latest concern of technological skeptics is biotechnology.

    a. On the one hand, the ability to create new forms of life holds out incredible potential for advances in medicine, food production, and other fields.

    b. On the other hand, detractors claim that, without moral and political decisions based on a firm sociological understanding of who benefits and suffers from these new techniques, the application of biotechnology may be a greater threat to our well being than any other technology developed.

5. These concerns with biotechnology suggest five questions:

    a. Is technology *the* great driving force of historical and social change?

    b. If some people do control technology, then exactly *who* are they?

    c. What are the most dangerous spin-offs of technology, and how is risk distributed among various social groups?

    d. How can we overcome the dangers of environmental degradation?

    e. <u>Underlying all the others</u>: of what use is sociology in helping us solve the world's technological and environmental problems?

D. Technology *and* People Make History

1. Russian economist Nikolai Kondratiev subscribed to a form of **technological determinism**, the belief that technology is the major force shaping human society and history. Technology does indeed shape society and history, as in the examples of:

    a. The steam engine;

    b. The internal combustion engine; and

    c. The computer.

2. However, these inventions did not become engines of economic growth until *social* conditions allowed them to do so.

    a. The original steam engine was not adopted until 1,700 years later when the Industrial Revolution began, and after the social need for it emerged.

    b. The car and petroleum industries grew out of the internal combustion engine only because an ingenious entrepreneur efficiently organized work in a new way.

    c. After stopping soon after the outbreak of World War II, the development of the computer resumed once its military potential became evident.

E. How High Tech Became Big Tech

1. The 19th Century

    a. In the 19th century, gaining technological advantage was still inexpensive. In contrast, feats of the 20th—and 21st—century technology require enormous capital investment, detailed attention to the organization of work, and legions of technical experts.

   b.   It was already clear in the last quarter of the 19th century that turning scientific principles into technological innovations was going to require not just genius but substantial resources, especially money and organization.

   c.   Thomas Edison established the first "invention factory" in the late 1870s.

   d.   Edison inspired the phonograph and the electric light bulb.

2.   The 20th and 21st Centuries

   a.   By the beginning of the 20th century, the scientific or engineering genius operating in isolation was rarely able to contribute to technological innovation. By mid-century, most technological innovation was organized along industrial lines.

   b.   There seemed to be no upper limit to the amount that could be spent on research development as the 20th century ended.

      i.   The United States has more than a million research scientists today.

      ii.   In 2003, American research and development spending reached $259 billion.

   c.   The time lag between new scientific discoveries and their technological application is continuously shrinking, because large multinational corporations invest astronomical sums in research and development to be first to bring innovations to market.

   d.   The military and profit-making considerations govern the direction of most research and development. However personal interests, individual creativity, and the state of the field's intellectual development sill influence the direction of inquiry, especially for theoretical work in colleges. Nonetheless, the connection between practicality and research is even more evident today.

F.   Environmental Degradation

1.   Global Warming

   a.   The **greenhouse effect** is a consequence of humans burning increasing quantities of fossil fuels since the Industrial Revolution, and the accumulation of carbon dioxide in the atmosphere that allows more solar radiation to enter the atmosphere and less solar radiation to escape.

   b.   The greenhouse effect is believed to contribute to **global warming**, a gradual increase in the world's average surface temperature.

      i.   Between 1866 and 1965, average surface air temperature rose at a rate of 0.25 degree Celsius per century.

      ii.   From 1966 to 2004, average surface air temperature rose at a rate of 1.76 degrees Celsius.

   c.   Many scientists believe global warming is already producing serious climatic change.

      i.   As temperatures rise, more water evaporates, causing more rainfall and bigger storms, which lead to more flooding and soil erosion, and less cultivable land. People suffer and die along the causal chain.

      ii.   Example: Hurricane Mitch in 1998.

2.   Industrial Pollution

   a.   Industrial pollution is the emission of various impurities into the air, water, and soil due to industrial processes.

b. More common ingredients include household trash, scrap automobiles, residue from processed ores, agricultural runoff containing dangerous chemicals, lead, carbon monoxide, carbon dioxide, sulfur dioxide, ozone, nitrogen oxide, various volatile organic compounds, chlorofluorocarbons (CFCs), and various solids mixed with liquid droplets floating in the air.

c. Pollutants may affect us directly (drinking water and the air we breathe), and indirectly (sulfur dioxide and other gases emitted by coal-burning power plants, pulp and paper mills, and motor-vehicle exhaust).

    i. They form **acid rain**, a form of precipitation whose acidity eats away at, and even eventually destroys, forests and the ecosystems of lakes.

    ii. The use of CFCs in industry and by consumers, notably refrigeration equipment, contain chlorine, which is responsible for the depletion of the **ozone layer** 5-25 miles above the earth's surface.

d. Radioactive waste results from more than 100 nuclear reactors that generate commercial electricity in the United States, which run on enriched uranium or plutonium fuel rods. When fuel rods decay beyond their usefulness in the reactor, they become highly radioactive waste material, and must decay about 10,000 years before humans can be safely exposed to it without special protective equipment. Decaying fuel rods need to be placed in watertight copper canisters and buried deep in granite bedrock. However, most Americans do not want nuclear waste facilities near their families. Spent fuel rods have been accumulating since the 1950s in "temporary" facilities, mainly pools of water near nuclear reactors. The first long-term nuclear waste repository in the United States, deep inside Nevada's Yucca Mountain, is scheduled to open in 2010, but delays are expected due to numerous regulatory and legal battles.

3. The Decline of Biodiversity

a. The third main form of environmental degradation is the decline in **biodiversity**, the enormous variety of plant and animal species inhabiting the earth.

b. As part of the normal evolutionary process, biodiversity changes as new species emerge and old species die off because they cannot adapt to their environment. In recent decades, the environment has become so inhospitable that the rate of extinction has greatly accelerated.

    i. For millions of years, an average of one to three species became extinct annually.

    ii. About 1,000 species are becoming extinct annually today.

c. Although the extinction of species is impoverishing in itself, it also has practical consequences for humans.

    i. Each species of animal and plant has unique properties that are medically useful.

    ii. The single richest source of genetic material with pharmaceutical value is found in the world's rain forests, particularly in Brazil, where more than 30 million species of life exist.

    iii. Strip mining, the construction of huge pulp and paper mills and hydroelectric projects, and the deforestation of land are destroying rain forests by farmers and cattle grazers.

    d. Fleets of trawlers from highly industrialized countries, equipped with sonar, increase their catch of fish using mesh nets, which has depleted fish stocks, one of the world's important sources of protein.

  4. Genetic Pollution

    a. **Genetic pollution** refers to the health and ecological dangers that may result from artificially splicing genes together.

    b. While the genetic information of all living things is coded in a chemical called DNA, **recombinant DNA** is a technique developed by molecular biologists in the last few decades, and involves artificially joining bits of DNA from a donor to the DNA of a host. In effect, a new form of life is created.

    c. Since 1990, governments and corporations have been engaged in a multibillion-dollar international effort to create a complete genetic map of humans and various plants, microorganisms, and animal species. By 2000, scientists had identified the location and chemical structure of every one of the approximately 40,000 human genes.

    d. The potential health and economic benefits to humankind of recombinant DNA are truly startling.

      i. It is possible to design what some people regard as more useful animals and plants and superior humans.

      ii. Hereditary propensities to a range of diseases can be detected and eliminated.

      iii. Farmers could grow disease—and frost-resistant crops with higher yields.

      iv. The dangers of mining could be greatly reduced, with the use of ore-eating microbes.

      v. Companies could grow plants that produce cheap biodegradable plastic and microorganisms that consume oil spills and absorb radioactivity.

    e. The dangers of genetic pollution poses to human health and the stability of ecosystems, are also startling.

    f. Global warming, industrial pollution, the decline of biodiversity, and genetic pollution threaten everyone.

II. THE SOCIAL CONSTRUCTION OF ENVIRONMENTAL PROBLEMS

  A. *Social constructionism* is a sociological approach to studying social problems such as environmental degradation, which emphasizes that social problems do not emerge spontaneously. Rather, they are contested phenomena whose prominence depends on the ability of supporters and detractors to make the public aware of them.

  B. The Case of Global Warming

    1. The controversy over global warming illustrates how people create and contest definitions of environmental problems.

    2. As the social constructionists suggest, the power of competing interests to get their definition of reality accepted as the truth will continue to influence public perceptions of the seriousness of global warming.

  C. The Social Distribution of Environmental Risk

    1. Whenever disaster strikes, economically and politically disadvantaged people almost always suffer most, because their circumstances render them most vulnerable.

2. Environmental Racism
   a. The advantaged often consciously put the disadvantaged in harm's way to avoid risk to themselves.
      i. Oil refineries, chemical plants, toxic dumps, garbage incinerators, and other environmentally dangerous installations are more likely to be built in poor communities with a high percentage of African Americans or Hispanic Americans than in more affluent, mainly white communities. Disadvantaged people are often too politically weak to oppose and some may value the jobs created.
      ii. Some poor Native American reservations have been targeted as possible interim nuclear waste sites, partly because states lack jurisdiction over reservations, so the usual state protests are less likely to be effective. In addition, the Goshute tribes in Utah and the Mescalero Apaches in New Mexico have expressed economic interest in the project.
   b. Some analysts refer to this recurrent pattern as **environmental racism**, the tendency to heap environmental dangers on the disadvantaged, and especially disadvantaged racial minorities.
3. Environmental Risk and the Less Developed Countries
   a. Environmental disadvantage also holds true for the world's less developed countries.
      i. Although, people in less developed countries are more concerned about the environment, their countries cannot afford much pollution control.
      ii. Currently, the rich countries do most of the world's environmental damage, because their inhabitants earn and consume more.
   b. Social inequalities also exist in biotechnology.
      i. Because large multinational companies dominate the pharmaceutical, seed, and agrochemical industries, and work for patents when they discover genetic material with commercial value, they have exclusive right to manufacture and sell genetic material without compensating the donors.
4. Environmental Risk and Social Class
   a. The possible consequence of having their babies genetically engineered, which is likely on a wide scale in 10 to 20 years is compelling.
      i. In spite of the health benefits, the well to do are more likely to afford genetically engineered babies.
      ii. Consequently, this could introduce an era of increased social inequality. The economically underprivileged would bear a risk of genetic inferiority, perhaps creating genetic stratification.

III. WHAT IS TO BE DONE?

A. The Market and High-Tech Solutions
   1. Some people believe the market and high technology will resolve the environmental crisis. For example: oil, rice, and wheat production.
   2. Some evidence supports this optimistic scenario.
      a. The replacement of brain-damaging leaded gas with unleaded gas.
      b. Development of environmentally friendly refrigerants.

    c. Subsidizing the cost of replacing CFCs in developing countries by rich countries.

    d. Windmills and solar panels.

    e. High-tech pollution control devices.

    f. Development of new methods for eliminating carbon dioxide emissions.

    g. Diesel-electric hybrid cars and electric cars.

3. Problems with the Market and High-Tech Solutions

    a. Three factors suggest market forces cannot solve environmental problems on their own.

        i. Imperfect price signals

        ii. Political pressure

        iii. The slow pace of change

B. The Cooperative Alternative

1. The alternative involves people cooperating to greatly reduce their overconsumption of everything, which would require renewed commitment to voluntary efforts, new laws and enforcement bodies, increased environmentally related research and development by industry and government, more environmentally directed foreign aid, hefty new taxes, careful risk assessment of biotechnology in consultation with the public, and equitable sharing of profit from genetic engineering.

2. Before such policies could be adopted, four conditions would have to be met. To be politically acceptable, three conditions have to be met. The broad public in North America, Western Europe, and Japan would have to be:

    a. Aware of the gravity of environmental problems;

    b. Confident in the capacity of citizens, their governments, and corporations to solve the problem;

    c. Willing to make substantial economic sacrifices to get the job done; and

    d. Able to overcome resistance to change on the part of interest groups with a deep stake in things as they are.

3. Based on the General Social Survey, in the United States:

    a. Nearly all Americans are aware of the environmental problem;

    b. A majority (56%) think they can do something about environmental issues themselves;

    c. A huge majority (89%) believe the government should pass more laws to protect the environment; and

    d. Most Americans (61%) say too little is being spent on environmental cleanup.

4. However, biting the bullet is another story. Most Americans are prepared to protect the environment if it does not inconvenience them too much. The numbers drop sharply when it is inconvenient.

    a. Fewer than half of Americans (47%) are willing to pay much higher prices to protect the environment.

    b. Fewer than a third (32%) are willing to accept cuts to their standard of living.

    c. Barely a third (34%) are willing to pay much higher taxes.

5. History teaches us that people are not usually prepared to make big personal sacrifices for seemingly remote and abstract goals, but are prepared to sacrifice if the goals become more

close to home (i.e. Pearl Harbor). Larger environmental catastrophes may have to occur before people are willing to take action, but there may still be time to act.

6. The Resistance of Powerful Interest Groups
   a. The final obstacle to implementing the cooperative alternative is the resistance of interest groups who benefit from things the way they are. Oil and automobile industries in the United States are two of the most powerful interest groups.
   b. The United States imports about one-half the petroleum it uses. About a third of its imports come from the Middle East, especially Saudi Arabia, Iraq, and Kuwait. We pay a heavy price economically, environmentally, and politically for this dependence. We went to war with Iraq in 1991 and 2003 partly (some would say mainly) to protect our oil interests in the region. Meanwhile the United States is the second largest producer of automobiles in the world.
   c. Given the extent and depth of the interests of the oil and automobile industries in the status quo, proposals to adopt cleaner and less expensive alternatives to gas-burning cars have been met with little response from government or industry.

7. Contrasting Cases: Japan and Iceland
   a. A company owned partly by Shell Oil opened Japan's first hydrogen filling station in 2003. Iceland also opened a hydrogen filling station in 2003.
      i. Cooperation among citizens, governments, and corporations can make progress in the fight against environmental degradation.
      ii. Cooperative strategy is not popular in the United States. What could change that?

IV. EVOLUTION AND SOCIOLOGY

A. We have two survival strategies to cope with future challenges: *competition* and *cooperation*.

B. Competition takes place when members of the same species compete for limited resources.
   1. Based on Charles Darwin's *The Origin of Species*, members of each species struggle against each other and other species to survive.
   2. The most fit live long enough to bear offspring, while most of the rest are killed off.
   3. The traits passed to offspring are those most valuable for survival. Ruthless competition is a key survival strategy of all species.

C. Cooperation occurs when species members struggle against adverse environmental circumstances.
   1. Based on Darwin's *The Descent of Man*, the species members who flourish are those who best learn to help each other.
   2. According to Russian geographer and naturalist Petr Kropotkin, survival in the face of environmental threat is best assured if species members help each other.
   3. Kropotkin shows that the most advanced species in any group are the most cooperative.

D. A strictly competitive approach to dealing with the environmental crisis seems inadequate. It requires more cooperation and self-sacrifice.

E. Sociology-yours and ours.
   1. Conceived at its broadest, sociology promises to help in the rational and equitable evolution of humankind.

a. "Calvin and Hobbes": Look down the road;-)

Study Activity: Applying The Sociological Compass

As noted in Chapter 18, "The Mass Media", there has been a proliferation of films that offer fictitious glimpses into a world driven by, controlled, and created through technology. The movies discussed in Box 22.1 – 'The Matrix' series and 'The Terminator' series - are cases in point. For this exercise, re-read Box 22.1, Sociology at the Movies. Write down the questions and major points posed by Brym and Lie in the box, and utilize these questions as a framework of analysis. View at least one of the films with a sociological eye, whether you've seen the movie before or not. In the process, take notes on the sociological implications of the movie in light of what you learned about social constructionism in Chapter 22. Although these films are action-packed, entertaining, and seemingly far-fetched, consider some social parallels between the movie and U.S. society. What are some of the parallel constructions of reality that "have taken over the seemingly stable reality of the physical world"? Explain these examples from the social constructionist approach.

Major Questions and Points Raised in Box 22.1:

1.

2.

3.

4.

Parallels of the movie and U.S. Society:

1.

2.

3.

4.

Applying Social Constructionism

Infotrac College Edition Online Exercises

For the following exercises, log on to the online library of InfoTrac College Edition at http://www.infotrac-college.com/. Make note that InfoTrac has implemented a new registration system that will allow easier access to InfoTrac through the use of a personalized username and password. Once you've created your username and password you may proceed directly to the Log On page. To create an account, register your passcode packaged with your textbook, and create a username and password, by following the online prompt. After you are logged in, click on "Infotrac College Edition." You will arrive at a screen that enables you to search topics.

Keyword: **environmental racism**. Choose two articles that discuss or highlight recent findings about the issue. After reading the articles: Summarize the major arguments of each source. Are the articles consistent, or do they diverge? Which article seems more consistent with a conflict perspective, and explain why?

Keyword: **global warming**. Identify two relevant articles. Summarize the major findings. What are the positions of each article? How do the articles frame the issue? From a social constructionist perspective, what kinds of realities are constructed? Why?

Keyword: **cellular phones**. An increasing number of people are using cellular phones. Find an article that discusses the negative aspects of cellular phone usage. Find an article that discusses legislation about cell phones. What are the key negative aspects of having so many people use cellular phones? Is there any legislation under review that can affect cellular phone usage?

Keyword: **global warming**. Environmentalists are currently more concerned with global warming than the rest of the population. Find an article that discusses solutions to this problem. What are some solutions suggested to this potentially hazardous problem? How do these suggestions compare to the discussion in Chapter 22?

Keyword: **recycling**. Find an article that discusses the financial costs of supporting a recycling program. What are the long-range costs and benefits for supporting a recycling program? Write a short paper discussing your findings.

Keyword: **Internet**. The Internet is a growing source of information and information management. Find an article that discusses how the information provided in the Internet can be assessed. Is the information provided on the Internet reliable?

Internet Exercises

The Global Network of Environment and Technology website at: http://www.gnet.com/

shows the interplay between environmental and technological issues. Browse the site and write down what you perceive to be the top five issues of concern. .

Environmental Racism is increasingly becoming an issue of concern to humans. Go to: http://egj.lib.uidaho.edu/egj01/weint01.html and find a document called *Fighting Environmental Racism: A Selected Annotated Bibliography*. This is a great resource for finding information on environmental racism. Look up three different resources and compare their approach to environmental racism. Write a short paper on your findings.

The United States Environment Protection Agency (EPA) sponsors a search engine located at: http://www.epa.gov/epahome/comm.htm. Look up environmental issues in specific communities. Pick a community from your home state and a community that you are not familiar with. Make a list for each community describing specific environmental issues and concerns of each community. Compare and contrast your lists.

Practice Tests

Fill-In-The-Blank

Fill in the blank with the appropriate term from the above list of "key terms."

1. _____ refers to the enormous variety of plant and animal species inhabiting the earth.

2. _____ is the tendency to heap environmental dangers on the disadvantaged, especially racial minorities.

3. A(An) _____ is a postmodern society defined by the way risk is distributed as a side effect of technology.

4. _____ is the gradual worldwide increase in average surface temperature.

5. _____ is the practical application of scientific principles.

6. _____ refers to the potential dangers of mixing genes of one species with those of another.

7. _____ is the belief that technology is the main factor shaping human society and history.

Multiple-Choice Questions

Select the response that best answers the question or completes the statement:

1. The bombing of Hiroshima by the United States on August 6, 1945, is believed to have divided the 20th century into two distinct periods. According to the textbook, the first period may be called:
    a. the era of mass destruction
    b. the era of naïve optimism
    c. the era of technological optimism
    d. the new dark age of industry
    e. the new renaissance of technological wonder

2. J. Robert Oppenheimer, the "father" of the atom bomb:
    a. supported the development of the nuclear bomb after witnessing the dominance of the United States following Hiroshima.
    b. directly organized one of the largest technological projects in the former Soviet Union toward developing their nuclear arsenal.
    c. wanted the United States and the Soviet Union to halt thermonuclear research and development of the hydrogen bomb.
    d. All of these are correct.
    e. None of the above.

3. In 2000, _____ of Americans agreed that science and technology do more good than harm.
    a. 70%
    b. 93%
    c. about 50%
    d. 30%
    e. 12%

4. _____ is a term that recognizes that the very complexity of modern technologies ensures they will inevitably fail, though in predictable ways.
    a. Risk society
    b. Normal accidents
    c. Technopoly
    d. Technological lag
    e. None of the above.

5. _____ is the belief that technology is the major force shaping human society and history.
    a. Technological determinism
    b. Technological lag
    c. Technological functionalism
    d. Technological ideology
    e. None of the above.

6. Which is an example of technological determinism?
   a. the steam engine
   b. the internal combustion engine
   c. the computer
   d. All of these are correct.
   e. None of the above.

7. Which type of environmental degradation is a form of precipitation that eats away and eventually destroys forests, and the ecosystems of lakes?
   a. the greenhouse effect
   b. global warming
   c. depletion of the ozone layer
   d. All of these are correct.
   e. None of the above.

8. Which two forms of environmental degradation are scientifically most closely related?
   a. species extinction and genetic engineering
   b. genetic pollution and depletion of the ozone layer
   c. the greenhouse effect and global warming
   d. radiation and toxic waste
   e. carbon dioxide and CFCs

9. Which is NOT a benefit of recombinant DNA?
   a. It is possible to design what some people regard as more useful animals and plants and superior humans.
   b. Hereditary propensities to a range of diseases can be detected and eliminated.
   c. Farmers could grow disease—and frost-resistant crops with higher yields.
   d. Companies could grow plants that produce cheap biodegradable plastic and microorganisms that consume oil spills and absorb radioactivity.
   e. All of the answers are benefits of recombinant DNA.

10. Oil refineries, chemical plants, toxic dumps, and garbage incinerators are more likely built:
    a. in middle-class industrial communities.
    b. in poor communities.
    c. near maximum correction facilities.
    d. near beaches, lakes, and streams.
    e. None of the above.

11. Currently, _____ countries do most the world's environmental damage.
    a. unindustrialized
    b. semiperipheral
    c. rich
    d. peripheral
    e. None of the above.

12. The idea of people inventing machines that destroy their creators:
    a. is a new idea.
    b. is a unique idea portrayed in *The Matrix* and *The Terminator* movies.
    c. first surfaced in the 1980s.
    d. All of these are correct.
    e. None of the above.

13. Accidents that occur inevitably due to the very complexity of modern technologies is referred to as:
    a. spurious accidents.
    b. normal accidents.
    c. risk accidents.
    d. technological accidents.
    e. environmental accidents.

14. As in the case of an early computer called "ORDVAC," developed at the University of Illinois, technology typically advances when there is:
    a. an urgent social need.
    b. economic crisis.
    c. economic and political prosperity.
    d. All of these are correct.
    e. None of the above.

15. Since its invention, which technological commodity penetrated the market most sharply and dramatically?
    a. VCR
    b. cell phone
    c. World Wide Web
    d. Microwave
    e. PC

16. Due to global warming:
    a. glaciers are melting.
    b. the sea level is rising.
    c. extreme weather events are becoming more frequent.
    d. All of these are correct.
    e. None of the above.

17. Even in the late 1980s, nearly _____ of the biotechnology scientists who belonged to the prestigious National Academy of Science had industry affiliations?
    a. 75%
    b. 40%
    c. 25%
    d. 10%
    e. 90%

18. Worldwide, what year did "natural" disasters take its greatest toll in terms of monetary damage in billions of dollars?
    a. 1978
    b. 1998
    c. 2004
    d. 1992
    e. 2003

19. The precipitation whose acidity destroys forests and the ecosystems of lakes is referred to as:
    a. biodiversity.
    b. ozone depletion.
    c. acid rain
    d. environmental racism.
    e. the greenhouse effect.

20. In terms of genetic pollution, the potential of ecological catastrophe has:
    a. multiplied.
    b. declined.
    c. remained relatively constant over that last 50 years.
    d. All of these are correct.
    e. None of the above.

21. Petrochemical plants between New Orleans and Baton Rouge form what local residents call:
    a. "Exxonville."
    b. "the Oily Way."
    c. "Cancer Alley."
    d. "The Black Gold Coast."
    e. "Poison Point."

22. Environmental racism tends to endanger:
    a. racial minorities exclusively.
    b. factory workers in particular.
    c. racial minority factory workers specifically.
    d. the disadvantaged, especially disadvantaged racial minorities.
    e. women across the social class hierarchy.

23. The first long-term nuclear waste repository in the United States, located _____, is scheduled to open in 2010, but delays are expected due to numerous regulatory and legal battles.
    a. deep inside Nevada's Yucca Mountain
    b. in the Alaskan wilderness
    c. in offshore containment rigs in the Gulf of Mexico
    d. on remote federal land in Tennessee
    e. in a series of deep, interconnected tunnels in North Dakota

24. According to the textbook, what is the final and perhaps most powerful source of resistance to the cooperative alternative to cleaning up the environment?
    a. Political and economic enemies of the United States.
    b. powerful U.S. interest groups in the oil and automobile industries.
    c. physical scientists.
    d. social scientists.
    e. the masses.

True False

1. Today, technology is overwhelmingly regarded as the application of scientific principles for the improvement of human life.

   TRUE or FALSE

2. The United States is by far the world leader in scientific research, publications, and elite achievements.

   TRUE or FALSE

3. The outbreak of "Legionnaires Disease" in a Philadelphia hotel in 1976 is among the most infamous technological disasters of the 1970s and 1980s.

   TRUE or FALSE

4. Recombinant DNA is a technique developed by molecular biologists in the last few decades, and involves artificially joining bits of DNA from a donor to the DNA of a host.

   TRUE or FALSE

5. Although research indicates the existence of environmental racism, the advantaged seldom put the disadvantaged in harm's way to avoid environmental risk.

   TRUE or FALSE

6. According to Charles Darwin, the species members who flourish are those who best learn to help each other.

   TRUE or FALSE

Short Answer

1. Define technology. (p. 633)

   _____

   _____

2. Explain the connection between the greenhouse effect and global warming. Be sure to provide a precise definition for each term. (p. 642-643, 647-648)

_____

_____

3. Explain the potential for ecological catastrophe by the testing of genetically altered plants in the field. (p. 646-647)

_____

_____

4. Define the term "environmental racism". Provide an example from Chapter 22. (p. 649)

_____

_____

5. List and describe three factors that suggest market forces alone cannot solve environmental problems. (p. 651-656)

_____

_____

Essay Questions

1. In the 19th century, gaining a technological advantage was inexpensive, and relied much on creativity and imagination. What changes occurred in the 20th century that led to the requirement of enormous capital investment in the high-tech industry? (p. 638-642)

2. Explain the major benefits and threats of genetic engineering and pollution. How do you see society keeping checks and balances on the progress and application of these technologies, so that the potential benefits can be shared democratically with all of humankind? (p. 646-647)

3. Explain the idea of environmental problems as social constructions. What are the implications of social constructionism for alleviating environmental degradation and crisis in the future? (p. 647-651)

4. Describe technological determinism and the theories of Nikolai Kondratiev. To what degree do you think technology shapes society and history, and do you see technology's influence changing in the future? (p. 635-642)

5. Compare and contrast the two survival strategies to cope with the environmental challenges in future. How can we save our planet by "looking down the road"? (p. 656-657)

Solutions

Practice Tests

Fill-In-The-Blank

1. Biodiversity
2. Environmental racism
3. risk society
4. Global warming
5. Technology
6. Genetic pollution
7. Technological determinism

Multiple-Choice Questions

1. B, (p. 633)
2. C, (p. 633-634)
3. A, (p. 634)
4. B, (p. 635)
5. A, (p. 638)
6. D, (p. 638-639)
7. E, (p. 642-647)
8. C, (p. 642-647)
9. E, (p. 646-647)
10. B, (p. 648-649)
11. C, (p 649-650)
12. E, (p. 637)
13. B, (p. 635)
14. A, (p. 640)
15. C, (p. 641)
16. D, (p. 642-643)
17. B, (p. 642)
18. C, (p. 644)
19. C, (p. 644).
20. A, (p. 646-647)
21. C, (p. 649)
22. D, (p. 649)
23. A, (p. 645)
24. B, (p. 655-656)

True False

1. F, (p. 633-634)
2. T, (p. 634)
3. T, (p. 634)
4. T, (p. 646)
5. F, (p. 648-649)
6. T, (p. 657)